Progress in Scientific Computing
Vol. 3

Edited by
S. Abarbanel
R. Glowinski
G. Golub
P. Henrici
H.-O. Kreiss

Birkhäuser
Boston · Basel · Stuttgart

Jane K. Cullum
Ralph A. Willoughby

Lanczos Algorithms for Large Symmetric Eigenvalue Computations Vol. I Theory

1985

Birkhäuser
Boston · Basel · Stuttgart

Authors:
Jane K. Cullum
Ralph A. Willoughby
IBM T. J. Watson Research Center
Yorktown Heights, NY 10598 (USA)

Library of Congress Cataloging in Publication Data

Cullum, Jane K., 1938–
  Lanczos algorithms for large symmetric eigenvalue
computations.
  (Progress in scientific computing ; v. 3)
  Bibliography: p.
  Includes indexes.
  Contents: v. 1. Theory.
  1. Symmetric matrices – – Data processing.   2. Eigenvalues
– – Data processing.   I. Willoughby, Ralph A.   II. Title.
III. Series.
QA193.C84   1985   512.9'434   81-38450
ISBN 978-1-4684-9192-0 (v. 1)

CIP-Kurztitelaufnahme der Deutschen Bibliothek

Cullum, Jane K.:
Lanczos algorithms for large symmetric eigenvalue
computations / Jane K. Cullum ; Ralph A. Willoughby.
– Boston ; Basel ; Stuttgart : Birkhäuser
  ISBN 978-1-4684-9192-0

NE: Willoughby, Ralph A.:
Vol. 1. Theory. – 1985.
  (Progress in scientific computing ; Vol. 3)
  ISBN 978-1-4684-9192-0

NE: GT

© 1985 Birkhäuser Boston, Inc.
Softcover reprint of the hardcover 1st edition 1985

ISBN 978-1-4684-9192-0   ISBN 978-1-4684-9190-6 (eBook)
DOI 10.1007/978-1-4684-9190-6

Table of Contents

PREFACE

INTRODUCTION

CHAPTER 0

PRELIMINARIES: NOTATION AND DEFINITIONS

0.1	Notation	1
0.2	Special Types of Matrices	3
0.3	Spectral Quantities	7
0.4	Types of Matrix Transformations	10
0.5	Subspaces, Projections, and Ritz Vectors	14
0.6	Miscellaneous Definitions	15

CHAPTER 1

REAL 'SYMMETRIC' PROBLEMS

1.1	Real Symmetric Matrices	17
1.2	Perturbation Theory	19
1.3	Residual Estimates of Errors	20
1.4	Eigenvalue Interlacing and Sturm Sequencing	22
1.5	Hermitian Matrices	23
1.6	Real Symmetric Generalized Eigenvalue Problems	24
1.7	Singular Value Problems	26
1.8	Sparse Matrices	28
1.9	Reorderings and Factorization of Matrices	29

CHAPTER 2

LANCZOS PROCEDURES, REAL SYMMETRIC PROBLEMS

2.1 Definition, Basic Lanczos Procedure   32

2.2 Basic Lanczos Recursion, Exact Arithmetic   35

2.3 Basic Lanczos Recursion, Finite Precision Arithmetic   44

2.4 Types of Practical Lanczos Procedures   53

2.5 Recent Research on Lanczos Procedures   58

CHAPTER 3

TRIDIAGONAL MATRICES

3.1 Introduction   76

3.2 Adjoint and Eigenvector Formulas   79

3.3 Complex Symmetric or Hermitian Tridiagonal   84

3.4 Eigenvectors, Using Inverse Iteration   86

3.5 Eigenvalues, Using Sturm Sequencing   89

CHAPTER 4

LANCZOS PROCEDURES WITH NO REORTHOGONALIZATION
FOR REAL SYMMETRIC PROBLEMS

4.1 Introduction   92

4.2 An Equivalence, Exact Arithmetic   95

4.3 An Equivalence, Finite Precision Arithmetic   101

4.4 The Lanczos Phenomenon   119

4.5 An Identification Test, 'Good' versus 'Spurious' Eigenvalues 121

4.6. Example, Tracking Spurious Eigenvalues   130

4.7 Lanczos Procedures, Eigenvalues   136

4.8 Lanczos Procedures, Eigenvectors   151

4.9 Lanczos Procedure, Hermitian, Generalized Symmetric   157

CHAPTER 5

REAL RECTANGULAR MATRICES

5.1	Introduction	164
5.2	Relationships With Eigenvalues	166
5.3	Applications	172
5.4	Lanczos Procedure, Singular Values and Vectors	178

CHAPTER 6

NONDEFECTIVE COMPLEX SYMMETRIC MATRICES

6.1	Introduction	194
6.2	Properties of Complex Symmetric Matrices	196
6.3	Lanczos Procedure, Nondefective Matrices	199
6.4	QL Algorithm, Complex Symmetric Tridiagonal Matrices	204

CHAPTER 7

BLOCK LANCZOS PROCEDURES, REAL SYMMETRIC MATRICES

7.1	Introduction	210
7.2	Iterative Single-vector, Optimization Interpretation	214
7.3	Iterative Block, Optimization Interpretation	220
7.4	Iterative Block, A Practical Implementation	232
7.5	A Hybrid Lanczos Procedure	244

REFERENCES

AUTHOR AND SUBJECT INDICES

PREFACE

Energy levels, resonances, vibrations, feature extraction, factor analysis - the names vary from discipline to discipline; however, all involve eigenvalue/eigenvector computations. An engineer or physicist who is modeling a physical process, structure, or device is constrained to select a model for which the subsequently-required computations can be performed. This constraint often leads to reduced order or reduced size models which may or may not preserve all of the important characteristics of the system being modeled. Ideally, the modeler should not be forced to make such a priori reductions. It is our intention to provide here procedures which will allow the direct and successful solution of many large 'symmetric' eigenvalue problems, so that at least in problems where the computations are of this type there will be no need for model reduction.

Matrix eigenelement computations can be classified as small, medium, or large scale, in terms of their relative degrees of difficulty as measured by the amount of computer storage and time required to complete the desired computations. A matrix eigenvalue problem is said to be small scale if the given matrix has order smaller than 100. Well-documented and reliable FORTRAN programs exist for small scale eigenelement computations, see in particular EIS-PACK [1976,1977]. Typically those programs explicitly transform the given matrix into a simpler canonical form. The eigenelement computations are then performed on the canonical form. For the EISPACK programs the storage requirements grow as the square of the order of the matrix being processed and the operation counts grow cubically with the order.

A matrix eigenvalue problem is said to be medium scale if it is real symmetric, and if it is computationally feasible to compute the eigenelements by using Sturm sequencing and bisection in combination with inverse iteration directly on the original matrix. For example, a band matrix with a reasonable band width will be said to be of medium scale.

Matrix eigenvalue computations are said to be large scale if the size of the matrix and the pattern of the nonzeros in the matrix preclude the use of EISPACK-type procedures or of a Sturm sequencing/bisection/inverse iteration approach. For example, if the given matrix has order larger than 200 and is not banded, then we would classify the associated eigenvalue computation for that matrix as large scale. In most of our experiments, large scale has meant of order greater than 500.

We focus on large scale, 'symmetric' matrix eigenelement computations. Symmetric is in quotes because we include a procedure for computing singular values and vectors of real rectangular matrices. In addition we also include a procedure for computing eigenelements of nondefective complex symmetric matrices. Such matrices do not possess the desirable properties of real symmetric matrices, and the amount of computation required to process them can be significantly more than that required in the real symmetric case. We will not address the general nonsymmetric eigenvalue problem.

This is a research monograph intended for engineers, scientists and mathematicians who are interested in computational procedures for large matrix eigenvalue problems. The discussion focuses on one particular subset of one particular family of procedures for large matrix eigenvalue computations. We are interested in Lanczos procedures.

Lanczos procedures derive their name from a famous (fifteen years ago some people may have been inclined to say infamous) 3-term recursion that was originally proposed as a means of transforming a general real symmetric matrix into a real symmetric tridiagonal matrix. Within the scientific community we find two basic approaches to Lanczos procedures. One approach maintains that 'global orthogonality is crucial' and the other approach maintains that 'local orthogonality is sufficient'. This terminology is explained in Chapter 2, Sections 2.4 and 2.5 where we briefly survey the literature on Lanczos procedures for eigenelement computations.

Our emphasis is on computation. We focus primarily on our research on single-vector Lanczos procedures with no reorthogonalization. These procedures belong to the class of 'local orthogonality is sufficient' procedures. The material is organized into two volumes. This volume contains the material necessary for understanding the proposed Lanczos procedures. The second volume contains the FORTRAN codes and documentation for each of the Lanczos procedures discussed in this volume.

Users with large problems are concerned about the amounts of computer storage and time required by the procedures which they have to use. Our single-vector Lanczos procedures are storage efficient. In most cases they are also time efficient if the matrix, whose eigenvalues (singular values) are to be computed, is such that matrix-vector multiplies can be computed rapidly and accurately. Typically if the given matrix is sparse, in the sense that there are only a few nonzero entries in each row and column, then this can be achieved.

Some of what is presented is new and has not yet been published elsewhere. Much of what is presented has appeared at least in preliminary form in papers and reports published in various places. It is hoped that by bringing all of this material together in one place, that these results will prove useful to a wide variety of users in the engineering and scientific community.

Jane K. Cullum

Ralph A. Willoughby

July 1984

INTRODUCTION

We consider the question of the computation of eigenvalues and eigenvectors of large, 'symmetric' matrices. While in a strictly mathematical sense, the scope of this book is very narrow, the potential applications for the material which is included are important and numerous. Perhaps the most familiar application of eigenvalue and eigenvector computations is to structural analysis, studies of the responses of aircraft, of bridges, or of buildings when they are subjected to different types of disturbances such as air turbulence, various types of loadings, or earthquakes. In each case, the physical system being analyzed varies continuously with time, and its true motion is described by one or more differential equations. Matrix eigenvalue problems and approximations to this motion are obtained by discretizing the system equations in some appropriate way.

For a given matrix A, the 'simple' eigenvalue-eigenvector problem is to determine a scalar λ and a vector $x \neq 0$ such that $Ax = \lambda x$. In structural problems, one typically encounters the generalized eigenvalue problem $Kx = \lambda Mx$, involving 2 different matrices, a mass matrix M and a stiffness matrix K. In fact the problems there can be nonlinear quadratic eigenvalue problems, see for example Abo-Hamd and Utku [1978]. However, the 'solution' of a linearization of one of these quadratic problems is often used as a basis for reducing the given nonlinear eigenvalue problem to a much smaller but dense generalized eigenvalue problem. In such a problem a few of the smallest eigenvalues and corresponding eigenvectors may be required or in some cases in order to determine the response of a given structure to external disturbances it may be necessary to compute eigenvalues and corresponding eigenvectors on some interior interval of the spectrum of the given matrix. In structures the matrices used are typically banded, that is all of the nonzero entries are clustered around the main diagonal of the system matrices. For many years simultaneous iteration techniques have been applied successfully to shifted and inverted matrices $(K - \mu M)$, using equation solving techniques designed for band matrices.

Very large matrix eigenvalue problems also arise in studies in quantum physics and chemistry, see for example Kirkpatrick [1972] and Gehring [1975]. The matrices generated are large and sparse. Typically significant numbers of the eigenvalues of these matrices are required. An entirely different kind of application is the use of eigenvectors in heuristic partitioning algorithms, see for example Barnes [1982]. For the particular application which Barnes considered, the placement of electrical circuits on silicon chips, the goal was to position a large number of circuits on a given number of chips in such a way that the resulting number of external connections between circuits on different chips was minimized.

Other applications for eigenvalue/eigenvector computations occur in quantum chemistry, see for example Nesbet [1981]; in power system analysis, see for example Van Ness [1980]; in oceanography, see for example Winant [1975] and Platzman [1978]; in magnetohydrodynamics, see for example Gerlakh [1978]; in nuclear reactor studies, see for example Geogakis [1977]; in helicopter stability studies, see for example Hodges [1979]; and in geophysics, see for example Kupchinov [1973].

As we said earlier we are considering the question of computing eigenvalues and eigenvectors of large 'symmetric' matrices which arise in various applications. The word symmetric is in quotes because we also present procedures for two types of matrix computations which are not symmetric in the ordinary sense. The basic ideas which we discuss are equally applicable to any matrix problem which is equivalent to a real symmetric eigenvalue/eigenvector problem. We consider several such equivalences. These include Hermitian matrices, certain real symmetric generalized eigenvalue problems, and singular value and singular vector computations for real, rectangular matrices. We also consider complex symmetric matrices which are not equivalent to real symmetric matrices.

The actual scope of this book is limited to a particular family of algorithms for large scale eigenvalue problems, the Lanczos procedures. Other types of eigenelement procedures suitable for large matrices exist, most of which are based upon either simultaneous iterations or upon Rayleigh quotient iterations, see Bathe and Wilson [1976] and Jennings [1977] for complete and very readable discussions of simultaneous iteration procedures. The research on Rayleigh quotient iteration procedures is scattered. Parlett [1980, Chapter 4] discusses the theoretical properties of such procedures and gives references for interested readers. We do not cover any of the non-Lanczos procedures in our discussions.

The research on Lanczos procedures for eigenelement computations (and for solving systems of equations) continues. Although many interesting results have been obtained, many of the theoretical questions concerning Lanczos procedures have not been satisfactorily resolved. Much of the existing literature on Lanczos procedures has not adequately incorporated the effects of roundoff errors due to the inexactness of the computer arithmetic. Numerical experiments with various Lanczos procedures have however clearly demonstrated their advantages and capabilities. Many different people have contributed to this research, and we apologize if we have neglected to mention one or more of these authors in our discussions or if for the authors we do mention we have not referenced all of their papers on this subject.

The demonstrated computational efficiencies and excellent convergence properties which can be achieved by Lanczos procedures, have generated much interest in the scientific and engineering communities. Parlett [1980] is primarily devoted to discussions of small to medium size real symmetric eigenvalue problems where other type of eigenelement procedures are applicable. However, Chapter 13 of that book is devoted to Lanczos procedures for large matrices, but that discussion focuses on the 'global orthogonality is crucial' approach to Lanczos procedures and there is not much discussion of the 'local orthogonality is sufficient' approach which we use in our single-vector Lanczos procedures. Specific comments regarding the differences between these two approaches are given in Sections 2.4 and 2.5 of Chapter 2. We focus primarily on one subset of the Lanczos eigenelement procedures, the single-vector Lanczos procedures which do not use any reorthogonalization. Iterative block Lanczos procedures with limited reorthogonalization are also discussed but to a lesser extent.

This book is divided into two volumes. This volume provides the background material necessary for understanding the Lanczos procedures which we have developed and gives some perspective of the existing research on Lanczos procedures for eigenvalue or singular value computations. The second volume contains FORTRAN programs for each of the Lanczos procedures discussed in this volume. We have tried to make these volumes self-contained by including the material from matrix theory which is necessary for following the arguments given. Both volumes of this book should be accessible to engineers and scientists who have some knowledge of matrix eigenvalue problems. References are given to other books and papers where the interested reader can pursue various topics discussed.

Chapter 0 is intended as a reference chapter for the reader. Basic definitions and concepts from matrix theory which are used throughout the book are listed. Our notation is specified and special types of matrices are defined, along with special types of matrix transformations and projections.

Chapter 1 contains brief summaries of fundamental results from matrix theory which are needed in later chapters. Properties of real symmetric matrices, of Hermitian matrices, and of real symmetric generalized eigenvalue problems are summarized. Sparse matrices are discussed along with sparse matrix factorizations.

Chapter 2 begins with a description of a basic single-vector Lanczos procedure for computing eigenelements of real symmetric matrices. Properties of this procedure are derived, assuming that the computations are being performed in exact arithmetic. However, we are interested in Lanczos procedures which do not use any reorthogonalization and must therefore be concerned with what happens in finite precision arithmetic. In Section 2.3 we summarize the results obtained by Paige [1971,1972,1976,1980], assuming finite precision arithmetic. These results are the basis for the arguments which are given in Chapter 4 to justify our Lanczos procedures with no reorthogonalization. In Section 2.4 we discuss the question of constructing practical Lanczos procedures, that is, procedures which are numerically-stable in finite precision arithmetic. Section 2.5 consists of a survey of the literature on Lanczos procedures.

Chapter 3 contains proofs of several basic properties of general tridiagonal matrices, including determinant recursions and formulas for computing eigenvectors from the determinants of the matrix, along with comments on inverse iteration computations. We need these properties in Chapters 4, 5, and 6.

Chapter 4 is the main chapter of this volume. Here we develop the single-vector Lanczos procedure with no reorthogonalization for real symmetric matrices. Included is a discussion of the relationships between Lanczos tridiagonalization and the conjugate gradient method for solving systems of equations. This relationship is used to construct a plausibility argument for the belief that the 'local orthogonality is sufficient' approach is legitimate. The key to the success of these types of eigenvalue procedures, an identification test which sorts the 'good' eigenvalues from the 'spurious' ones, is developed in Section 4.5. This test is justified heuristically using the connection of the Lanczos recursion with conjugate gradient iterations. Results of numerical experiments are used to demonstrate the performance of this procedure on different types of matrices. FORTRAN code for this procedure is given in Chapter 2 of Volume 2. Chapters 3, 4, and 5 of Volume 2 contain respectively, FORTRAN codes for corresponding Lanczos procedures for Hermitian matrices, for factored inverses of real symmetric matrices, and for certain real symmetric generalized eigenvalue problems.

Chapter 5 addresses the question of constructing a single-vector Lanczos procedure for computing singular values and singular vectors of real rectangular matrices. A general discussion of basic properties of singular values and singular vectors and of the relationships between singular values and eigenvalues is given. Section 5.3 contains a very brief discussion of several applications of singular values and vectors. Section 5.4 centers on our single-vector Lanczos procedure with no reorthogonalization. Results of numerical experiments are included to demonstrate the performance of this procedure. FORTRAN code for this procedure is given in Chapter 6 of Volume 2.

Chapter 6 addresses the question of constructing a single-vector Lanczos procedure for diagonalizable complex symmetric matrices. This class of matrices is genuinely nonsymmetric, possessing none of the desirable properties of real symmetric matrices. Relevant properties of complex symmetric matrices are included. The Lanczos procedure which is proposed maps the given complex symmetric matrix into a family of complex symmetric tridiagonal matrices. FORTRAN code for this procedure is given in Chapter 7 of Volume 2.

Chapter 7 addresses the question of iterative block Lanczos procedures. First a practical implementation of the block Lanczos procedure given in Cullum and Donath [1974] is discussed. We then describe a recently-developed hybrid procedure which combines ideas from the single-vector Lanczos procedures and from the iterative block procedure. FORTRAN code for this hybrid procedure is given in Chapter 8 of Volume 2.

CHAPTER 0

PRELIMINARIES: NOTATION AND DEFINITIONS

In this preliminary chapter we have included a summary of the notation and of the basic mathematical quantities which are used repeatedly throughout this book.   Many of these definitions will also be given somewhere else in this book.   However they are provided here in a contiguous fashion as an aid to the reader.   Additional information on these quantities may be found in basic textbooks on matrix theory, for example Stewart [1973] and Parlett [1980].   No attempt is made to provide complete coverage of the basic notions of matrix theory.   Only those concepts which are actually used in the discussions in this book are included.

SECTION 0.1   NOTATION

We are dealing with finite-dimensional matrices and therefore all of the vectors and subspaces which are used are also finite-dimensional.

SCALARS   Scalar integers will be denoted by lower case Roman letters such as i, j, k, ℓ, m, and n.   Non-integer scalars will be denoted by either lower case Roman letters other than i, j, k, ℓ, m, or n, or by lower case Greek letters such as α, β, ε, λ, and σ.   The complex conjugate of a complex scalar b will be denoted by $\bar{b}$.

VECTORS   Vectors will be denoted by lower case Roman letters such as x and v.   If the vector x is a column vector, then x^T denotes the corresponding row vector and x^H denotes the corresponding row vector whose components are complex conjugates of the components of x. We call x^H the complex conjugate transpose of x.   For a real vector $x^T = x^H$.   The ith component of a vector x will be denoted by x(i).

The real and the Hermitian vector norm will be defined by

$$\| x \|_2 \equiv (\sum_{k=1}^{n} | x(k) |^2)^{1/2} .$$ (0.1.1)

$| x(k) |$ denotes the absolute value of the kth component x(k).   If x is a real vector then we have that

$$\| x \|_2^2 \equiv x^T x \equiv x^H x \equiv \sum_{k=1}^{n} [x(k)]^2.$$ (0.1.2)

If x is a complex vector then we have that

$$\| x \|_2^2 \equiv x^H x \equiv \sum_{k=1}^{n} \bar{x}(k)x(k)$$ (0.1.3)

1

Thus, we use the same symbol to denote the vector norm for both real and for complex vectors. If x is a complex vector then clearly

$$x^H x \neq x^T x. \tag{0.1.4}$$

The inner product of two real vectors x and y will be defined by $x^T y$. The inner product of two vectors x and y, at least one of which is complex, will be defined by $x^H y$. We note however that in Chapter 6 where we discuss complex symmetric matrices, we will be using the real inner product when x and y are complex vectors. In this case of course it is no longer an inner product since there are complex vectors $x \neq 0$ such that $x^T x = 0$.

We denote an ordered set of vectors $\{x_1,...,x_k\}$ by X_k and denote the span of this set of vectors by $\mathscr{R}^k \equiv sp\{X_k\}$. A vector $y \, \varepsilon \, \mathscr{R}^k$ if and only if $y = X_k u$ for some vector u. A set of vectors $X_k \equiv \{x_1, ..., x_k\}$ will be said to be an orthogonal set of vectors if and only if $x_j^T x_\ell = 0$ for $j \neq \ell$. We denote Euclidean n-space by E^n, and the corresponding complex n-space by $\mathscr{C}^n$.

MATRICES Matrices are denoted by capital Roman letters such as A, T, and X except for diagonal matrices whose diagonal entries are eigenvalues or singular values. Such diagonal matrices will be denoted by capital Greek letters such as Σ or Λ. Diagonal matrices may also be denoted by $diag\{d_1, ..., d_n\}$.

The notation

$$A = (a_{ij}) \quad \text{where} \quad 1 \leq i \leq m, \, 1 \leq j \leq n \tag{0.1.5}$$

denotes a rectangular matrix with m rows and n columns whose (i,j)-th entry is a_{ij}. The entries a_{ij} may be real-valued or complex-valued. Hermitian and complex symmetric matrices will have one or more entries which are complex. Except in our discussions of singular value/vector computations we will always be talking about square matrices. That is, m = n. In some cases we use A(i,j) to denote the (i,j)-th element in a matrix A. Thus, for any matrix A we have the following notational equivalence.

$$a_{ij} \equiv A(i,j), \quad 1 \leq i \leq m, \, 1 \leq j \leq n. \tag{0.1.6}$$

The determinant of a square matrix A will be denoted by det(A).

The matrix norms corresponding to the vector norms in Eqn(0.1.1) are defined by

$$\| A \|_2 \equiv \max_{u \neq 0} \| Au \|_2 / \| u \|_2. \tag{0.1.7}$$

We note that if the vector u is real then $\| u \|_2$ denotes $u^T u$, and if u is complex, it denotes $u^H u$. For any A we have that $\| A \|_2 = \sigma_1(A)$ where $\sigma_1(A)$ denotes the largest singular value of A.

The transpose of an mxn matrix A is denoted by A^T. If $A \equiv (a_{ij})$ then the transpose is the nxm matrix

$$A^T \equiv (a_{ji}) \tag{0.1.8}$$

obtained from A by interchanging the rows and the columns of A. The Hermitian transpose of an mxn matrix A is denoted by A^H and is defined by

$$A^H \equiv (\bar{a}_{ji}) \tag{0.1.9}$$

where the bar denotes the complex conjugate of a_{ji}. We denote the nxn identity matrix by I_n.

APPROXIMATIONS Given a function $f(z)$ where z is a vector and $f(z)$ may be a scalar or a vector-valued function, we use the notation $f(z) = O(\varepsilon^q)$ to mean that there exists a constant C such that $\| f(z) \|_2 \leq C\varepsilon^q$. If f is a scalar-valued function the norm reduces to the absolute value of f. Similarly, we use the notation $f(z) = o(\varepsilon^q)$ to mean that there exists a $\delta > 0$ such that $f(z) = O(\varepsilon^{q+\delta})$.

SECTION 0.2   SPECIAL TYPES OF MATRICES

Throughout the discussions in this book we refer repeatedly to several different types of matrices. In this section definitions of each of these types of matrices are provided.

SPARSE MATRIX An mxn matrix A will be said to be a sparse matrix if and only if each row of the matrix contains at most a few entries which are not zero. Typically in the applications, the matrices are very large, several hundred to several thousand or more in size, and sparse.

REAL SYMMETRIC MATRIX An nxn matrix A is real symmetric if and only if A is real and it is equal to its transpose. That is, $A^T = A$.

COMPLEX SYMMETRIC MATRIX An nxn matrix A is complex symmetric if and only if A is complex and symmetric. That is, $A^T = A$ and at least one of the entries in A is complex-valued. The main topic of discussion in this book is real 'symmetric' problems. However, Chapter 6 of Volume 1 and Chapter 7 of Volume 2 deal with complex symmetric matrices and single-vector Lanczos procedures with no reorthogonalization for such matrices. The spectral properties of real symmetric matrices and those of complex symmetric matrices bear little resemblance to each other as we will see in Chapter 6 of this volume.

HERMITIAN MATRICES An nxn complex matrix A is Hermitian if and only if $A^H = A$. That is, for each $1 \leq i, j \leq n$, $a_{ij} = \bar{a}_{ji}$. Therefore, the diagonal entries of any Hermitian matrix must be real. Hermitian matrices possess essentially the same spectral properties as real symmetric matrices. More will be said about this in Section 1.5 of Chapter 1.

TRIDIAGONAL MATRICES A nxn matrix $T \equiv (t_{ij})$ is tridiagonal if and only if

$$t_{ij} \equiv 0 \quad \text{for all} \quad 1 \leq i,j \leq n \quad \text{and} \quad |i - j| > 1. \tag{0.2.1}$$

That is, only the main diagonal, the first superdiagonal and the first subdiagonal may contain nonzero entries. The single-vector Lanczos procedures which we discuss will replace the given matrix problem by 'equivalent' tridiagonal problems. If T is an mxm symmetric tridiagonal matrix we denote its nonzero entries as follows. Define $\beta_1 \equiv 0$. Then define

$$\begin{aligned}
\alpha_k &\equiv T(k,k), \quad \text{for} \quad 1 \leq k \leq m \quad \text{and} \\
\beta_{k+1} &\equiv T(k+1,k) \equiv T(k,k+1) \quad \text{for} \quad 1 \leq k \leq m-1
\end{aligned} \tag{0.2.2}$$

UNITARY MATRICES A nxn matrix W is a unitary matrix if and only if W is complex and the Hermitian transpose of W is the inverse of W. That is,

$$W \text{ is complex and } W^H W = I_n. \tag{0.2.3}$$

If we write a unitary matrix W as $W = (w_1, w_2, ..., w_n)$ where w_j is the jth column of W, then we have the following orthogonality relationships. For $1 \leq j,k \leq n$,

$$w_j^H w_k = \delta_{jk} \quad \text{where} \quad \delta_{jk} = 0 \text{ if } j \neq k \text{ and } \delta_{jj} = 1. \tag{0.2.4}$$

We call δ_{jk} the Kronecker delta. Typically whenever we discuss orthogonal sets of vectors we will use the Kronecker delta notation.

ORTHOGONAL MATRIX A nxn matrix Q is an orthogonal matrix if and only if Q is real and the transpose of Q is its inverse. That is,

$$Q \text{ is real and } Q^T Q = I_n. \tag{0.2.5}$$

REAL ORTHOGONAL MATRIX A nxn matrix Q is a real orthogonal matrix if and only if Q is a complex matrix and its transpose is its inverse. That is,

$$Q \text{ is complex and } Q^T Q = I. \tag{0.2.6}$$

The class of real orthogonal matrices plays a fundamental role in the construction of our Lanczos procedures for complex symmetric matrices. These procedures (see Chapter 6) generate complex symmetric tridiagonal matrices. Real orthogonal transformations are used to obtain a storage efficient procedure for computing the eigenvalues of these complex symmetric tridiagonal matrices.

PERMUTATION MATRIX A nxn matrix P is a permutation matrix if and only if each column of P is some column of the identity matrix I_n, and every column of I_n appears as some column of P. Specifically, there exists a permutation p(i) of the integers i=1,2,...,n such that for each i

$$Pe_i = e_{p(i)}, \tag{0.2.7}$$

where e_i is the ith column of I_n and $\{p(1),p(2),.,p(n)\}$ is a permutation of the numbers $\{1,2,...,n\}$.

Permutation matrices will be useful in our discussions of generalized eigenvalue problems. Generalized eigenvalue problems involve two matrices. Our Lanczos procedures will require the factorization of one of the two matrices involved. In many cases the sparsity of a given matrix can be preserved in its factorization if the matrix is permuted prior to its factorization. Permutation matrices have the property that

$$P^T \equiv P^{-1}. \tag{0.2.8}$$

Therefore, the eigenvalues of a given matrix are preserved under the mapping of A into P^TAP. The eigenvectors of the permuted matrix are simply permutations of the eigenvectors of A.

UPPER HESSENBERG MATRICES   A nxn matrix H is an upper Hessenberg matrix if and only if all of its entries below the main subdiagonal are zero. That is, if $H \equiv (h_{ij})$ then

$$h_{ij} \equiv 0, \text{ for } i > j + 1 \text{ and } 1 \leq i, j \leq n. \tag{0.2.9}$$

In our discussions we will not have any direct need for Hessenberg matrices because even in the complex symmetric case the Lanczos matrices which we generate will be tridiagonal. In this latter situation however the EISPACK Library [1976,1977] uses Hessenberg matrices, and we will be referring to this fact in our discussions.

UPPER AND LOWER TRIANGULAR MATRICES   An nxn matrix A is upper triangular if and only if all the nonzero entries of A are on or above the main diagonal of A. An nxn matrix A is lower triangular if and only if all the nonzero entries of A are on or below the main diagonal of A. We will encounter such matrices when we are discussing factorizations of matrices. An upper or lower triangular matrix is unit upper or unit lower triangular if and only if all of the diagonal entries are 1.

DIAGONAL MATRIX   A nxn matrix D is diagonal if and only if all of the nonzero entries of D are on the main diagonal of D. That is, if

$$D \equiv (d_{ij}), \text{ then for all } 1 \leq i \neq j \leq n, \quad d_{ij} \equiv 0. \tag{0.2.10}$$

We may denote a diagonal matrix D by $D = \text{diag}\{d_1, d_2,...,d_n\}$ where d_k is used here to denote the d_{kk} entry of D. This may be somewhat confusing because we also may use the notation $A = (a_1,...,a_n)$ to denote a matrix A. Typically however, the diagonal matrices used in our discussions will be matrices whose entries are either eigenvalues or singular values. In these cases we use lower case Greek letters to denote the nonzero entries in the matrix which itself would be denoted by a capital Greek letter, and no confusion should result. For example the diagonal matrix could be denoted by Λ and its nonzero entries by λ_k. Thus, $\Lambda = \text{diag}\{\lambda_1,...,\lambda_n\}$.

SKEW SYMMETRIC MATRIX An nxn matrix $A \equiv (a_{ij})$ is skew symmetric if and only if A is real and $A^T = -A$. That is,

$$\text{for } 1 \leq i,j \leq n, \quad a_{ij} = -a_{ji}. \tag{0.2.11}$$

In particular, the diagonal entries of any skew symmetric matrix must all be zero, and any skew symmetric matrix of odd order must be singular. The nonzero eigenvalues of a skew symmetric matrix are pure imaginary pairs, i.e. of the form $\pm ir$ where r is real and positive. Therefore the related matrix iA has all real roots; in fact it is a Hermitian matrix.

ADJOINT OF A MATRIX If A is a nxn matrix then its adjoint Adj(A) is defined by the following relationships.

$$\text{Adj}(A)A = A\text{Adj}(A) \equiv \det(A)I_n. \tag{0.2.12}$$

If A is nonsingular so that $\det(A) \neq 0$, then $A^{-1} \equiv \text{Adj}(A)/\det(A)$. If the matrix A is real symmetric and has rank (n-1) and u is a n-vector such that $u \neq 0$ but $Au = 0$, then $\text{Adj}(A) = \gamma uu^T$ for some scalar multiple γ. In particular if λ is an eigenvalue of A and the matrix $(A - \lambda I)$ has rank (n-1), then the corresponding eigenvector x of A is in the null space of this matrix. Thus, we have that

$$\text{Adj}(A - \lambda I) = \gamma xx^T. \tag{0.2.13}$$

We will need this property in Chapter 3 in our discussions of tridiagonal matrices. It will prove to be very valuable in some of our arguments justifying our approach to single-vector Lanczos procedures with no reorthogonalization.

PRINCIPAL SUBMATRIX OF A MATRIX Given a nxn matrix A, a kxk matrix B is a principal submatrix of A of order k if and only if B is the submatrix which consists of all of the elements of A which are simultaneously in both the rows and the columns $i_1 < i_2 < ... < i_k$ of A.

LEADING PRINCIPAL SUBMATRIX OF A MATRIX Given a nxn matrix A, a kxk matrix B is the leading principal submatrix of A of order k if and only if B is the submatrix which consists of all of those elements of A which are simultaneously in both the first k rows and the first k columns of A. Leading principal submatrices are central to the Sturm sequencing computations used to compute eigenvalues of real symmetric tridiagonal matrices.

SECTION 0.3   SPECTRAL QUANTITIES

In this section we summarize the basic definitions needed in discussing eigenvalue and eigenvector problems for matrices, as well as those needed for the discussions of singular value and vector problems given in Chapter 5 of Volume 1 and Chapter 6 of Volume 2.

EIGENVALUES AND EIGENVECTORS   Given any nxn matrix A, a scalar λ and nonzero vector x are called an eigenvalue and corresponding right eigenvector of A if

$$Ax = \lambda x. \tag{0.3.1}$$

Similarly, a nonzero vector y is called a left eigenvector associated with the eigenvalue λ if

$$A^T y = \lambda y. \tag{0.3.2}$$

Any nxn matrix has n eigenvalues; they may not all be distinct. However, a matrix need not have a full set of right and left eigenvectors. If A is a real symmetric matrix, then its left and right eigenvectors are identical and A has a full complement of eigenvectors. In fact for any real symmetric matrix there is an orthonormal set of eigenvectors which is a basis for the entire space E^n.

Given any two nxn matrices A and B we can define the generalized eigenvalue problem as follows. Determine scalars λ and nonzero vectors x such that

$$Ax \equiv \lambda Bx. \tag{0.3.3}$$

Such problems are considered briefly in Chapter 1 and Chapter 4 but only for the case when A and B are both real symmetric matrices and either A or B is a positive definite matrix. Volume 2 Chapter 5 contains a single-vector Lanczos procedure for such problems which is applicable whenever the Cholesky factorization $B = LL^T$ is available.

SPECTRUM OF A MATRIX   The set of eigenvalues of any given nxn matrix A is called the spectrum of A. If a matrix has all real eigenvalues, then the extreme eigenvalues of A are defined as those eigenvalues which are the algebraically-largest or the algebraically-smallest. Typically for such a matrix we write the eigenvalues as

$$\{\lambda_1 \geq \lambda_2 \geq \dots \geq \lambda_n\} \tag{0.3.4}$$

with the algebraically-largest eigenvalues assigned low indices and the algebraically-smallest eigenvalues assigned the highest indices.

POSITIVE DEFINITE MATRICES   An nxn real symmetric or Hermitian matrix A is said to be positive definite if and only if all of its eigenvalues are positive. If all of its eigenvalues are nonnegative then it is said to be positive semidefinite. There are analogous definitions for negative definite and negative semidefinite matrices. Matrices which are not definite or semidefinite are called indefinite.

NONDEFECTIVE MATRIX An nxn matrix A is said to be a nondefective matrix if and only if it is diagonalizable. That is there exists a nonsingular matrix X and a diagonal matrix $\Lambda = \text{diag}\{\lambda_1, \lambda_2, ., \lambda_n\}$ such that

$$A \equiv X\Lambda X^{-1}. \tag{0.3.5}$$

If this is the case then clearly $AX = X\Lambda$ and $A^T X^{-T} = X^{-T}\Lambda$. Therefore, the columns of X must be right eigenvectors of A, the columns of X^{-T} must be left eigenvectors of A, and the λ_i are the eigenvalues of A. Any real symmetric, Hermitian, or skew symmetric matrix is nondefective. Furthermore, any nxn matrix with n distinct eigenvalues is nondefective (diagonalizable).

SPECTRAL DECOMPOSITION OF A MATRIX If A is a nxn nondefective matrix then from Eqn(0.3.5) we have the following decomposition of A into a sum of rank 1 matrices.

$$A \equiv \sum_{k=1}^{n} \lambda_k x_k y_k^T \text{ where } Y^T = X^{-1}. \tag{0.3.6}$$

If the matrix A is real symmetric, then X can be taken to be an orthogonal matrix and $A = X\Lambda X^T$. If the matrix A is Hermitian, then X can be taken to be a unitary matrix and $A = X\Lambda X^H$.

RAYLEIGH QUOTIENTS Let A be a nxn real symmetric (Hermitian) matrix with eigenvalues $\{\lambda_1, ..., \lambda_n\}$ and let $X \equiv \{x_1, ..., x_n\}$ be a corresponding orthonormal (unitary) basis of eigenvectors. For a real symmetric matrix A and any nonzero vector x we define the Rayleigh quotient $\rho(x,A)$ of A and x as follows.

$$\rho(x,A) \equiv (x^T Ax)/(x^T x). \tag{0.3.7}$$

If A is Hermitian we replace the real inner products in Eqn(0.3.7) by the corresponding complex inner product. We may use $\rho(x)$ to denote the Rayleigh quotient when it is clear from the context which matrix A is being used. For any nonzero vector x we have that

$$\lambda_n(A) \leq \rho(x,A) \leq \lambda_1(A). \tag{0.3.8}$$

For any eigenvalue-eigenvector pair (λ_j, x_j), $\rho(x_j) = \lambda_j$.

The eigenvalues of a given real symmetric or Hermitian matrix A have the following minmax characterization as the minimums of the maximums of Rayleigh quotients over certain size subspaces. In Eqn(0.3.9) $z \neq 0$ and $1 \leq j \leq n$.

$$\lambda_{n-j+1} = \min_{\mathbf{Z}^j} \max_{z \in \mathbf{Z}^j} \rho(z) \tag{0.3.9}$$

where $\mathbf{Z}^j$ denotes an arbitrary j-dimensional subspace. There is an analogous characterization using the max-min and subspaces $\mathbf{Z}^{(n-j+1)}$. See for example, Stewart [Chapter 6, 1973] for a proof of this well-known result.

If for some vector y and some eigenvector x_j corresponding to an eigenvalue λ_j we have that

$$y = x_j + O(\varepsilon) \text{ then } \rho(y) = \lambda_j + O(\varepsilon^2). \tag{0.3.10}$$

Eqn(0.3.10) states that if we have an eigenvector approximation y of order ε, then the corresponding Rayleigh quotient is an approximation to the corresponding eigenvalue of order ε^2. This property has been used to justify Rayleigh-Ritz methods for computing eigenvalues and eigenvectors of real symmetric or Hermitian matrices. Such methods are iterative and on each iteration include the computation of an approximate eigenvector (typically obtained using some form of inverse iteration with the current eigenvalue approximation) followed by the subsequent computation of the corresponding Rayleigh quotient which then becomes the updated approximation to the eigenvalue. The single-vector Lanczos procedures which we describe in Chapters 4, 5 and 6 are not Rayleigh-Ritz type algorithms. However, the iterative block Lanczos procedures which we describe in Chapter 7 can be interpreted as block analogs of Rayleigh-Ritz procedures.

INVERSE ITERATION Let A be a nxn matrix. Given any scalar λ which is a good approximation to a simple eigenvalue of A, the method of inverse iteration provides a mechanism for computing an approximation to the corresponding eigenvector of A. For any such λ and any given random n-vector w, the following ill-conditioned set of equations is solved for z.

$$(A - \lambda I)z = w. \tag{0.3.11}$$

The starting vector w is scaled so that $\sum_{k=1}^{n} |w(k)| = \gamma\varepsilon_{mach}$ where ε_{mach} is the machine epsilon for the computer being used and γ is a scaling obtained from A. (The machine espilon is the smallest positive number such that $1.0 + \varepsilon_{mach} \neq 1.0$). If the computed vector z satisfies the condition $\sum_{k=1}^{n} |z(k)| \geq 1$, then z is accepted as an approximate eigenvector associated with λ and the approximate eigenvector $y \equiv z/\|z\|_2$. Otherwise z replaces w in Eqn(0.3.11), normalized as before, and another step of inverse iteration is performed. Only a small number of iterations is allowed. Typically one iteration is sufficient.

INVARIANT SUBSPACES Let $\mathcal{X}^k \equiv \text{sp}\{X_k\}$, then $\mathcal{X}^k$ is an invariant subspace of a given matrix A if and only if for any $x \in \mathcal{X}^k$ and the corresponding $y = Ax$, we have that $y \in \mathcal{X}^k$. For any set of eigenvectors $X_k \equiv \{x_1, ..., x_k\}$ of a matrix A the corresponding subspace $\mathcal{X}^k$ is an invariant subspace of A.

SINGULAR VALUES OF MATRICES Let A be a real ℓxn matrix with $\ell \geq n$. A scalar σ and two nonzero vectors x and y are respectively a singular value, a right singular vector and a left singular vector of A if and only if

$$Ax = \sigma y, \text{ and } A^T y = \sigma x. \tag{0.3.12}$$

If A is complex then A^T has to be replaced by A^H in Eqn(0.3.12). Typically we label the singular values as follows $\sigma_1 \geq \sigma_2 \geq ... \geq \sigma_n$, and denote corresponding right and left singular vectors by $X \equiv \{x_1, ..., x_n\}$ and $Y \equiv \{y_1, ..., y_\ell\}$.

We have the following singular value decomposition for any ℓxn matrix. There exists an ℓxℓ unitary matrix Y of left eigenvectors and a nxn unitary matrix X of right eigenvectors such that

$$A \equiv Y\Sigma X^H \text{ where } \Sigma = \begin{bmatrix} \Sigma_1 \\ 0 \end{bmatrix} \qquad (0.3.13)$$

with $\Sigma_1 = \text{diag}\{\sigma_1, ..., \sigma_n\}$. When A is a real matrix X and Y are orthogonal matrices. This is the generalization of the spectral decomposition for diagonalizable matrices given in Eqn(0.3.5) to an arbitrary matrix. Observe that $\|A\|_2 \equiv \sigma_1(A)$.

ILL-CONDITIONED MATRIX An nxn nonsingular matrix A is said to be ill-conditioned if small changes in a vector b cause relatively large changes in the solution x of the corresponding system of equations Ax = b. One commonly-used measure of ill-conditioning is the ratio of the largest to the smallest singular value of A, $\sigma_1(A)/\sigma_n(A)$. This measure applies equally well to a mxn rectangular matrix where this ratio uses the smallest nonzero singular value.

SECTION 0.4   TYPES OF MATRIX TRANSFORMATIONS

In this section we briefly describe each type of matrix transformation which will be used somewhere in this book.

UNITARY TRANSFORMATION Given a nxn matrix A and a nxn unitary matrix U, the matrix

$$A' \equiv U^H A U \qquad (0.4.1)$$

obtained from A is called a unitary transformation of A. Unitary transformations preserve the norms of the vectors they are transforming. This property is very desirable numerically because it controls the sizes of the quantities encountered. Many of the eigenelement procedures in the EISPACK Library [1976,1977] are primarily sequences of unitary or orthogonal transformations.

ORTHOGONAL TRANSFORMATION Given a nxn matrix A and a nxn orthogonal matrix W, the matrix

$$A' \equiv W^T A W \qquad (0.4.2)$$

obtained from A is called an orthogonal transformation of A. Unitary and orthogonal transformations preserve eigenvalues and singular values. In either case the resulting transformed matrix A' has the same eigenvalues and singular values as the original matrix. This is important for both numerical and theoretical considerations.

SIMILARITY TRANSFORMATION Given a nxn matrix A and a nxn nonsingular matrix S, the matrix

$$A' \equiv S^{-1}AS \tag{0.4.3}$$

obtained from A is called a similarity transformation of A. Similarity transformations preserve the eigenvalues of the given matrix. Furthermore, we have that for any eigenvalue and eigenvector pair (λ, x) of A that $(\lambda, S^{-1}x)$ is an eigenvalue and eigenvector pair of A'.

ROW PERMUTATION Given a nxn matrix A and a nxn permutation matrix P, the matrix

$$A' \equiv PA \tag{0.4.4}$$

obtained from A is called a row permutation of A. That is, the rows of A' are simply a permutation of the rows of A. Specifically, P defines some permutation p(i) of the integers $1 \le i \le n$, and the ith row of A' is simply the p(i)-th row of A.

COLUMN PERMUTATION Given a nxn matrix A and a nxn permutation matrix P, the matrix

$$A' \equiv AP \tag{0.4.5}$$

obtained from A is called a column permutation of A. That is, the columns of A' are simply a permutation of the columns of A. Specifically, P defines some permutation p(i) of the integers $1 \le i \le n$ and the ith column of A' is simply the p(i)-th column of A.

PLANE ROTATION Let A be a nxn real symmetric matrix. Select real scalars c and s such that $c^2 + s^2 = 1$. Select integers k and ℓ such that $1 \le k < \ell \le n$. Define the associated plane rotation matrix $R \equiv (r_{ij})$ as follows.

$$\begin{aligned} r_{kk} &\equiv r_{\ell\ell} \equiv c, \\ r_{\ell k} &\equiv -r_{k\ell} \equiv s, \\ r_{ij} &\equiv \delta_{ij}, \text{ for } i, j \ne k, \ell. \end{aligned} \tag{0.4.6}$$

The matrix

$$A' = R^T A R, \tag{0.4.7}$$

is called a plane rotation of the matrix A. Observe that only rows and columns k and ℓ of A are modified. Specifically we have that

$$a'_{ik} = a'_{ki} = ca_{ki} + sa_{\ell i}$$
$$a'_{i\ell} = a'_{\ell i} = ca_{\ell i} - sa_{ki}$$
$$a'_{kk} = a_{kk} + sb, \quad a'_{\ell\ell} = a_{\ell\ell} - sb \qquad (0.4.8)$$
$$a'_{k\ell} = a'_{\ell k} = -a_{k\ell} + cb$$
$$\text{where} \quad b = 2ca_{k\ell} + s(a_{\ell\ell} - a_{kk}).$$

Any plane rotation matrix R is real symmetric and orthogonal. In Chapter 6 in our discussion of Lanczos methods for complex symmetric matrices we find it necessary to use a complex generalization of these real plane rotation matrices. There we consider 'rotation' matrices defined by the following equations.

$$r_{\ell\ell} \equiv -r_{kk} \equiv c,$$
$$r_{\ell k} \equiv r_{k\ell} \equiv s, \qquad (0.4.9)$$
$$r_{ij} \equiv \delta_{ij}, \quad \text{for} \quad i, j \neq k, \ell.$$

where the scalars c and s are complex-valued. These matrices are complex symmetric and real orthogonal. See Chapter 6 for more details.

TRIANGULAR FACTORIZATION   Given a nxn matrix A we define the triangular factorization A = LU of A where L is a lower triangular matrix and U is a unit upper triangular matrix. The entries in L and U are defined by the following recursions. For each k=1,2,...,n compute

$$\ell_{ik} \equiv a_{ik} - \sum_{j=1}^{k-1} \ell_{ij}u_{jk}, \quad \text{for} \quad k \leq i \leq n$$
$$u_{ki} \equiv (a_{ki} - \sum_{j=1}^{k-1} \ell_{kj}u_{ji}) / \ell_{kk}, \quad \text{for} \quad k+1 \leq i \leq n. \qquad (0.4.10)$$

There are matrices for which this factorization is not well-defined. However, if we assume that all of the leading principal submatrices of A are nonsingular, then this triangularization is well-defined. If for each k = 1,...,n we denote the kxk leading principal submatrix of A, L, and U by respectively, A_k, L_k, and U_k, then we have that for each k, $A_k = L_k U_k$. That is, the factors of the leading principal submatrices of A are obtained from the leading principal submatrices of the factors of A. An alternative form for the factorization is A = $\bar{L}$DU where $\bar{L}$ is a unit lower triangular matrix. This form can be obtained by replacing ℓ_{ik} in Eqn(0.4.10) by ℓ_{ik}/ℓ_{kk} and defining D $\equiv$ diag$\{\ell_{11}, ..., \ell_{nn}\}$. If A is a symmetric matrix then U = $\bar{L}^T$.

CHOLESKY FACTORIZATION   For any positive definite symmetric matrix A the triangular factorization of A is called the Cholesky factorization. In this case the formulas in Eqn(0.4.10) reduce to the following. For each $k=1,2,...,n$ compute

$$\ell_{kk} \equiv (a_{kk} - \sum_{j=1}^{k-1} \ell_{kj}^2)^{1/2} > 0 \text{ and}$$

$$\ell_{ik} \equiv (a_{ik} - \sum_{j=1}^{k-1} \ell_{ij}\ell_{kj}) / \ell_{kk}, \text{ for } k<i\leq n. \tag{0.4.11}$$

The associated factorization is $A = LL^T$ where $L \equiv (\ell_{ik})$ defined in Eqn(0.4.11).

HOUSEHOLDER TRANSFORMATION   Let u be a n-vector such that $u^H u = 1$. Then

$$Q \equiv I - 2uu^H \tag{0.4.12}$$

is called a Householder transformation matrix.   Q is a Hermitian unitary matrix.   That is, $Q^H = Q^{-1} = Q$. If u is a real vector then u^T is used and Q is real symmetric and orthogonal. For any such Q, the corresponding matrix $A' \equiv QAQ$ has the same eigenvalues as A.   Given any complex nxn matrix A, a sequence of n Householder transformations can be used to reduce A to upper Hessenberg form.   If the matrix A is Hermitian or real symmetric, then the resulting matrix can be made to be real symmetric and tridiagonal.   Householder transformations are used in many of the EISPACK Library programs [1976,1977].

GRAM-SCHMIDT ORTHOGONALIZATION   Let $Z \equiv \{z_1, z_2, ..., z_n\}$ be a set of linearly independent mx1 real vectors.   The Gram-Schmidt orthogonalization procedure generates a corresponding orthogonal set of vectors $Q = \{q_1, q_2, ..., q_n\}$ such that

$$sp\{Q\} \equiv sp\{Z\}, \tag{0.4.13}$$

and $Z = QR$ where $R \equiv (r_{ij})$ is a unit upper triangular matrix.   The vectors q_k and the matrix R are defined recursively by the following equations.   Define $q_1 \equiv z_1$. Then for each $k=2,...,n$ define

$$q_k = z_k - \sum_{i=1}^{k-1} r_{ki}q_i, \text{ where}$$

$$r_{ki} \equiv z_k^T q_i / d_i, \text{ and } d_i \equiv q_i^T q_i. \tag{0.4.14}$$

Observe that $Q^T Q = D \equiv diag\{d_1, ..., d_n\}$.   The error propagation which occurs when Eqns(0.4.14) are implemented on a computer is such that they are seldom used in practice. See for example Bjorck [1967] for a discussion of this problem.   Typically the modified form of the Gram-Schmidt orthogonalization which is defined in the next paragraph is used.

MODIFIED GRAM-SCHMIDT   The following modification of the basic Gram-Schmidt procedure yields an orthogonalization procedure with better numerical stability than the Gram-Schmidt procedure as defined in Eqns(0.4.14).   Let $Z = \{z_1, ..., z_n\}$ be a set of linearly independent mx1 real vectors.   The modified Gram-Schmidt orthogonalization procedure generates a corresponding orthogonal set of vectors $Q = \{q_1, q_2, ..., q_n\}$ such that

$sp\{Q\} \equiv sp\{Z\}$. using the following recursions. The vectors q_k and the matrix R are defined recursively by the following equations.

For each k define $q_k = z_k^{(k)}$ where

$$z_j^{(k+1)} \equiv z_j^{(k)} - r_{kj}q_k, \text{ for } j>k. \tag{0.4.15}$$

$$r_{kj} \equiv q_k^T z_j^{(k)}/d_k \text{ and } d_k \equiv q_k^T q_k.$$

Column interchanges (pivoting) can be used in the orthogonalization process. In particular at any stage in the process the vector in the remaining subset of vectors which has maximal norm can be normalized. That is at stage k, choose vector $z_s^{(k)}$ such that

$$\| z_s^{(k)} \|_2 = \max_{k \leq j \leq n} \| z_j^{(k)} \|_2. \tag{0.4.16}$$

SECTION 0.5   SUBSPACES, PROJECTIONS, AND RITZ VECTORS

The terms Lanczos recursion, Krylov subspaces, projections of matrices, and Ritz vectors and values are used repeatedly throughout this book. In this section we give basic definitions for each of these concepts.

KRYLOV SUBSPACES   Given a nxn matrix A and a vector v_1 we define the family of Krylov subspaces $\mathcal{K}^k$, $k = 1,2,...,n$, as follows.

$$\mathcal{K}^k \equiv sp\{v_1, Av_1, A^2v_1, ..., A^{k-1}v_1\}. \tag{0.5.1}$$

As we will see in Chapter 2, theoretically for any given real symmetric matrix A, the real symmetric Lanczos recursion defined below in Eqn(0.5.3) generates orthonormal bases for the Krylov subspaces corresponding to the starting vector v_1 used in the Lanczos recursion. (In practical implementations of this recursion this will not be true. We discuss the numerical problems in Section 2.3 in Chapter 2.)

ORTHOGONAL PROJECTIONS OF MATRICES   Let A be a real nxn matrix. Let V be a nxm matrix whose columns are orthonormal. Then the matrix

$$B \equiv V^T A V \tag{0.5.2}$$

is an orthogonal projection of A onto the space spanned by the columns of V. Theoretically, for a given real symmetric matrix A the Lanczos recursion in Eqn(0.5.3) generates tridiagonal matrices which are the orthogonal projections of A onto the corresponding subspaces $sp\{V_m\}$, spanned by the Lanczos vectors generated. These subspaces are in fact Krylov subspaces. In particular $sp\{V_m\} = \mathcal{K}^m(v_1,A)$. See Eqns(0.5.3) below.

BASIC LANCZOS RECURSION   Let A be a nxn real symmetric matrix. Let v_1 be a starting vector. Define Lanczos vectors v_j, $j = 2,...,m$ and Lanczos matrices T_m for $m = 1,2,...,M$ as follows. Define $\beta_1 \equiv 0$, and $v_0 \equiv 0$. Then for $k=1,2,...,M$

$$\beta_{k+1}v_{k+1} = Av_k - \alpha_k v_k - \beta_k v_{k-1},$$
$$\alpha_k = v_k^T Av_k \ \text{and} \ \beta_{k+1} = v_{k+1}^T Av_k \tag{0.5.3}$$

Let $V_m \equiv (v_1, v_2, ... v_m)$ be the matrices of Lanczos vectors generated. Then in matrix form Eqn(0.5.3) becomes

$$AV_m = V_m T_m + \beta_{m+1}v_{m+1}e_m^T \tag{0.5.4}$$

where T_m is the real symmetric, tridiagonal Lanczos matrix of order m generated. We have that

$$T_m(k,k) \equiv \alpha_k, \ \text{for} \ 1 \leq k \leq m$$
$$T_m(k,k+1) \equiv T_m(k+1,k) \equiv \beta_{k+1} \ \text{for} \ 1 \leq k \leq m-1.$$

LANCZOS/RITZ VECTORS Let A be a nxn real symmetric matrix. Refer to Eqns(0.5.3) for the definition of the basic single-vector Lanczos recursion. If the basic Lanczos recursion for $k = 1, ..., m$ or a variant of it is used to generate a Lanczos tridiagonal matrix T_m, then for any eigenvalue-eigenvector pair (μ, u) of T_m we define a corresponding Ritz vector

$$y \equiv V_m u. \tag{0.5.5}$$

Eigenvalues of any Lanczos matrix will be called Ritz values for the corresponding matrix A and the vectors obtained as in Eqn(0.5.5) from the corresponding eigenvectors of the Lanczos matrix will be called Ritz vectors for A. It is important to note that for the single-vector Lanczos procedures presented in Chapter 4 it will not be the case that every eigenvalue-eigenvector pair of every Lanczos matrix generated yields a Ritz value and Ritz vector which are approximations to eigenvalues and eigenvectors of A. We will see that in general it is only appropriate to use certain subsets of the eigenvalues of the Lanczos matrices as approximations to eigenvalues of A. Please see Chapter 4 for detailed comments.

SECTION 0.6   MISCELLANEOUS DEFINITIONS

In this section we include a definition which does not fit naturally into the earlier sections.

CHEBYSHEV POLYNOMIALS Error estimates for various Lanczos procedures can be obtained (at least in the case when we assume that we are using infinite precision arithmetic) by using the special properties of the Krylov subspaces generated by the Lanczos recursions. These estimates utilize the Chebyshev polynomials, $\mathscr{T}_k$, $k = 1, ..., m$. These polynomials are defined as follows. For any $k > 0$,

$$\mathscr{T}_k(x) \equiv \cos(k \ \text{arccos} x), \ \text{for} \ |x| \leq 1,$$
$$\mathscr{T}_k(x) \equiv \cosh(k \ \text{arccosh} x) \ \text{for} \ |x| \geq 1. \tag{0.6.1}$$

Chebyshev polynomials satisfy the following recursion. (This is actually a trigonometric identity.)

$$\mathscr{T}_{k+1}(x) \;=\; 2x\mathscr{T}_k(x) \,-\, \mathscr{T}_{k-1}(x). \tag{0.6.2}$$

These polynomials have the special property that if we define the corresponding polynomials $p_k(x) \equiv \mathscr{T}_k(x)/2^{k-1}$, then $p_k(x)$ is the monic polynomial of degree k with the smallest maximum absolute value over the interval $[-1,1]$. This maximum is simply $1/2^{k-1}$. The error estimates obtained using these polynomials are so-called 'worst' case estimates. That is, they provide estimates of the maximum error which can occur and do not necessarily give indications of the average error incurred. For more details about Chebyshev polynomials, see for example, Hamming [Chapter 28, 1973] and Parlett [Appendix B, 1980].

CHAPTER 1

REAL 'SYMMETRIC' PROBLEMS

SECTION 1.1   REAL SYMMETRIC MATRICES

This book focuses on the solution of real symmetric eigenvalue problems. This includes not only real symmetric matrices but also includes other classes of problems which are equivalent to eigenvalue problems for real symmetric matrices. One primary example of such an equivalence is the computation of singular values and vectors of real rectangular matrices. Lanczos [1961] suggested a simple mechanism for obtaining a real symmetric eigenvalue problem from any real rectangular singular value problem, and we take advantage of this mechanism to obtain a single-vector Lanczos procedure for computing singular values and vectors of such matrices. This is discussed in Chapter 5. Hermitian matrices provide another example. As the reader will see in Chapter 4, it is possible to define a Lanczos procedure which reduces the required Hermitian computations to computations on real symmetric tridiagonal matrices.

A is a real symmetric nxn matrix if and only if it is real and symmetric. That is $A^T = A$. This class of matrices is the most well-behaved and thus the 'easiest' to handle numerically. All of the eigenvalues of a real symmetric matrix are real and all of the eigenvectors are real. We denote the eigenvalues of A by $\{\lambda_1 \geq \lambda_1 \geq ... \geq \lambda_n\}$ and denote any corresponding set of eigenvectors by $X = \{x_1, x_2, ..., x_n\}$. For any real symmetric (or Hermitian) matrix $\|A\|_2 = \max\{|\lambda_1|, |\lambda_n|\}$.

Real symmetric matrices are discussed in detail in Stewart [1973] and Parlett [1980]. There are three principal properties of real symmetric matrices which we will need in our discussions. First, real symmetric matrices have complete eigensystems. The dimension of the eigenspace corresponding to each eigenvalue of the matrix A is the same as the multiplicity of that eigenvalue as a root of the characteristic polynomial of A. Second, for any two distinct eigenvalues λ and μ of A and corresponding eigenvectors x and y, $x^T y = 0$. Thus, eigenvectors corresponding to different eigenvalues are orthogonal, and we can therefore construct an eigenvector basis which is orthonormal. Third, small symmetric perturbations in any real symmetric matrix cause only small perturbations in the eigenvalues. Corresponding perturbations in the eigenvectors must be viewed in terms of perturbations on eigenspaces not as perturbations on individual vectors. If there exists a decent gap between two eigenvalues λ_j and λ_{j+1} of A, then small symmetric perturbations in A will yield small perturbations in the eigenspaces corresponding to the two sets of eigenvalues $\Lambda_1 \equiv \{\lambda_1, ..., \lambda_j\}$ and $\Lambda_2 \equiv \{\lambda_{j+1}, ..., \lambda_n\}$.

We are interested in very large real symmetric matrices. Such matrices arise in a variety of applications. The following incomplete list indicates how diverse the application areas are.

17

CLUSTERING ANALYSIS

1. Computation of eigenvectors for use in constructing a model of a power system for use in transient stability studies. See for example, Chow et al [1984].

2. Calculation of eigenvectors for use in the partitioning of graphs which represent the interconnections between electrical circuits which are to be placed upon a silicon chip for use in an electronic device. The objective is to partition the graph (i.e. circuits) in such a way as to minimize the number of edges (connections between the circuits) which connect the resulting subgraphs of the partition to each other. See for example Barnes [1982].

PHYSICS

1. Computation of eigenvalues and eigenvectors of the Schrodinger equation for use in the comparison of various theories of quantum mechanical interactions with experimental observations. See for example, Haydock [1980].

2. Eigenvalue and eigenvector computations for matrices which model the arrangement of atoms in a disordered material. The objective is to determine various physical properties of such materials. See for example Kirkpatrick [1979], Dean [1972], and Stein and Krey [1979].

3. Eigenvalue and eigenvector computations used in attempts to understand the Jahn-Teller effect. This effect depends upon the coupling between electronic and nuclear (vibrational) motion within the nucleus of an atom. See for example, Sears et al [1981].

4. Eigenvalue and eigenvector computations used in studies of the many-body nuclear system of an atom. See for example Whitehead et al [1977].

5. Eigenvalue and eigenvector computations used to estimate the atomic densities of states near the surfaces in semiconductors. See for example Pandey [1983].

CHEMISTRY

1. Eigenvalue computations for the calculation of bond energies in molecules. See for example, Nesbet [1981] and Davidson [1983].

STRUCTURAL MECHANICS

1. Eigenvalue and eigenvector computations for the vibrational or the buckling analysis of structures. Typically these problems are generalized eigenvalue problems. See for example Jennings [1977] and NASTRAN [1977].

OCEANOGRAPHY

1. Eigenvalue computations for the analysis of models of ocean tides.   See for example Platzman [1978] and Cline et al [1976].

The power systems application in Chow et al [1984] requires the computation of a number of the algebraically-smallest eigenvalues and corresponding eigenvectors of a large sparse real symmetric matrix which describes the power system being analyzed.   The computed eigenvectors are used to partition the network of electric generators and buses in such a way that the connections between the resulting disjoint subsets of machines and buses are weak.   The objective is to obtain a subdivision such that if an electrical fault occurs within one of the subgroups, the effects of this fault can be studied by modelling that subgroup in detail but treating each of the other subgroups (in terms of their interactions with the subgroup where the fault occurred) as simple average 'machines'.   In Chow et al [1984] a matrix of size 2161 was used to model a hypothetical but realistic grid of generators and buses for the Western United States.   The eigenvectors required for the analysis in Chow et al [1984] were computed using the version of our single-vector Lanczos procedure given in Volume 2, Chapter 4.

SECTION 1.2   PERTURBATION THEORY

We simply restate three well-known results.   See Stewart [Chapter 6, 1973] for more details.

THEOREM 1.2.1   Stewart [Chapter 6, 1973]   Let A be a nxn Hermitian (real symmetric) matrix.   Let $A' \equiv A + E$ where E is a Hermitian (real symmetric) perturbation of A.   Define the eigenvalues of A as $\{\lambda_1 \geq ... \geq \lambda_n\}$ and the eigenvalues of A' by $\{\lambda'_1 \geq ... \geq \lambda'_n\}$.   Then for each i=1,2,...,n we have that

$$\lambda_i - \|E\|_2 \leq \lambda'_i \leq \lambda_i + \|E\|_2. \qquad (1.2.1)$$

We have that $\|E\|_2 = \max\{|\lambda_1(E)|, |\lambda_n(E)|\}$ for any Hermitian or real symmetric matrix E.   Therefore, the eigenvalues of a matrix resulting from such a perturbation of a Hermitian or of a real symmetric matrix cannot differ from the eigenvalues of the original matrix by more than the largest eigenvalue of the perturbation matrix E.   Thus if the norm of the perturbation matrix is small compared to the norm of A, the changes in the eigenvalues must also be small.   Note that these are absolute changes not relative changes.   If a matrix has both very large and very small eigenvalues, then the relative changes in the smaller eigenvalues may be large.

Stewart [Chapter 6, 1973] gives the following theorem for the perturbations in the corresponding eigenvectors.   Davis and Kahan [1970] provide generalizations of this result to subspaces of eigenvectors of A and A'.   These generalizations do not look at changes in individual eigenvectors but look at the angles between subspaces defined by the eigenvectors of A and subspaces defined by the corresponding eigenvectors of the perturbed matrix A'. Such generalizations are included in the discussions in Section 2.2 of Chapter 2.

THEOREM 1.2.2 Stewart [Chapter 6, 1973] If under the hypotheses of Theorem 1.2.1, λ and x denote an eigenvalue-eigenvector pair of A, then for any Hermitian (real symmetric) perturbation matrix E, there is an eigenvalue-eigenvector pair λ' and x' of $A' \equiv A + E$ which satisfies the following inequality.

$$\| x - x' \|_2 \leq \| E \|_2 / \delta + O(\| E \|_2^2),$$
$$\text{where } \delta \equiv \min_{\lambda_i \neq \lambda} | \lambda - \lambda_i |$$

(1.2.2)

Theorem 1.2.2 states that for any simple isolated eigenvalue λ of a Hermitian (real symmetric) matrix A with a corresponding eigenvector x, a Hermitian (real symmetric) perturbation E of A of size $\| E \|_2$ yields a matrix A' with a corresponding eigenvector x' which varies from x by at most as much as the ratio of the largest eigenvalue of E to the gap between that eigenvalue and the other eigenvalues of A. We state one more perturbation result which is useful in practice.

THEOREM 1.2.3 Hoffman and Wielandt [1953] If under the hypotheses of Theorem 1.2.1 $\{\gamma_1 \geq ..., \geq \gamma_n\}$ denote the eigenvalues of the perturbation matrix E, then we have that

$$\sum_{k=1}^{n} (\lambda_k - \lambda'_k)^2 \leq \sum_{k=1}^{n} \gamma_k^2.$$

(1.2.3)

SECTION 1.3   RESIDUAL ESTIMATES OF ERRORS

For each of the Lanczos procedures we discuss it is necessary to compute error estimates for the eigenvalue and eigenvector approximations in order to determine when the desired convergence has been achieved. These error estimates are obtained by either computing the norms of various residuals which arise in the computations or by computing quantities which are good estimates of the norms of residuals. For the single-vector Lanczos procedures which we discuss in Chapters 4, 5 and 6, error estimates are obtained by computing eigenvectors of certain Lanczos matrices and using the sizes of the last components of these eigenvectors as estimates on the sizes of the associated residuals. For details see Section 4.7 in Chapter 4. In the block Lanczos procedures discussed in Chapter 7, norms of residuals are computed as a natural part of the block generation on each iteration. See Chapter 7 for details. The theorems in this section relate the sizes of the norms of certain residuals to errors in associated eigenvalue and eigenvector approximations.

THEOREM 1.3.1 Parlett [Chapter 11, 1980] Let A be a nxn Hermitian or real symmetric matrix. Let $Q \equiv (q_1, ..., q_k)$ be a nxk matrix of unitary vectors. That is, $Q^H Q = I_k$. Define the kxk projection matrix $H \equiv Q^H A Q$ and let $\{\gamma_1 \geq ... \geq \gamma_k\}$ denote the eigenvalues of H. Then if we define the corresponding block residual matrix $R \equiv AQ - QH$, we have the following relationship between the eigenvalues of A and those of H as measured by the spectral norm of the residual matrix R. There are k of the eigenvalues of A, $\{\lambda_{i_1} \geq .... \geq \lambda_{i_k}\}$, such that

$$| \lambda_{i_j} - \gamma_j | \leq \| R \|_2, \quad \text{for } 1 \leq j \leq k. \tag{1.3.1}$$

Thus, if we have a projection matrix H, as we will have in our iterative block Lanczos procedures, and the norm of the corresponding block residual R is small, then the eigenvalues of that projection matrix are close to eigenvalues of the original matrix. Note that when Q is a single vector, then H is just the Rayleigh quotient of that vector. Thus, we call the matrix H in Theorem 1.3.1 a generalized Rayleigh quotient matrix of A corresponding to the given vectors Q.

When the eigenvalue being considered is a simple isolated eigenvalue of A, then the error estimate for the corresponding Ritz vector obtained from the projection matrix H is typically expressed in terms of the angle between the Ritz vector and the corresponding eigenvector of A. When the eigenvalue being considered is not isolated or not simple the error estimates are expressed in terms of a generalization of the notion of the angle between two vectors to an angle between two subspaces. This concept is defined in Chapter 2, Section 2.2 where we use it to describe theoretical estimates for the accuracy of the Ritz values and vectors obtained using the Lanczos recursions. In this chapter we include only an estimate for a single eigenvector corresponding to a simple, isolated eigenvalue. The generalization of this statement to the case of subspaces uses straight-forward generalizations of both the notion of a gap between two eigenvalues and of the angle between two vectors, to that of a gap between two sets of eigenvalues and of an angle between two subspaces. The interested reader can refer to Section 2.2 of this book for brief comments and to Chapter 11 of Parlett [1980] for details.

THEOREM 1.3.2 Parlett [Chapter 11, 1980]. Under the hypotheses of Theorem 1.3.1, let μ and u be an eigenvalue and a unit eigenvector pair of the projection matrix H. Define the corresponding Ritz vector $y \equiv Qu$. Let λ be the eigenvalue of A closest to μ and let x be a unit eigenvector of A for λ. Define the associated gap $\gamma \equiv \min_{\lambda_i \neq \lambda} | \lambda_i - \mu |$. Let Θ denote the angle between the Ritz vector y and the A-eigenvector x. Then

$$| \sin \Theta | \leq \| r(y) \|_2 / \gamma \quad \text{where } r(y) \equiv \| Ay - \mu y \|_2. \tag{1.3.2}$$

Parlett [Chapter 11, 1980] contains a generalization of Theorem 1.3.1 to the case when the set of orthogonal vectors Q is replaced by a set $Z \equiv \{z_1, ..., z_k\}$ of unit vectors which are not orthogonal but which are linearly independent. This generalization states essentially that the eigenvalues of the nonorthogonal projection matrix $H \equiv Z^T A Z$ can be paired with eigenvalues of A in such a way that the differences between any two paired eigenvalues is bounded by $\{\sqrt{2} \|R\|_2 / \sigma_k(Z)\}$. Here $\sigma_k(Z)$ denotes the smallest singular value of the matrix Z and R denotes the corresponding block residual matrix $R \equiv AZ - ZH$.

Thus, the relationship in Theorem 1.3.1 persists as long as the independence of the columns of Z is maintained. We will see however that for the single-vector Lanczos procedures which we propose that even this generalization is not sufficient. The projection matrices which we will encounter, namely the Lanczos matrices which we will generate, will be obtained from vectors Z which are not only not necessarily orthogonal but in fact may not even be linearly independent. Thus, Theorem 1.3.1 cannot be used to justify our single-vector Lanczos procedures. We will instead use the following basic inequality, see Wilkinson [1965, p.171]. Let A be Hermitian or real symmetric. Let a scalar μ and a nonzero vector Vu be given. Then there is an eigenvalue λ of A such that

$$| \mu - \lambda | \leq \frac{\| AVu - \mu Vu \|}{\| Vu \|}. \tag{1.3.3}$$

Theorems 1.3.1 and 1.3.2 are however applicable to our iterative block Lanczos procedures which we discuss in Chapter 7 because in those procedures we maintain near-orthogonality of the Lanczos vectors.

SECTION 1.4   EIGENVALUE INTERLACING AND STURM SEQUENCING

LEMMA 1.4.1 Jennings [1977] (Interlacing Property) Let A be a nxn real symmetric matrix A with eigenvalues $\lambda_1 \geq \lambda_2 \geq ... \geq \lambda_n$. Let A_{n-1} denote the $(n-1) \times (n-1)$ principal submatrix obtained from A by deleting the last row and column of A. Let $\lambda_1^{n-1} \geq \lambda_2^{n-1} \geq ... \geq \lambda_{n-1}^{n-1}$ denote the eigenvalues of A_{n-1}. Then

$$\lambda_k \geq \lambda_k^{n-1} \geq \lambda_{k+1} \quad \text{for} \quad 1 \leq k \leq n-1. \tag{1.4.1}$$

That is the eigenvalues of the submatrix A_{n-1} interlace the eigenvalues of A. Interlacing gives rise to the important Sturm sequence property for real symmetric matrices.

LEMMA 1.4.2 Jennings [1977] (Sturm Sequencing Property) Under the hypotheses of Lemma 1.4.1 set $d_0 \equiv 1$ and for $k \geq 1$ define

$$d_k \equiv \det(\mu I - A_k), \tag{1.4.2}$$

the determinant of the the kxk leading principal submatrix of $(\mu I - A)$. Then the number of alternations in sign in the sequence $\{d_0, d_1, ... , d_n\}$ is equal to the number of eigenvalues of A which are greater than the specified scalar μ.

Given a real symmetric matrix A we can use the Sturm Sequencing Property to compute the eigenvalues of A in any specified interval by evaluating sequentially the determinants given in Eqn(1.4.2) for well-chosen values of μ. There are however, numerical difficulties with this approach. Depending upon the matrix and upon its size these determinants can range from being exorbitantly large to being excessively small, exceeding the ranges of the floating point numbers on the computer being used. This difficulty can be overcome if we use ratios of determinants. In particular define

$$r_k(\mu) \equiv \det(\mu I - A_k)/\det(\mu I - A_{k-1}). \tag{1.4.3}$$

For a detailed discussion of interlacing and of the Sturm sequencing property please see for example, Jennings [1977, Chapter 9].

In this monograph we are interested in the Sturm sequencing property only as it applies to real symmetric tridiagonal matrices. For such matrices it is easy to compute the determinants needed to use the Sturm sequencing property in a practical algorithm for computing the eigenvalues of such matrices. In particular if T is a real symmetric tridiagonal matrix defined by

$$T \equiv (t_{ij}), \quad \text{where} \quad t_{kk} \equiv \alpha_k, \text{ and } t_{k+1,k} \equiv \beta_{k+1}, \tag{1.4.4}$$

then the determinant ratios defined in Eqn(1.4.3) satisfy the following recursion.

$$\begin{aligned}
r_1 &= (\mu - \alpha_1) \text{ and for } k = 2, \dots \\
r_k(\mu) &= (\mu - \alpha_k) - \beta_k^2/r_{k-1}(\mu).
\end{aligned} \tag{1.4.5}$$

See Section 3.5 in Chapter 3 for a discussion of Sturm sequencing as it applies to the computation of eigenvalues of a real symmetric tridiagonal matrix. See Section 1.6 of Chapter 1 for the generalization of Sturm sequencing to certain real symmetric generalized eigenvalue problems.

SECTION 1.5   HERMITIAN MATRICES

DEFINITION 1.5.1. A complex square matrix A of order n, $A \equiv (a_{ij})$, $1 \le i, j \le n$, is a Hermitian matrix if and only if for every i and j, $a_{ij} = \bar{a}_{ji}$. The overbar denotes the complex conjugate of a_{ij}.

It is straight-forward to demonstrate from Definition 1.5.1 that for any Hermitian matrix $A = B + Ci$, where B and C are real matrices and $i = \sqrt{-1}$, that B must be a real symmetric matrix and C must be skew symmetric. That is, $B^T = B$ and $C^T = -C$. Furthermore, it is not difficult to see that Hermitian matrices must have real diagonal entries and real eigenvalues. However, the eigenvectors are complex vectors. Any Hermitian matrix can be transformed into a real symmetric tridiagonal matrix for the purposes of computing the eigenvalues of the Hermitian matrix, Stewart [1973]. In fact the Lanczos recursion which we use in our Lanczos procedure for Hermitian matrices, see Chapter 4, transforms the given Hermitian matrix into a family of real symmetric tridiagonal matrices, thereby reducing the eigenvalue computations for Hermitian matrices to computations on real symmetric tridiagonal matrices.

Hermitian matrices possess the 'same' properties as real symmetric matrices do, except that these properties are defined with respect to the complex or Hermitian norm, rather than with respect to the Euclidean norm, see Stewart [1973]. The Hermitian norm of a given complex-valued vector x is defined as $\| x \|_2^2 \equiv \sum_{i=1}^{n} \bar{x}(i)x(i)$. There are three principal properties of Hermitian matrices which we need in our discussions. First, Hermitian matrices have complete eigensystems. That is, the dimension of the eigenspace corresponding to any eigenvalue of a Hermitian matrix is the same as the multiplicity of that eigenvalue as a root of the characteristic polynomial of that matrix. Second given any two distinct eigenvalues λ and μ, and corresponding eigenvectors x and y, we have that $x^H y = 0$. Thus such eigenvectors are orthogonal with respect to the Hermitian inner product. Third, small Hermitian perturbations in a Hermitian matrix cause only small perturbations in the eigenvalues. Similarly, perturbations in the eigenspaces are small whenever the sets of eigenvalues being considered are separated by a reasonable gap.

Hermitian eigenvalue problems arise in quantum physics and quantum chemistry. One important application is the analysis of surface states in semiconductor physics; see for example Pandey [1983]. Another application is shell-model calculations in nuclear physics; see for example Whitehead et al [1977]. An application in quantum chemistry is the analysis of bonding energies in molecules. See for example Nesbet [1981] and Davidson [1983].

Estimates of the accuracy of approximations to eigenvalues and eigenvectors of Hermitian matrices are based upon the sizes of norms of residuals and are the 'same' as those for the real symmetric case. See Section 1.3. The eigenvalues of a Hermitian matrix A interlace the eigenvalues of any $(n-1)x(n-1)$ principal submatrix of A. For any real λ the determinant of $B - \lambda I$, where B is any kxk principal submatrix of A, is real and the Sturm sequence property applies to $A - \lambda I$. There is a unitary matrix of eigenvectors U with $U = (u_1,u_2,...,u_n)$, $u_k^H u_j = \delta_{kj}$ such that $A = U\Lambda U^H$. Lanczos eigenvalue (and eigenvector) procedures for Hermitian matrices are discussed briefly in Chapter 4. FORTRAN code and documentation are given in Chapter 3 of Volume 2.

SECTION 1.6   REAL SYMMETRIC GENERALIZED EIGENVALUE PROBLEMS

Let A and B be real symmetric matrices then the real symmetric generalized eigenvalue problem is to find scalars λ and vectors $x \neq 0$ such that

$$Ax = \lambda Bx. \tag{1.6.1}$$

If we assume further that the matrix B is positive definite and $B = LL^T$ is the Cholesky factorization of B, then this problem is equivalent to the following real symmetric problem and therefore has all the same desirable properties as a standard real symmetric problem. Determine scalars λ and vectors $y \neq 0$ such that

$$L^{-1}AL^{-T}y = \lambda y, \text{ where } y = L^{T}x. \tag{1.6.2}$$

Therefore, in this case we could solve the generalized problem by applying the real symmetric Lanczos procedure directly to the composite matrix $C \equiv L^{-1}AL^{-T}$ given in Eqn(1.6.2). However, in our Lanczos procedure we will choose to work directly with the generalized problem and use a real symmetric generalized form of the Lanczos recursion which yields real symmetric Lanczos matrices. For details see Section 4.9 in Chapter 4 and also see Chapter 5 in Volume 2. Observe that the B-orthogonality of the eigenvectors is an immediate consequence of the orthogonality of the eigenvectors of the real symmetric problem in Eqn(1.6.2).

Generalized eigenvalue problems arise primarily in structural mechanics applications. See for example Bathe and Wilson [1976], Jennings [1977], and NASTRAN [1977]. In these situations typically either A or B is at least positive semi-definite, and a factorization is available. The following example clearly illustrates the difficulties one can encounter if both the A and the B matrices are indefinite. In that situation there is no equivalent real symmetric formulation. This case is actually analogous to what one encounters with nonsymmetric matrices. In particular we can have generalized eigenvalue problems with deficient eigenspaces. In this situation the eigenvalue and eigenvector computations become very difficult.

EXAMPLE 1.6.1   Let A and B be the following 2x2 matrices.

$$A \equiv \begin{bmatrix} 1 & 2/a \\ 2/a & -1 \end{bmatrix} \text{ and } B \equiv \begin{bmatrix} a & 1 \\ 1 & -1/a \end{bmatrix}. \tag{1.6.3}$$

The eigenvalues of the corresponding generalized eigenvalue problem given by Eqn(1.6.1) are equal to the eigenvalues of the nonsymmetric matrix $M \equiv B^{-1}A$ and these two problems also have the same eigenvectors.

Computing the matrix M explicitly we obtain

$$M \equiv 0.5 \begin{bmatrix} 3/a & -1 + 2/a^2 \\ -1 & a + 2/a \end{bmatrix}. \tag{1.6.4}$$

Consider the case when $a^2 = (\sqrt{8} - 1)$, then the matrix M has the double root $\lambda = .25(a + 5/a)$. The eigenvalues of any matrix are preserved under any similarity transformation so we know that for any nonsingular matrix T that the corresponding matrix $T^{-1}MT$ has the same eigenvalues as M. But observe that if we set $\mu \equiv .5[(1/a) - a]$ then

$$T \equiv \begin{bmatrix} \mu & \mu + 2 \\ -1 & -1 \end{bmatrix} \text{ and } T^{-1}MT = \begin{bmatrix} \lambda & 1 \\ 0 & \lambda \end{bmatrix}. \tag{1.6.5}$$

Eqn(1.6.5) tells us that λ is a deficient eigenvalue of the original generalized eigenvalue problem. A deficient eigenvalue with deficiency q can be computed only to within the qth root of the accuracy achievable for nondeficient eigenvalues. Deficiencies can cause havoc numerically. Therefore, we restrict our considerations to the case when at least one of the two matrices is positive definite.

It is interesting to note that if B is positive definite then the real symmetric generalized problem also possesses the Sturm Sequencing property. The matrix $C(\mu) \equiv (A - \mu B)$ is used directly; the factors of B are not required. The following discussion is taken from Peters and Wilkinson [1969]. It is included only for completeness. The Lanczos programs provided in Volume 2 for real symmetric generalized problems transform the given matrix into real symmetric tridiagonal matrices and do not use this generalization of Sturm Sequencing.

To see that the Sturm sequence property is applicable to such generalized problems consider the following argument. Since B is positive definite it has a Cholesky factorization $B = LL^T$. Then for any given μ we have that

$$C(\mu) = L(S - \mu I)L^T \text{ where } S \equiv L^{-1}AL^{-T}. \tag{1.6.6}$$

Therefore the eigenvalues of the generalized problem in Eqn(1.6.1) are the same as the eigenvalues of the standard real symmetric eigenvalue problem $Sx = \lambda x$.

Let $S(\mu)$ denote $(S - \mu I)$, and let B_k, $S_k(\mu)$, and $C_k(\mu)$ $(k = 1,2,...,n)$ denote respectively the $k \times k$ principal minors of B, $S(\mu)$, and $C(\mu)$. Since B is positive definite, no interchanges of rows are required in the factorization of B. Therefore, we have that for each k, $B_k = L_k L_k^T$. From this we obtain

$$C_k(\mu) \equiv (A_k - \mu B_k) = L_k(S_k - \mu I_k)L_k^T.$$

Let $\sigma_k \equiv \text{sign}[\det(S_k - \mu I)]$. Then by the Sturm Sequence Property we know that the eigenvalues of S can be determined by keeping track of σ_k for various choices of μ. However it is clear that $\sigma_k = \text{sign}[\det(C_k)]$ since the determinant of L_k is positive. Therefore these two determinants always have the same sign, and therefore the determinants of $C_k(\mu)$ can be used to count the number of eigenvalues of S in any given interval. The eigenvalues of S are however just the eigenvalues of the given generalized problem.

SECTION 1.7   SINGULAR VALUE PROBLEMS

The eigenvector decomposition of an arbitrary real Hermitian matrix A consists of an unitary matrix X of eigenvectors and a simple diagonal matrix Λ of eigenvalues. That is, $A = X\Lambda X^H$. Using such a decomposition, one can easily predict the behavior of various kinds of operations and algorithms when they are applied to Hermitian matrices.

It is therefore natural to ask if an analogous decomposition exists for any general ℓxn rectangular matrix A. If $\ell = n$ but A is not Hermitian, then in general the eigenvectors of A do not form a basis for $\mathscr{C}^n$. Even if they do form such a basis, it may be very nonunitary. General non-Hermitian matrices A have two sets of eigenvectors, the left eigenvectors satisfying $A^T y = \lambda y$ and the right eigenvectors satisfying $Ax = \lambda x$. If A is square and nondefective then we have that $A = X\Lambda X^{-1}$, however the X matrix may be very ill-conditioned. If A is rectangular with $\ell \neq n$, then the range of A, that is the space spanned by the columns of A, is in $\mathscr{C}^\ell$ whereas the range of A^H, that is the space spanned by the complex conjugates of the rows of A is in $\mathscr{C}^n$. Thus, it is clear that if we want to obtain an analog of the Hermitian eigenvector decomposition for a general matrix, then we must deal with at least two different sets of vectors, which we will choose to denote by Y and by X.

The analog for a general matrix of the eigenvector expansion for Hermitian matrices is the singular value decomposition. We assume w.l.o.g. that $\ell \geq n$.

$$A = Y\Sigma X^H \text{ where } \Sigma = \begin{bmatrix} \Sigma_1 \\ 0 \end{bmatrix}, \tag{1.7.1}$$

Y is a ℓxℓ unitary matrix, X is a nxn unitary matrix, and Σ_1 is an nxn diagonal matrix with nonnegative entries and 0 denotes the $(\ell-n)$xn matrix of zeros. To simplify the discussion we assume that $\ell \geq n$ unless it is explicitly stated otherwise. This is more typical in the applications. The corresponding statements for the case $\ell \leq n$ can be obtained directly by applying the arguments given to A^H instead of A. If A is real then Y and X are orthogonal matrices.

From Eqn(1.7.1) we have that for each σ_i and the two corresponding vectors x_i and y_i that

$$Ax_i = \sigma_i y_i \text{ and } A^T y_i = \sigma_i x_i. \tag{1.7.2}$$

DEFINITION 1.7.1 Let A be a ℓxn matrix ($\ell \geq n$) with the singular value decomposition given in Eqn(1.7.1). Then the scalars σ_j, $1 \leq j \leq n$, are called the singular values of A. The columns of Y are called left singular vectors of A and the columns of X are called right singular vectors of A.

As defined the problem of computing singular values and vectors does not appear to be equivalent to a real symmetric problem. However, Lanczos [1961] noted that an 'equivalent' real symmetric problem could be defined as follows. Define

$$B \equiv \begin{bmatrix} 0 & A \\ A^T & 0 \end{bmatrix}. \tag{1.7.3}$$

It is easy to prove and this is done in Chapter 5, that B has pairs of eigenvalues $\pm\sigma_i$, $1 \leq i \leq n$, corresponding to the singular values of A. In addition if $\ell \neq n$, then there are $|\ell-n|$ additional zero eigenvalues. Similarly, there is a very straight-forward relationship between the eigenvectors of B and the left and right singular vectors of A. Thus, the singular values and corresponding singular vectors of A can be computed by computing eigenvalues and corresponding eigenvectors of the associated matrix B.

We could apply our real symmetric version of the Lanczos recursion directly to the matrix B in Eqn(1.7.3). However, due to the fact that this matrix may be considerably larger than the A-matrix, we use instead a modification which reduces the amount of computation to essentially that required for a nonsymmetric matrix of the size of A. This idea is due to Golub and Kahan [1965] and is discussed in detail in Chapter 5.

Singular values and vectors are useful tools in many applications, providing for example, measures of the sensitivity of solutions of given problems to errors in the data provided (see e.g. Lawson and Hanson [1974]), ways of stabilizing computational methods for ill-posed problems (see e.g. Hanson and Phillips [1975]), and data reduction schemes useful for example in pattern recognition problems (see e.g. Andrews and Hunt [1977]). Reliable programs for computing singular values and vectors of matrices of moderate size are available in the LINPACK Library [1979]. The amount of computer storage for these programs is proportional to the product ℓn and the corresponding arithmetic count is proportional to $\ell^2 n$. (Here we assumed that $\ell \geq n$.)

SECTION 1.8   SPARSE MATRICES

We are interested in very large matrices. Typically such matrices are sparse; that is they have very few nonzero entries. Since the matrices are very large, it is essential that the nonzero entries be stored in the most economical manner. This can be done using sparse matrix formats, Gustavson [1972]. We consider two classes of matrices, real symmetric matrices and general real nonsymmetric matrices. We encounter nonsymmetric matrices in our treatment of singular value problems. For simplicity we restrict ourselves to real matrices but corresponding complex matrices may be stored in similar formats with the obvious changes in the data types.

Let A be a nxn sparse, real symmetric matrix. The sparse matrix format requires the number NZS of nonzero subdiagonal entries in A and the number NZL, the index of the last column which contains at least one nonzero subdiagonal element. Two integer vector arrays are required , one of length NZL which in our programs we call ICOL and the other of length NZS which we call IROW. Two real, single or double precision, arrays are required, one of length n which we call AD and the other of length NZS which we call ASD. For $1 \leq j \leq NZL$, ICOL(j) contains the number of nonzero subdiagonal entries in column j of A. For $1 \leq j \leq n$, AD(j) contains the jth diagonal entry of A including any entries which are zero. For $1 \leq i \leq NZS$, ASD contains the nonzero subdiagonal entries of A, in column order. The array IROW contains the corresponding row indices for the entries in ASD.

The sparse matrix format for a nonsymmetric ℓxn matrix A is very similar except that we do not store the diagonal entries in a separate array. Without loss of generality assume that

$\ell \geq n$ and that every column of A contains at least one nonzero entry. Let NZ denote the total number of nonzero entries in A. We use two integer arrays ICOL and IROW and one real, single or double precision, array which we also call A. Then for each $1 \leq j \leq n$, ICOL(j) contains the number of nonzeros of A in column j. The nonzeros of A are stored in the A-array in increasing column order and within each column in increasing row order. The corresponding row index for each A(j) is contained in IROW(j). Thus, IROW and the A-array have length NZ. ICOL has length n.

Use of the above sparse matrix formats minimizes the amount of storage required to store a very large but sparse matrix. Since the Lanczos procedures do not explicitly modify the given matrix A, this storage advantage is maintained throughout the entire Lanczos computations. Lanczos procedures require the repeated computation of matrix-vector products Ax and in the singular value case also of $A^T y$. Users must provide subroutines which compute these quantities in the most economical, accurate and rapid way possible. Accuracy and consistency are keys to the successful implementation of these Lanczos procedures. The values of the matrix-vector products Ax (and of $A^T y$) provided to the Lanczos procedures must consistently represent the given matrix whose eigenvalues are to be determined. Sample matrix-vector multiply subroutines for general real symmetric matrices and for general real rectangular matrices are provided in Chapters 2 and 6 of Volume 2.

SECTION 1.9   REORDERINGS AND FACTORIZATIONS OF MATRICES

Matrix factorization is an important tool in the solution of many eigenvalue problems. The rates of convergence of the Lanczos procedures for solving eigenvalue problems depend upon the distribution of the eigenvalues of the matrix used in the procedure. If a given matrix A has an eigenvalue distribution which is unfavorable for the procedure being used, it may be possible to accelerate the rate of convergence of the computations significantly by applying the procedure to the inverse of A. When A is very large, using A^{-1} is feasible only if A can be factored in such a way that the factors are sparse. In particular if A is real symmetric, positive definite and sparse, and a sparse Cholesky factorization is available, then for any given vector z, the vector $A^{-1}z$ can be evaluated efficiently by solving sequentially the triangular systems of equations Lu = z and $L^T w = u$. The eigenvalue distribution for A^{-1} may be radically better than that of A and we give an example of this below.

The proposed method of solution of the generalized eigenvalue problem (see Sections 1.6 and 4.9) requires the factorization of the positive definite matrix B (or A). Ericsson and Ruhe [1980] present a Lanczos procedure which requires the factorization of the matrix B and of matrices of the form $(A-\mu B)$ where μ is a shift which is changed repeatedly during the eigenvalue computations. That procedure is described briefly in Chapter 2, Section 2.5.

Typically a straight-forward factorization of a given positive definite matrix B yields factors which are not sparse. However by judiciously reordering the rows and columns of the given matrix, in many important cases we can obtain sparse factors. One reordering rule which is commonly used and which has proved to be successful in many situations is the minimum degree rule. Using this rule at each step in the factorization of a positive definite symmetric matrix we take as the next pivot a diagonal element residing in a column of the reduced matrix with the minimum number of nondiagonal nonzero entries. The resulting factorization corresponds to a permutation B' of the original matrix B. In particular we have the following relationships.

$$B' \equiv PBP^T = LL^T \tag{1.9.1}$$

where P is a permutation martix.

Chow et al [1984] provides an example of the effective use of matrix factorization in Lanczos eigenvalue computations. A real symmetric matrix A of order n=2161 was constructed which represented a hypothetical but realistic network of electric power generators and buses. The fifteen eigenvalues smallest in magnitude and their corresponding eigenvectors had to be computed. The A-matrix had only 2828 subdiagonal nonzeros. It was real symmetric and negative definite with eigenvalues on the interval $[-.10368346 \times 10^6, -.84836988 \times 10^{-6}]$. Observe that the condition number of A is 1.2×10^{11}.

Because of the very large negative eigenvalues, the approximations to the desired eigenvalues obtained using our real symmetric Lanczos procedure directly on A converged very slowly. The eigenvalue gaps and gap ratio in A strongly favored the convergence of the numerically large eigenvalues. This lead us to consider using a factorization of A. Since it is not wise to work directly with such a poorly-conditioned matrix, we used the better-conditioned, positive definite, scaled and shifted matrix $C \equiv (-A + I/100)$. The sparsity of C is the same as that of A and the eigenvalues of A can be easily obtained from those of C.

The sparse matrix package SPARSPAK [1979] developed by Alan George et al at Waterloo University was used. This package is designed to completely handle the problem of solving the system of equations Bu = v for any u where B is a sparse, real symmetric and positive definite matrix. Subroutines are included for determining an appropriate reordering for which the factors of the reordered matrix are also sparse. Several algorithms for reordering matrices such as bandwidth minimization and the minimum degree algorithm are included in SPARSPAK. Chow et al [1984] used the minimum degree algorithm to determine an appropriate permutation P, and then the factorization of the permuted matrix $B \equiv PCP^T = LL^T$ was obtained. In this case the resulting factor L had 5023 subdiagonal nonzeros. This should be compared with the 2828 subdiagonal nonzeros in the given matrix.

Our single-vector Lanczos procedure for real symmetric matrices was applied to B^{-1}. The vectors $u = B^{-1}v_i$ were computed by solving the two triangular systems $Lw = v_i$ and $L^Tu = w$. Since the factor L was very sparse, these triangular systems could be solved very rapidly. The solution of each triangular system required NZS multiply-adds and n multiplies where NZS was the total number of nonzero subdiagonal entries in L and n was the order of A.

Our Lanczos procedure computed all of the 15 desired eigenvalues and corresponding eigenvectors in only 125 steps of the Lanczos recursion. The direct use of A would have required several thousand steps. Each step using B^{-1} was more expensive than a corresponding step using A. However, the large decrease in the number of Lanczos steps required and the fact that the factors were also sparse, yielded overall large gains in storage and time. In addition the subsequent Lanczos eigenvalue-eigenvector computations required were vastly cheaper than those which would have been required using A directly.

In general a crude test of the resulting accuracy achieved in computing $B^{-1}v_i$ can be obtained by generating a random unit vector w, setting $z = Bw$, and then solving $Bu = z$ using the factorization. The resulting u is compared with the known solution w using $\max_k |u(k) - w(k)|$. If this indicated accuracy is not satisfactory, then iterative refinement can be applied. However, typically it is sufficient to work with a shifted matrix, and iterative refinement is not necessary. Iterative refinement consists of the following two steps. (1) For the given approximation u to $B^{-1}v$ compute the corresponding residual $r \equiv v - Bu$. This residual must be calculated in a higher precision than that used for the original computations. (2) Solve the equations $B\delta = r$ for δ and set the corrected solution $\bar{u} = u + \delta$. In the example cited above iterative refinement was not required. We should note that it is not always advantageous to use A^{-1} to compute the small eigenvalues of A. The eigenvalue distribution in A^{-1} corresponding to the desired eigenvalues may actually be worse than the corresponding one in A. However, in many cases significant gains can be achieved.

CHAPTER 2

LANCZOS PROCEDURES, REAL SYMMETRIC PROBLEMS

SECTION 2.1   DEFINITION, BASIC LANCZOS PROCEDURE

Lanczos [1950] introduced what has since become known as the Lanczos recursion or Lanczos tridiagonalization. Lanczos procedures for computing eigenvalues and eigenvectors of real symmetric matrices (and also those for solving real symmetric systems of linear equations) are based upon one or more variants of the basic Lanczos recursion for tridiagonalizing a real symmetric matrix.

BASIC (SINGLE-VECTOR) LANCZOS RECURSION

Let A be a nxn real symmetric matrix and let v_1, be a unit starting vector, typically generated randomly. For $j=1,2,...,m$ define corresponding Lanczos matrices T_j using the following recursion. Define $\beta_1 \equiv 0$, and $v_0 \equiv 0$. Then for $i=1,2,...,m$ define Lanczos vectors v_i and scalars α_i and β_{i+1} where

$$\beta_{i+1}v_{i+1} = Av_i - \alpha_i v_i - \beta_i v_{i-1} \text{ and} \qquad (2.1.1)$$

$$\alpha_i \equiv v_i^T Av_i \text{ and } \beta_{i+1} \equiv v_{i+1}^T Av_i. \qquad (2.1.2)$$

For each j, the corresponding Lanczos matrix T_j is defined as the real symmetric and tridiagonal matrix with diagonal entries α_i, $1 \leq i \leq j$, and subdiagonal (superdiagonal) entries β_{i+1}, $1 \leq i \leq (j-1)$. Therefore for each j,

$$T_j \equiv \begin{bmatrix} \alpha_1 & \beta_2 & & & \\ \beta_2 & \alpha_2 & & & \\ & & \ddots & & \\ & & & \beta_j & \\ & & & \beta_j & \alpha_j \end{bmatrix} \qquad (2.1.3)$$

By definition the vectors $\alpha_i v_i$ and $\beta_i v_{i-1}$ in Eqn(2.1.1) are respectively, the orthogonal projections of the vector Av_i onto the two most recently generated Lanczos vectors v_i and v_{i-1}. Thus for each i, the next Lanczos vector v_{i+1} is determined by orthogonalizing the vector Av_i with respect to v_i and v_{i-1}. The resulting collection of scalar coefficients α_i, β_{i+1} obtained in these orthogonalizations defines the corresponding Lanczos matrices.

Rewriting Eqn(2.1.1) in matrix form, for each j we obtain the matrix equation

$$AV_j = V_j T_j + \beta_{j+1} v_{j+1} e_j^T \qquad (2.1.4)$$

where $V_j \equiv \{v_1,...,v_j\}$ is the nxj matrix whose kth column is the kth Lanczos vector, and e_j is the coordinate vector whose jth component is 1 and whose other components are 0. Thus, given a real symmetric matrix A and a starting vector v_1, the Lanczos recursion generates a family of real symmetric tridiagonal matrices related to A and to v_1 through Eqn(2.1.4). In the next few sections we will discuss some of the relationships between a given matrix A (and a given v_1) and its associated Lanczos matrices. In this section we will only give the reader some feeling for what a Lanczos procedure is, in a purely mechanical sense, without justifying or explaining why such a procedure should be expected to work.

The basic idea in any Lanczos procedure is to replace the eigenelement problem for the given matrix A by eigenelement computations on one or more of the simpler Lanczos matrices T_j. Lanczos procedures use the Lanczos recursion to transform a general real symmetric matrix eigenvalue problem into a highly-structured (in fact simple tridiagonal) real symmetric problem. Real symmetric tridiagonal matrices, can be viewed as computationally optimal matrices. They have minimal storage requirements, as do the associated algorithms for eigenvalue and eigenvector computations and for solving tridiagonal systems of equations. Also the arithmetic operations required are small. Furthermore, the Sturm sequencing property (see Chapter 3, Section 3.5) allows us to readily compute any portions of the eigenvalue spectrum without having to compute the rest of it. This latter property will allow us to work with very large tridiagonal matrices, and therefore with very large general real symmetric matrices. Many of these computational advantages or analogs of them also apply to general symmetric tridiagonal matrices and this will be important in Chapter 6 where we will discuss the class of diagonalizable complex symmetric matrices.

BASIC LANCZOS EIGENELEMENT PROCEDURE.

Step 1.   Use a variant of the Lanczos recursion in Eqns(2.1.1) - (2.1.2) to transform the given real symmetric matrix A into a family of real symmetric tridiagonal matrices, T_j, $j = 1,2,...,M$.

Step 2.   For some $m \leq M$, compute the relevant eigenvalues of the Lanczos matrix T_m.

Step 3.   Select some or all of these eigenvalues as approximations to eigenvalues of the given matrix A.

Step 4.   If eigenvectors of A are desired for one or more of the eigenvalues selected in Step 3, then for each such eigenvalue μ compute a corresponding unit eigenvector u such that $T_m u = \mu u$. Map such vectors into corresponding Ritz vectors $y \equiv V_M u$, which are then used as approximations to the desired eigenvectors of A.

The basic Lanczos recursion given in Eqns(2.1.1) - (2.1.2) has two properties which make it particularly attractive for dealing with very large but sparse matrices. First, the matrix A enters the recursion only through the matrix-vector multiply terms Av_i. Thus, the given matrix A is not modified by the computations. A Lanczos procedure requires only that a user provide a subroutine which for any given x, computes Ax. If the given matrix is sparse, then this computation can be performed using an amount of storage which is only linear in the size n of the matrix. Second, at each iteration the basic Lanczos recursion requires only the two most recently-generated Lanczos vectors. The other Lanczos vectors need not be stored for future use by the recursion. We note however, that some proposed modifications of this basic Lanczos procedure do require that these vectors be available on some readily accessible storage for use repeatedly throughout the Lanczos procedure, see for example Parlett and Scott [1979]. We further note that our single-vector Lanczos procedures, as the reader will see in Chapters 4,5 and 6, do not need these Lanczos vectors (at least not for the eigenvalue computations) so that the storage requirements for our procedures are minimal.

It is easy to prove (at least theoretically) that for the basic Lanczos recursion that orthogo-nalization with respect to only the two most recently-generated Lanczos vectors is sufficient to guarantee that each succeeding Lanczos vector is orthogonal with respect to all previously-generated Lanczos vectors. (See Theorem 2.2.1 in Section 2.2). The basic Lanczos procedure can be viewed as the Gram-Schmidt orthogonalization of the set of Krylov vectors $\{v_1, Av_1, ..., A^{m-1}v_1\}$. We note that the Gram-Schmidt orthogonalization of an arbitrary set of vectors would require that at each stage in the process that all of the vectors which have already been orthogonalized be available for orthogonalizing each additional vector as it is considered. The storage required for implementing the basic Lanczos procedure is minimal. The basic storage requirements are some small multiple of n, the order of the given matrix A, plus whatever storage is required to compute the matrix-vector multiplies Ax. Thus in the case when A is sparse, the total storage requirements are in fact a small multiple of n.

A simple comparison of the requirements of a basic Lanczos eigenelement procedure with those of the corresponding procedure in the EISPACK Library [1976,1977] quickly demon-strates the reasons for considering Lanczos procedures. The eigenelement procedures for real symmetric matrices which are in the EISPACK library [1976,1977] also transform the given matrix A into an equivalent real symmetric tridiagonal matrix. However, they do this by using orthogonal transformations which explicitly alter the original matrix. Fill-in can and almost surely does occur during this process. Zero elements become nonzero, and one cannot predict where this fill-in will occur. See the example in Chapter 3, Section 3.1. Thus, the EISPACK subroutines have a priori computer storage requirements of more than $[n(n + 1)/2]$ words. This limits the size of the matrix which can be handled. For example, if n = 200 then 160,000 bytes of storage are required. If n = 1000 then 4,000,000 bytes of storage are required. For a sparse matrix this of course compares very unfavorably with the storage requirements of the basic single-vector Lanczos procedure which are a linear (not quadratic) function of the size of the matrix.

The number of arithmetic operations required by the EISPACK procedures as a function of the size n of the given matrix grows like n^3. This is actually a reflection of the cost of the transformation of the given matrix A to tridiagonal form. On the other hand, for a sparse matrix the number of arithmetic operations required to generate the Lanczos matrices using the basic Lanczos recursion is at most proportional to n^2. The actual number of arithmetic operations required by a Lanczos procedure can often be much less than this if only a few eigenvalues are required. The eigenelement computations required for the resulting real symmetric tridiagonal matrices are feasible even for very large tridiagonal matrices. The storage requirements for an eigenvalue computation of any real symmetric tridiagonal matrix depend only linearly upon the order of the matrix, and the number of arithmetic operations required to compute even all of the eigenvalues of such a matrix is proportional to the square of the order of the matrix. Any required eigenvectors of the Lanczos matrices can be computed economically using inverse iteration Peters and Wilkinson [1979], once good approximations to the corresponding eigenvalues are known. Thus, a Lanczos approach offers the possibility of obtaining an eigenvalue/eigenvector algorithm which can be applied to very large real symmetric matrices.

In the next section we summarize some of the attractive theoretical properties of the basic single-vector Lanczos recursion. We note in passing, more will be said about this in Section 2.3, that a basic Lanczos procedure which has not been properly modified will not behave in accordance with the theoretical properties summarized in the next section. The reasons for this will be made clear in Section 2.3 where we consider the effect of the finite precision of the computer arithmetic upon the quantities generated by the Lanczos recursion. First however, in the next section we look at the properties of this recursion assuming that we are using infinite precision arithmetic. This is the way in which the Lanczos recursion was viewed by Lanczos. We want to emphasize however that in the main discussion of our Lanczos procedures in Chapter 4, we will be primarily interested in the theorems from Section 2.3 and not in those from Section 2.2.

SECTION 2.2   BASIC LANCZOS RECURSION, EXACT ARITHMETIC

In this section we summarize the classical arguments supporting the use of the basic Lanczos recursion in an eigenelement procedure. All of the arguments in this section assume that all of the required numerical computations are performed using infinite precision arithmetic. This is obviously not the case when these computations are performed on a computer. For some types of mathematical computations the properties and theorems which can be derived assuming exact arithmetic are still essentially valid when finite precision arithmetic is used. However, this is not the case for the computations defined by the basic Lanczos recursion. The practical situation, finite precision arithmetic, is discussed in Section 2.3.

Obviously the Lanczos recursion is a mechanism for replacing a general real symmetric matrix by a family of much simpler matrices, namely the real symmetric tridiagonal Lanczos matrices. However, it is not immediately obvious from the definitions given in Section 2.1 that these Lanczos matrices possess any special properties which make them particularly suitable approximation matrices for A. To see that this is in fact the case, we first prove that the Lanczos vectors form an orthonormal set of vectors (in exact arithmetic).

THEOREM 2.2.1 (Exact Arithmetic) Let A be a nxn real symmetric matrix with n distinct eigenvalues. Let v_1 be a unit starting vector with a nonzero projection on every eigenvector of A. Use the basic single-vector Lanczos recursion defined by Eqns(2.1.1) and (2.1.2) to generate Lanczos matrices T_j and Lanczos vectors $V_j \equiv \{v_1,...,v_j\}$. Then for any $j \leq n$ we have that

$$V_j^T V_j = I_j . \tag{2.2.1}$$

Furthermore, for each such j,

$$T_j = V_j^T A V_j \tag{2.2.2}$$

is the orthogonal projection of A onto the subspace spanned by the V_j. In other words T_j represents the operator A restricted to $sp\{V_j\}$.

Proof. By construction $v_1^T v_1 = 1$ and $v_2^T v_1 = 0$. Assume Eqn(2.2.1) is true for $j \leq K$ and show it is true for $j=K+1$. We have that

$$v_j^T v_{K+1} \beta_{K+1} = v_j^T A v_K - \alpha_K v_j^T v_K - \beta_K v_j^T v_{K-1}. \tag{2.2.3}$$

But, by construction $v_K^T v_{K+1} = 0$ and $v_{K-1}^T v_{K+1} = 0$. So consider $j \leq K - 2$. By induction $v_j^T v_K = v_j^T v_{K-1} = 0$. Use Eqn(2.1.1) to replace $A v_j$ in Eqn(2.2.3) and then use the induction hypothesis to show that the resulting expression is zero. In particular we get that

$$v_j^T v_{K+1} \beta_{K+1} = v_K^T v_{j+1} \beta_{j+1} + v_K^T v_j \alpha_j + v_K^T v_{j-1} \beta_j.$$

But, clearly the right-hand side of this expression vanishes.

To prove Eqn(2.2.2) consider the following. Since each V_j is an orthonormal set of vectors the projection of A onto the space spanned by the V_j is simply the matrix $V_j^T A V_j$. But then for each j we get Eqn(2.2.2) from the global orthogonality of the V_j. Thus, (theoretically) for each $1 \leq j \leq n$, the symmetric tridiagonal matrices T_j are representations of the projections of the given matrix A onto the subspaces spanned by the Lanczos vectors V_j. □

Theorem 2.2.1 tells us that the eigenvalues of the Lanczos matrices are the eigenvalues of the A-matrix restricted to the Krylov subspaces $\mathcal{K}^j \equiv sp\{v_1, A v_1, A^2 v_1,..., A^{j-1} v_1\}$. Therefore if j is sufficiently large, we expect the eigenvalues of the T_j to provide good approximations to some of the eigenvalues of A. Clearly (at least theoretically), if we continue the Lanczos recursion until j=n, the order of the given matrix A, then the eigenvalues of T_n will be the eigenvalues of A. In this case T_n is simply an orthogonal similarity transformation of A and must therefore have the same eigenvalues as A. Moreover, any Ritz vector $V_j u$ obtained from an eigenvector u of a given T_j is an approximation to a corresponding eigenvector of A.

COROLLARY 2.2.1   Under the assumptions of Theorem 2.2.1 and for each $2 \leq j \leq n$, the basic single-vector Lanczos recursion given in Eqns(2.1.1) - (2.1.2) generates an orthonormal basis for each of the Krylov subspaces $\mathcal{K}^j \equiv sp\{v_1, Av_1,..., A^{j-1}v_1\}$. Furthermore, for each such j the Lanczos matrix T_j is the representation of the restriction of the matrix A to that Krylov space. Therefore, the eigenvalues of T_j are the eigenvalues of A restricted to $\mathcal{K}^j$.

COROLLARY 2.2.2   Under the assumptions of Theorem 2.2.1, the eigenvalues of T_n are the eigenvalues of A.

Theoretically a single-vector Lanczos procedure cannot determine the multiplicity of any multiple eigenvalue of the given matrix and can compute only one eigenvector for each such eigenvalue. In particular if the given matrix A has only r<n distinct eigenvalues then (in exact arithmetic) the Lanczos recursion would terminate after r steps, having computed one copy of each distinct eigenvalue and one eigenvector for each distinct eigenvalue. This property of single-vector Lanczos procedures has been used as one of the strong arguments for devising and using a block version of the basic Lanczos recursion. In practice we will see however that at least for Lanczos procedures that do not use any reorthgonalization, multiplicities can be determined if one is willing to do extra computation. The amount of extra computation required varies with the matrix and with the particular eigenvalue being considered.

Theorem 2.2.1 by itself does not totally answer the question, why is it reasonable to use a Lanczos procedure to compute approximations to the eigenvalues and eigenvectors of A. The key point in that argument however will be the fact that each subspace $sp\{V_j\}$ is a Krylov subspace for A. The Lanczos procedure is simply a mechanism for generating orthonormal bases for these Krylov subspaces and for computing the orthogonal projection of A onto these subspaces. Computing the eigenvalues of the Lanczos matrices T_j is then equivalent to computing the best approximations to the eigenvalues and eigenvectors of A restricted to the corresponding Krylov subspace. To see how well we can expect to approximate eigenvalues using Krylov subspaces, we consider the following theorem, Saad [1980]. The version given below is taken from Parlett [1980, Chapter 12]. As it indicates the Krylov subspaces are very desirable subspaces to use to obtain eigenvalue/eigenvector approximations. For more details the reader is referred to either Saad [1980] or Parlett [1980].

First we have to introduce some notation. Let $\{\mu_1 \geq ... \geq \mu_m\}$ denote the eigenvalues of the Lanczos matrix T_m and let $\{y_1,...,y_m\}$ denote a corresponding set of Ritz vectors. Let $\{\lambda_1 \geq \lambda_2 \geq ... \geq \lambda_n\}$ denote the eigenvalues of A and let $\{x_1,...,x_n\}$ denote a corresponding orthonormal set of eigenvectors of A. Let $\mathcal{X}^j$ denote the subspace spanned by the first j

A-eigenvectors. We need to define the angle $\angle[z,\mathcal{W}]$ between a vector z and a subspace $\mathcal{W}$. We use the definitions given in Davis and Kahan [1970] and summarized in Parlett [1980, Chapter 11].

DEFINITION 2.2.1 Given two subspaces $\mathcal{W}$ and $\mathcal{Z}$ we define the angle between these two subspaces to be the maximum of the minimum of the angles between any vector $w \epsilon \mathcal{W}$ and any vector $z \epsilon \mathcal{Z}$. Specifically,

$$\angle[\mathcal{W},\mathcal{Z}] \equiv \max_{w \epsilon \mathcal{W}} \min_{z \epsilon \mathcal{Z}} \angle[w,z]. \qquad (2.2.4)$$

In Eqn(2.2.4) $\cos(\angle[w,z]) \equiv w^T z / \|w\| \|z\|$, which is the usual definition for the angle between two vectors in a Euclidean space. The error bounds obtained will be stated in terms of angles between the starting vector v_1 used in the Krylov subspace generation and subspaces spanned by the eigenvectors of A corresponding to the eigenvalues $\{\lambda_1,...,\lambda_j\}$. The primary term in the bound, however, will be a Chebyshev polynomial. Chebyshev polynomials have the property that for a given k the Chebyshev polynomial $\mathcal{T}_k(1 + 2\varepsilon)$ grows like $e^{2k\sqrt{\varepsilon}}$ for $k\sqrt{\varepsilon} > 1$. See for example, Appendix B of Parlett [1980] for a brief summary of the properties of Chebyshev polynomials.

THEOREM 2.2.2 Saad [1980] Let A be a real symmetric nxn matrix. Let v_1 be a specified unit starting vector and apply the basic Lanczos recursion to A using v_1. For any $j = 1,2,...,m$ with $m \leq n$, let $\mu_1 \geq ... \geq \mu_m$ be the eigenvalues of the corresponding Lanczos matrix T_m. Let $\lambda_1 \geq ... \geq \lambda_m$ be the corresponding eigenvalues of A and let $\{x_1,...,x_m\}$ be a corresponding orthonormal set of eigenvectors of A. Then for each $j = 1,...,m$ the computed eigenvalue approximations satisfy the following inequality.

$$|\lambda_j - \mu_j| \leq |\lambda_j - \lambda_n| \left[\frac{\sin\angle[v_1,\mathcal{R}^j]}{\cos\angle[v_1,x_j]} \frac{\left(\prod_{\ell=1}^{j-1} \frac{(\mu_\ell - \lambda_n)}{(\mu_\ell - \lambda_j)} \right)}{\mathcal{T}_{m-j}(1 + 2\gamma_j)} \right]^2. \qquad (2.2.5)$$

where $\gamma_j \equiv |\lambda_j - \lambda_{j+1}| / |\lambda_{j+1} - \lambda_n|$. For each j, the tangent of the angle between the corresponding eigenvector of A and the Krylov subspace satisfies the following inequality.

$$\tan \angle[x_j,\mathcal{K}^m] \leq \frac{\sin\angle[v_1,\mathcal{R}^j]}{\cos\angle[v_1,x_j]} \frac{\left(\prod_{\ell=1}^{j-1} \frac{(\lambda_\ell - \lambda_n)}{(\lambda_\ell - \lambda_j)} \right)}{\mathcal{T}_{m-j}(1 + 2\gamma_j)} \qquad (2.2.6)$$

Observe that the starting vector enters these bounds through its projection on the eigenvector x_j which is being approximated and its projection on the subspace $\mathcal{R}^j$ corresponding to the first j eigenvectors of A. The product term in Eqn(2.2.5) is a complicated product of ratios of gaps between the computed eigenvalue approximations μ_k and the true eigenvalues λ_k. The key component in this bound is however the Chebyshev polynomial in the denominator. We know that if $2\gamma_j(m - j) > 1$ then this polynomial grows exponentially like $e^{2(m-j)\sqrt{2\gamma_j}}$, so that the error bound decays at this rate as we increase m. Observe that as would be expected that as we proceed further into the spectrum the bound increases.

The error bound given in Eqn(2.2.6) does not directly estimate the goodness of the corresponding Ritz vectors. Note also that it does not use the computed eigenvalue approximations. Instead it is a measure of how good the entire Krylov subspace is. As expected this bound on the eigenvector approximation indicates an accuracy which is essentially the square root of the indicated accuracy for the eigenvalue approximations. Parlett [1980] provides a detailed proof of Theorem 2.2.2 and we do not repeat that proof here. However, we will briefly summarize the basic ideas used in that proof. The following two generic lemmas are used. These are Lemmas 12.3.4 and 12.3.8 in Chapter 12 in Parlett [1980]. The Parlett numbering has been retained here.

LEMMA 12.3.4 Parlett [1980]   Given a real symmetric matrix A, then for any Rayleigh-Ritz approximation (μ,y) from a Krylov subspace $\mathcal{K}^m(A,v_1)$ and any polynomial p(t) of degree at most m-1, the corresponding vector $p(A)v_1$ is orthogonal to the Rayleigh-Ritz vector y if and only if $p(\mu) = 0$. That is, if and only if μ is a root of that polynomial.

The second lemma relates any Rayleigh quotient for a shifted A-matrix $A - \lambda_j I$ of a vector of the form $p(A)v_1$ where p(t) is a polynomial of degree not more than m-1, (these are obviously members of the Krylov subspace $\mathcal{K}^{m-1}$ corresponding to v_1) to the norm of the corresponding vector p(A)h where h is the normalized projection of the vector v_1 onto the orthogonal complement of the invariant subspace $\mathcal{X}^j$. To be more precise we write v_1 as $v_1 = g\cos\theta + h\sin\theta$ where $\theta \equiv \angle[v_1,\mathcal{X}^j]$. In the lemma below and elsewhere $\mathcal{P}^{m-1}$ denotes the family of scalar polynomials of degree $\leq m - 1$.

LEMMA 12.3.8 Parlett [1980]   For each polynomial $p(t)\epsilon\mathcal{P}^{m-1}$ and each $j\leq m$ the Rayleigh quotient ρ satisfies the following inequality.

$$\rho(p(A)v_1,\lambda_j I - A) \leq (\lambda_j - \lambda_n) \left[\frac{\sin\angle[v_1,\mathcal{X}^j]}{\cos\angle[v_1,x_j]} \frac{\|p(A)h\|}{p(\lambda_j)} \right]^2 \tag{2.2.8}$$

The notation $\rho(s,B)$ for a given matrix B and vector s stands for the Rayleigh quotient $(s^T Bs)/(s^T s)$. To prove Theorem 2.2.2 one first observes that if for a given j=1,2,..., m we pick a vector s_j which is in the orthogonal complement of the invariant subspace $\mathcal{X}^{j-1}$, then the following inequality holds

$$|\lambda_j - \mu_j| \leq \rho(s_j;\lambda_j I - A). \tag{2.2.9}$$

Thus if we construct a vector properly, we can use Lemma 12.3.8 in Eqn(2.2.9). Lemma 12.3.4 tells us how to construct such a vector. Lemma 12.3.8 is then invoked to obtain an upper bound on this Rayleigh quotient. The resulting ratio is then majorized by using results from Chebyshev polynomials. That is, the error estimation problem then reduces to finding a polynomial of degree not more than m-j with a minimum maximum over the interval $[\lambda_n, \lambda_{j+1}]$. This value of this min-max is given by the (m-j)th Chebyshev polynomial evaluated at $1 + 2\gamma_j$. Thus the key parts of the upper bounds on the convergence of μ_j given in Eqns(2.2.6) - (2.2.7), depend upon the gap between the jth eigenvalue and the closest eigenvalue which is smaller than it is and upon the effective gap, that is the gap between this closest eigenvalue and the smallest eigenvalue of A.

Saad's result, Theorem 2.2.2, is an improvement of a similar result due to Kaniel [1966]. Kaniel's error bounds incorporate the computed Ritz vectors explicitly, and the effect upon his bounds as one moves into the interior of the spectrum is much more obvious. Those bounds indicate specific decreases in accuracy of the computed eigenvalues and of the corresponding Ritz vectors as we move into the interior of the spectrum. We just note here that this need not happen in practice. Depending upon how the Lanczos procedure is constructed, interior eigenvalues of a matrix can be computed with the same accuracy as extreme eigenvalues can. This is the case for our single-vector Lanczos procedures. See Chapter 4. Kaniel's theorem is not stated below. The interested reader can find it in Parlett [1980, Chapter 12, p.245].

These error bounds indicate that for many matrices and for relatively small m, several of the extreme eigenvalues of A, that is several of the algebraically-largest or algebraically-smallest of the eigenvalues of A, are well approximated by eigenvalues of the corresponding Lanczos matrices. In practice it is not always true that both ends of the spectrum of the given matrix are equally well-approximated. However, it is generally true (but again not always) that at least one end of the spectrum is approximated well. See for example Example 4.7.6 in Chapter 4. Example 4.7.7 which is summarized in Table 4.7.9 in Chapter 4 illustrates however, that in some cases the Lanczos matrix must be as large as the original matrix before good eigenvalue approximations are obtained for any of the eigenvalues of A, including the extreme ones. Certainly Lanczos procedures tend to favor the extremes of the spectrum. However, depending upon the type of Lanczos procedure being used, the extremes of the spectrum need not be the first eigenvalues to converge.

Theorem 2.2.2 is applicable to a noniterative single-vector Lanczos procedure, and the assumption is made that all computations are in exact arithmetic. In some situations one is interested in computing only a few of the algebraically-largest (or algebraically-smallest) eigenvalues. Karush [1951] considered this latter question but only for the computation of the algebraically-largest or the algebraically-smallest eigenvalue of a given matrix A. We have the following basic iterative, single-vector scheme. We note that in practice block not single-vector versions of such an iterative procedure are used. Block Lanczos procedures are discussed in detail in Chapter 7.

BASIC SINGLE-VECTOR ITERATIVE LANCZOS PROCEDURE

Step 1. Let A be a real symmetric matrix of order n. Select a specified number of steps, $s \ll n$, and a unit starting vector v_1^0 and set $k = 0$.

Step 2. For each iteration k use the basic single-vector Lanczos recursion to generate a real symmetric tridiagonal Lanczos matrix of order s, T_s^k. For each k use the norm of the unnormalized second Lanczos vector $p_1^k \equiv Av_1^k - \alpha_1^k v_1^k$ to check for convergence of the iterations. If convergence is observed terminate.

Step 3. Compute the algebraically-largest eigenvalue of T_s^k and the corresponding eigenvector u^k.

Step 4. Compute the corresponding normalized Ritz vector of A, $y^k \equiv V_s^k u^k$ and return to Step 2, using this Ritz vector as the new starting vector for the next iteration. That is, $v_1^{k+1} \equiv y^k$.

This iterative Lanczos procedure is designed to compute only the algebraically-largest (or the algebraically-smallest) eigenvalue of A. Note that the size of the small Lanczos eigenelement problem to be solved on each iteration is fixed, namely size s. This is in contrast to the basic noniterative Lanczos procedure described earlier which for a given starting vector looks at what happens to the eigenvalue and eigenvector approximations as we increase the size of the Krylov subspace being used. Here in the basic iterative procedure we fix the size of the Krylov subspaces being considered, but on each iteration generate a new Krylov subspace using the best current approximation to the desired eigenvector as the starting vector in the Krylov subspace generation. In the iterative block Lanczos procedures which we will discuss in Chapter 7 we will be using a block analog of this basic idea for computing a few of the algebraically-largest or the algebraically-smallest eigenvalues of a given real symmetric matrix. There however the size of the small problem used on each iteration will be allowed to vary somewhat. Block analogs of Theorem 2.2.2 are given in Golub and Underwood [1977] and in Saad [1980].

Karush [1951], assuming all computations were done in exact arithmetic, proved that the procedure converged. Using the Lanczos recursions he derived an explicit formula for the small-dimensional eigenvector corresponding to the algebraically-largest eigenvalue of each of the Lanczos matrix iterates, T_s^k. He then used these formulas to obtain explicit expressions for the corresponding Ritz vectors for A. He then showed that these Ritz vectors converged to the desired eigenvector of A along with the corresponding eigenvalues, as the number of iterations $k \Rightarrow \infty$. Specifically, we have the following Theorem.

THEOREM 2.2.3 Karush [1951] Let A be a real symmetric nxn matrix. Let λ_1 denote the largest eigenvalue of A. Assume λ_1 is a simple eigenvalue and let x_1 denote the normalized eigenvector of A corresponding to λ_1. Given an initial vector $v_1^0 \neq 0$ with a nonzero projection on x_1 and a fixed dimension s ($1 < s < n$), construct a sequence of vectors $\{z^k\}$ as follows. Let z^{k+1} be the Ritz vector corresponding to the eigenvector of the Lanczos matrix T_s^{k+1} corresponding to the algebraically-largest eigenvalue. Then

$$y^k \equiv z^k / \| z^k \| \Rightarrow x_1 \text{ as } k \Rightarrow \infty. \qquad (2.2.11)$$

Furthermore, the corresponding algebraically-largest eigenvalues of the Lanczos matrices converge to λ_1.

Thus, we have restated error bounds or convergence results for two basic types of Lanczos procedures. The basic Kaniel-Saad bounds presented apply to noniterative single-pass single-vector Lanczos procedures. Convergence results were stated for an iterative single-vector Lanczos procedure designed to compute an extreme eigenvalue and corresponding eigenvector. Block Lanczos analogs of the Saad bounds are given in Saad [1980]. These analyses lead one to believe that the basic Lanczos recursion can provide a practical way of dealing with the computation of eigenelements of real symmetric matrices. However, that is not the case in general. In general, losses in the orthogonality of the Lanczos vectors occur as the recursion proceeds. These losses can occur after only a few steps of the Lanczos recursion or in some special cases may not occur at all. Typically, however they do occur at some point in the process, and when they do Theorem 2.2.2 can no longer be applied. Not only do the Lanczos vectors become non-orthogonal, they can even become linearly dependent. Thus, in this situation the corresponding Lanczos matrices T_m are not orthogonal projections of the given matrix A onto the associated Krylov spaces, and the theoretical error estimates for how well the eigenvalues of T_m should approximate eigenvalues of A are not valid.

That such losses in orthogonality would occur became apparent not long after Lanczos introduced his scheme in the 1950's. At that point in time it was assumed that such losses were caused by cancellation errors which occurred in the process of generating the Lanczos matrices when two numbers of similar size were subtracted. Thus, these errors were assumed to be unavoidable and not correctable. This behavior together with the research by Givens [1954] and by Householder [1958] on algorithms for the tridiagonalization of real symmetric matrices which do not suffer such losses in orthogonality, quelled the initial interest in the basic Lanczos recursion. Givens advocated the use of plane rotations to transform a real symmetric matrix A to tridiagonal form. Plane rotations are orthogonal matrices selected to zero out specified matrix entries. Such orthogonal transformations are numerically stable because norms of vectors are preserved under such transformations.

Householder countered with a proposal that elementary reflectors be used instead of plane rotations. Any real matrix of the form $H = I - 2xx^T$ where $x^Tx = 1$, is called an elementary reflector. An elementary reflector can be used to zero out simultaneously all of the elements below any subdiagonal entry. Both of these types of elementary orthogonal transformations, when implemented properly, provided numerically stable mechanisms for transforming a given real symmetric matrix into a symmetric tridiagonal matrix with the same eigenvalues. The eigenvalue algorithms which were subsequently developed used elementary reflectors because they were shown by Wilkinson [1960] to be superior with respect to speed and accuracy. See Wilkinson [1965, p.343], Stewart [1973] and EISPACK [1976,1977] for additional comments.

Interest in the Lanczos algorithm faded because most people viewed it primarily as a mechanism for tridiagonalization and that role was superseded by the Givens and Householder work. However, it should be noted that Lanczos had observed that even in cases where the global orthogonality of the Lanczos vectors was totally lost, some of the eigenvalues of the tridiagonal Lanczos matrices generated still approximated extreme eigenvalues of A. Thus, he suggested that in spite of all its numerically difficulties the Lanczos recursion could possibly be used to obtain a few extreme eigenvalues of a given real symmetric matrix.

With the advent of large computers, interest in solving large eigenvalue problems grew. Subsequently, primarily because of the low storage requirements, attempts were made to stabilize the Lanczos recursion so that it could be used for eigenelement computations on large symmetric matrices. Since initially no one understood the mechanism for the losses in orthogonality in the recursion, the first proposals to stabilize the Lanczos procedure were not very imaginative. A seemingly-obvious way to correct for the losses in orthogonality of the Lanczos vectors is to explicitly reorthogonalize each newly-generated Lanczos vector against each and every previously-generated Lanczos vector. This was in fact something that Lanczos had himself suggested. In practice this reorthogonalization worked, in the sense that when the Lanczos recursion was used together with this complete reorthogonalization of the Lanczos vectors, a few of the extreme eigenvalues of A could be reliably computed. The cost however of performing this total reorthogonalization is large, and the added storage requirements limit the number of eigenvalues which can be computed as well as the size of the matrix which can be handled reasonably. More will be said about such procedures in Section 2.5 where we briefly survey the recent research on Lanczos procedures.

Finally, a satisfactory explanation for the losses of orthogonality in the Lanczos vectors was provided by Paige [1971] in his Ph.D. thesis. This is the topic of discussion in the next section. Before proceeding to Section 2.3 we want to provide a simple example of the hazards of accepting theoretical results without checking them in a practical numerical situation. We consider simple real symmetric tridiagonal matrices. We have the following theorem which can be proved in a very straight-forward manner.

THEOREM 2.2.4 Let T be a real symmetric tridiagonal matrix with diagonal entries, $T(i,i) \equiv \alpha_i$ and with off-diagonal entries $T(i,i+1) = T(i+1,i) \equiv \beta_{i+1} \neq 0$, then the eigenvalues of T are distinct.

Proof. We need only show that each eigenvalue is simple. If not, then there would be an eigenvalue μ for which the matrix $T - \mu I$ has a null space of dimension two or greater. This can happen only if the rank of this matrix is less than n-1. But, the rank of a matrix is determined by the size of the largest nonsingular submatrix of T. But the determinant of the submatrix obtained by crossing out the first column and the last row of T is just $\prod_{k=2}^{n} \beta_k \neq 0$. Therefore, $T - \mu I$ has rank at least n-1. Thus, μ must be a simple eigenvalue. $\quad\square$

Theorem 2.2.4 states unequivocally that all of the eigenvalues of an irreducible real symmetric tridiagonal matrix are distinct. However, using the basic Lanczos recursion with no reorthogonalization of the Lanczos vectors, one can easily generate tridiagonal Lanczos matrices which simultaneously have uniformly 'large' off-diagonal entries β_{i+1} and many numerically-multiple eigenvalues (to 12-13 digits). In fact this will happen starting with any real symmetric matrix, as long as one chooses a large enough corresponding Lanczos matrix T_j. See Table 4.5.2 in Chapter 4 for an example of this behavior. This is in fact an observation of one of the difficulties of the basic single-vector Lanczos recursion with no reorthogonalization, namely that a simple eigenvalue of the given matrix A may appear as an eigenvalue of a corresponding Lanczos matrix with a (numerical) multiplicity greater than one.

SECTION 2.3   LANCZOS RECURSION, FINITE PRECISION ARITHMETIC

In practical situations the arithmetic is not exact and errors are incurred on every computation. Computers provide only a gridwork of numbers covering the continuous real number system. The fineness of this grid depends upon the number of digits allowed to represent a number and upon the base for the number system used, base 2, base 8, base 16, etc. The resulting arithmetic operations have finite precision, and therefore roundoff errors occur in any numerical computations. These roundoff errors make the computed quantities differ from their theoretical values. The question which will be addressed here is, what are the effects of these differences upon the relationships between the given matrix A, the Lanczos vectors V_j, and the Lanczos matrices T_j generated by the basic Lanczos recursion when there is no reorthogonalization of the Lanczos vectors.

The Lanczos vectors generated lose their orthogonality as j is increased. Once this global orthogonality is lost, the Lanczos matrices T_j are no longer simply orthogonal projections of the given matrix A onto the Lanczos subspaces $sp\{V_j\}$, and therefore the classical relationships which are summarized in Section 2.2 are not applicable. In fact typically one observes not only losses in the orthogonality of the Lanczos vectors, but the total loss of even the linear independence of these vectors. Thus, even a generalization of the classical estimates to the case of linearly independent Lanczos vectors (rather than orthogonal vectors) is not enough to obtain estimates and bounds which can then be applied to the practical situation.

For many years it was assumed that these losses in orthogonality were simply caused by the accumulation of cancellation and roundoff errors, and therefore were not controllable or correctable in any nice way other than by the incorporation of the total reorthogonalization of the Lanczos vectors. Paige [1971] showed that this was not the case. In his thesis and in several subsequent papers Paige [1972,1976,1980], he reexamined the Lanczos procedure, incorporating the effects of finite precision arithmetic. He found that the major losses in global orthogonality of the Lanczos vectors were not a simple consequence of the accumulation of the roundoff errors in the computation. Rather they were due to a combination of the effects of the roundoff errors together with the convergence of one or more of the eigenvalues of the Lanczos matrices T_j as j was increased. Such converged eigenvalues were shown to be good approximations to eigenvalues of the original A-matrix. Moreover, he showed that these losses in orthogonality were occurring along the directions of the corresponding Ritz vectors, the approximations to the eigenvectors of A. In particular he showed that for any given matrix A, for all j for which none of the eigenvalues of the associated Lanczos matrix T_j had converged, there were no significant losses in the orthogonality of the associated Lanczos vectors V_j.

A complementary and even more important observation (from our point of view in Chapter 4) was that although in general the global orthogonality of the Lanczos vectors was totally lost after perhaps very few many steps of the basic Lanczos recursion, a local nearest-neighbor, near-orthogonality of the Lanczos vectors persisted. In this section we restate each of these and several other key observations in the above papers by Paige. We basically consider these papers in chronological order. We note however that much of what is in Paige [1980] was actually contained in his 1971 thesis, but that since this 1980 paper is more readily accessible to the reader, we will refer to it rather than quoting from the thesis.

First we look at the question of which of several possible theoretically equivalent recursive formulas for the Lanczos recursion is the best, from a numerical standpoint. This question is discussed in Paige [1972]. After this discussion we consider the general error analysis of this best Lanczos recursion. This analysis is contained in Paige [1976] where he is asking the basic question, what are the relationships and bounds for the rounding errors produced in computing the Lanczos vectors and the Lanczos matrices, and in computing $AV = VT$, and what can we say about the losses in orthogonality, i.e. the differences $V_j^T V_j - I$. After we finish summarizing Paige [1976] we move on to Paige [1980]. We restate Paige's basic result relating the losses of orthogonality to the convergence of eigenvalues plus give some of the very fundamental properties of the Lanczos matrices generated. A key result given in Paige [1980] is that regardless of any losses in orthogonality, the eigenvalues of each of the Lanczos matrices generated must be (to within a small perturbation) contained in the spectral interval of the original matrix. We will need these results in Chapter 4 in the development of our Lanczos procedures with no reorthogonalization.

Paige [1972] considers several theoretically equivalent variants of the formulas used to compute the scalars α_i and β_{i+1} in the Lanczos recursions. We list these below. These are taken directly from Paige's paper. In each case select a starting vector v_1 with $\| v_1 \| = 1$. Then for each j=1,2,....,m define the Lanczos matrices T_j and the Lanczos vectors v_j by some combination of the following formulas. In particular, let $u_1 = Av_1$, then for each $j = 1,2,...$ set

$$\alpha_j = v_j^T A v_j \qquad (2.3.1)$$

or

$$\alpha_j = v_j^T u_j, \qquad (2.3.2)$$

$$w_j = u_j - \alpha_j v_j \qquad (2.3.3)$$

$$\beta_{j+1} = (w_j^T w_j)^{1/2} \qquad (2.3.4)$$

$$v_{j+1} = w_j / \beta_{j+1} \qquad (2.3.5)$$

$$\gamma_{j+1} = v_j^T A v_{j+1} \qquad (2.3.6)$$

or

$$\gamma_{j+1} = \beta_{j+1}. \qquad (2.3.7)$$

$$u_{j+1} = Av_{j+1} - \beta_{j+1} v_j. \qquad (2.3.8)$$

The four possible recursions will be denoted respectively by A(1,6), A(1,7), A(2,6) and A(2,7). This is the notation which Paige [1972] uses. Here we understand for example that A(2,6) means the Lanczos recursion which we obtain if we use formulas (2.3.2)-(2.3.6) together with (2.3.8). Thus, the A(2,6) and the A(2,7) recursions use a modified Gram-Schmidt formula for the α_i coefficients. Recursions A(1,6) and A(2,6) use the definition of γ_{j+1} which directly enforces the orthogonality of the (j+1)st Lanczos vector w.r.t the (j-1)st vector. For recursions A(1,7) and A(2,7), this 2-step orthogonality of v_{j+1} to v_{j-1} depends recursively upon the orthogonality of v_j to v_{j-2}.

Paige considers each of these 4 possible Lanczos recursions in the practical situation where the quantities in the recursions are computed with errors due to the use of finite precision arithmetic. The objective of the 1972 paper was to identify the best of these 4 recursions. The following lemma gives an indication of one type of difficulty identified by that analysis.

LEMMA 2.3.1 Paige [1972] Let 4 variations of the basic Lanczos recursion be defined as above and denoted by A(1,6), A(2,6), A(1,7), and A(2,7) respectively. Then the following statements are true. In recursion A(1,6) for each j we have that

$$\gamma_{j+1} v_j^T v_{j+1} = \left[\sum_{k=2}^{j} \prod_{\ell=k}^{j} \frac{\beta_\ell}{\gamma_\ell} + 1 \right] O(\varepsilon). \qquad (2.3.9)$$

For recursions A(2,6), A(2,7) and A(1,7) for each j we have that

$$\gamma_{j+1} v_{j+1}^T v_{j+1} = O(\varepsilon). \qquad (2.3.10)$$

In each case $O(\varepsilon)$ is bounded by a multiple of $\varepsilon \| A \|$, where ε denotes the machine epsilon.

Lemma 2.3.1 tells us that preservation of the nearest-neighbor orthogonality of the Lanczos vectors in recursion A(1,6) is sensitive to any differences in sizes of the off-diagonal entries β_{i+1} and γ_{i+1} in the corresponding Lanczos matrices. The other three recursions do not suffer from this sensitivity. Paige uses Lemma 2.3.1 together with an argument which shows that if γ_{j+1} is computed using Eqn(2.3.6) then the values of γ_{j+1} and β_{j+1} can diverge significantly to argue that therefore one can expect to encounter numerical problems with recursion A(1,6). (Note that theoretically these values should be identical.) He reports numerical experiments using the recursion A(1,6) for which the eigenvalues of the corresponding Lanczos matrices wandered aimlessly.

For both recursions A(1,6) and A(2,6) the resulting Lanczos matrices are not symmetric. Paige shows that this means that close eigenvalues of the original matrix A may therefore appear as eigenvalues of the Lanczos matrices with accuracies only of the order of $\sqrt{\varepsilon}$ instead of ε. He encountered such inaccuracies in his numerical experiments. On the other hand the Lanczos matrices for the Lanczos recursions A(1,7) and A(2,7) are symmetric and with these recursions the local orthogonality is preserved. From these and other arguments Paige concludes that A(1,7) and A(2,7) are the correct recursions to use in practice.

Paige [1972] contains a numerical comparison of these 4 variants of the basic Lanczos recursion for a test matrix of order n=8 with m=20 and for a test matrix of order n=1000 with m=600. For the 8x8 matrix at m=20, Paige gets all of the eigenvalues of this matrix with varying degrees of accuracy, plus he gets some extra eigenvalues which are due to the losses of orthogonality of the Lanczos vectors which have occurred. Thus, he provides an example of what we will call in Chapter 4, the Lanczos phenomenon. Namely, that if for a given real symmetric matrix we run the Lanczos recursion out far enough (obviously beyond m=q where q is the number of distinct eigenvalues of the given matrix), then for large enough m the Lanczos matrix T_m will have all of the distinct eigenvalues of the original matrix A among its eigenvalues. The 1000×1000 matrix was derived from the Laplace operator on a rectangular region with Dirichlet (zero) boundary conditions. At m=600 Paige obtained 184 converged eigenvalue approximations, each with absolute error less than 3×10^{-10}. Paige comments that this accuracy is remarkable, as it is. This type of accuracy and even better is achievable on many real symmetric problems.

In our tests and in tests by others it has further been demonstrated that for some large matrices there can be a significant difference between the performance of the two Lanczos recursions A(1,7) and A(2,7). These tests showed that it is important to use the modified

Gram-Schmidt version A(2,7). Therefore, this is the version which we use in our Lanczos algorithms which are discussed in Chapter 4, and in fact this is the version Paige focussed on in his later papers.

For the Lanczos recursion A(2,7), Paige [1976] gives relationships and bounds for the rounding errors produced in computing the Lanczos vectors and the Lanczos matrices and for the losses in orthogonality of the Lanczos vectors. Paige points out that if in the implementation of A(2,7) the vectors w_j are allowed to overwrite the vectors u_j, (see Eqn(2.3.3) and Eqn(2.3.8)) the vectors v_{j+1} overwrite the vectors w_j, and the vectors u_{j+1} overwrite the vectors v_j, then the basic recursions require computer storage for only two Lanczos vectors. Therefore, not only is the recursion A(2,7) the most numerically stable, it also requires the least amount of computer storage. In the error analysis provided, Paige ignores terms which are second and higher order in the machine epsilon ε. We restate his basic theorem below. Recall that for a given matrix A, $|A| \equiv (|a_{ij}|)$.

THEOREM 2.3.1 Paige [1976] Let A be a nxn real symmetric matrix with at most n_z nonzero elements in any row. Define $\sigma \equiv \|A\|$, and $\alpha\sigma \equiv \||A|\|$. Let ε be the relative precision of the computer arithmetic being used. Then if the basic Lanczos recursions A(2,7) are applied to A for j steps starting with a unit vector v_1, then the Lanczos matrices and the Lanczos vectors satisfy the following relationships for i = 1,2,...,j. We assume that each $\beta_{i+1} \neq 0$ and that for the ε_0 and the ε_1 defined in Eqn(2.3.13) that $4j(3\varepsilon_0 + \varepsilon_1) \ll 1$.

$$AV_j = V_jT_j + \beta_{j+1}v_{j+1}e_j^T + F_j \tag{2.3.11}$$

$$|v_{i+1}^Tv_{i+1} - 1| \leq \varepsilon_0 \tag{2.3.12}$$

$$\|f_i\| \leq \sigma\varepsilon_1, \text{ where } F_j \equiv \{f_1,...,f_j\} \tag{2.3.13}$$

$$\beta_{i+1}|v_i^Tv_{i+1}| \leq 2\sigma\varepsilon_0 \tag{2.3.14}$$

where we have used the notation

$$\varepsilon_0 \equiv (n + 4)\varepsilon, \quad \varepsilon_1 \equiv (7 + n_z\alpha)\varepsilon. \tag{2.3.15}$$

Furthermore, if we let R_j be the strictly upper triangular matrix such that

$$V_j^TV_j \equiv R_j^T + \text{diag}(v_i^Tv_i) + R_j \tag{2.3.16}$$

then

$$T_jR_j - R_jT_j = \beta_{j+1}V_j^Tv_{j+1}e_j^T + H_j \tag{2.3.17}$$

where H_j is an upper triangular matrix with elements $h_{\ell k}$ such that

$$|h_{\ell k}| \leq \sigma(4\varepsilon_0 + 2\varepsilon_1) \tag{2.3.18}$$

The proof is obtained by bounding all of the interesting quantities (for example the α_i and the β_{i+1}) and the rounding errors by expressions involving the norms $\|u_i\|$ of the vectors defined in Eqn(2.3.8), and then obtaining a bound for these norms. The determination of this bound rests upon the determination of an appropriate bound on the two-step orthogonality of each Lanczos vector v_{j+1} to the Lanczos vector v_{j-1}. This two-step orthogonality depends of course upon the one-step orthogonality stated in Eqn(2.3.14). Note that we have one-step orthogonality as long the β_{i+1} are not too small. We will use a related inequality in Chapter 4 so we repeat Paige's version here.

LEMMA 2.3.2. Paige [1976]. Under the assumptions and definitions in Theorem 2.3.1 we have the following bounds. Here we assume that $2j[7 + n_z\alpha + 3(n + 4)]\varepsilon << 1$. For each $i = 1,2,...,j$

$$\beta_i\beta_{i+1} |v_{i-1}^T v_{i+1}| \leq 2(i - 1)[3(n + 6)\mu_i + (1 + n_z\alpha)\sigma]\mu_i\varepsilon \qquad (2.3.19)$$

where

$$\mu_i \equiv \max_{1 \leq k \leq i}\{\|u_k\|\}, \quad \mu \equiv \max(\mu_j,\sigma), \text{ and} \qquad (2.3.20)$$

$$\mu^2 \leq \sigma^2(1 + 4j[7 + n_z\alpha + 3(n + 4)]\varepsilon). \qquad (2.3.21)$$

Inequality (2.3.21) is used repeatedly throughout the proof of Theorem 2.3.1. Thus, if the off-diagonal entries in the Lanczos matrices are not too small, two-step near-orthogonality of the Lanczos vectors is also preserved. This two-step orthogonality bound is, however, much much weaker than the bound for one-step orthogonality. The relationships Paige obtained in Theorem 2.3.1 can be applied to the analysis of specific single-vector Lanczos procedures which have been proposed in the literature. Paige [1980] has done this for the basic single-vector Lanczos eigenvalue procedure. We next summarize most of the key results in the 1980 paper.

Paige's analyses assume that no other rounding errors occur beyond those incurred in the generation of the Lanczos matrices. In particular, he assumes that the eigenvalues and the eigenvectors of the Lanczos matrices can be computed with no error. This assumption is acceptable because we know from past research that those computations are numerically stable. Before proceeding we need to recall some notation. For a given Lanczos matrix T_j we denote the eigenvalues of the Lanczos matrices by $\{\mu_1^j \geq \mu_2^j \geq ... \geq \mu_j^j\}$ and those of A by $\{\lambda_1 \geq \lambda_2 \geq ... \geq \lambda_n\}$. We let u_i^j denote corresponding orthonormal eigenvectors of T_j, and let $y_i^j \equiv V_j u_i^j$ denote the corresponding Ritz vectors. For any eigenvector u_i^j corresponding to the eigenvalue μ_i^j, $i = 1,2,...,j$ we define

$$\varepsilon_{ik}^j \equiv (u_i^j)^T H_j u_k^j \qquad (2.3.22)$$

where H_j is specified by Eqn(2.3.17). Paige [1980] contains a restatement of the fundamental result in his thesis concerning the observed losses in orthogonality of the Lanczos vectors. We restate that result first.

LEMMA 2.3.3 Paige [1971,1980] Under the hypotheses of Theorem 2.3.1 we have the following equality relating each Ritz vector y_i^j to the Lanczos vectors V_j. For each i=1,2,...,j we have

$$| v_{j+1}^T y_i^j | \ = \ \frac{\varepsilon_{ii}^j}{| \beta_{j+1} u_i^j(j) |} \tag{2.3.23}$$

where $u_i^j(j)$ denotes the jth component of the ith eigenvector of the Lanczos matrix T_j and for each i and k, $| \varepsilon_{ik}^j | \ \leq \ j\sigma\varepsilon_2$ with

$$\varepsilon_2 \ \equiv \ 2\sqrt{2}\max(6\varepsilon_0,\varepsilon_1). \tag{2.3.24}$$

Since the proof is not difficult to follow and this is a very important observation, we include the proof given by Paige.

Proof. Given an eigenvector u_i^j apply it to both sides of Eqn(2.3.17) to obtain upon simplification

$$(\mu_i - \mu_k)(u_i^T R_j u_k) \ = \ \beta_{j+1} y_i^T v_{j+1} u_k(j) \ + \ \varepsilon_{ik}.$$

Note that to simplify the notation we have deleted the superscript j. If we then set i=k, we obtain Eqn(2.3.23). □

Note that from Theorem 2.3.1 we have that for i,k = 1,2,...,j,

$$| \varepsilon_{ik} | \ \leq \ \sigma\varepsilon_2 \tag{2.3.25}$$

and that ε_2 is a function of j. In Lemma 2.3.6 we will state the relationship between significant losses in the orthogonality of the Lanczos vectors and the convergence of eigenvalues of the Lanczos matrices to eigenvalues of A which one obtains from Eqn(2.3.23). However, first we must state two other lemmas which we will need in order to make this connection. First we must know that the norms of the Ritz vectors being considered are not too small. The arguments that this is the case for any isolated eigenvalue of a Lanczos matrix and the corresponding generalization to clusters of eigenvalues are given in Paige [1971,1980]. Here we simply restate results from the 1980 paper. In the case of clusters of eigenvalues, the result says that at least one eigenvalue in any cluster must have a corresponding Ritz vector which has a reasonable size norm. Specifically in Paige [1980] we find the following statements.

LEMMA 2.3.4 Paige [1980] Let the hypotheses of Theorem 2.3.1 be satisfied. For a given j let y_i^j denote the Ritz vector obtained from the ith eigenvalue/eigenvector of T_j. Then we have the following inequality. If

$$\min_{k \neq i} |\mu_i^j - \mu_k^j| \geq j^{5/2}\sigma\varepsilon_2, \tag{2.3.26}$$

then

$$0.42 < \|y_i^j\| < 1.4. \tag{2.3.27}$$

Furthermore, if $\mu_i^j, ..., \mu_{i+s}^j$ are $s+1$ eigenvalues of T_j which form a cluster which is well-separated from the other eigenvalues of T_j, then for the corresponding Ritz vectors y_k^j we have that

$$\sum_{k=i}^{i+s} \|y_k^j\|^2 = s + 1. \tag{2.3.28}$$

That is, a Ritz vector of some member of this cluster must have a norm that is not small.

Paige [1980] makes the comment that the bounds obtained are weaker than they need to be. Lemma 2.3.4 tells us that even though the Lanczos vectors are not orthogonal or even linearly independent, the norms of the Ritz vectors, corresponding to isolated eigenvalues of the Lanczos matrices are not small. This has to be considered a very interesting result. Paige uses known eigenvalue perturbation theorems such as those summarized in Chapter 1, Lemma 2.3.4, along with modifications of Eqn(2.1.4) which include his error estimates to get the following lemma relating the computed eigenvalues to those of the original matrix.

LEMMA 2.3.5 Paige [1980] Under the hypotheses of Theorem 2.3.1, for any j and for any isolated eigenvalue μ_i^j of the Lanczos matrix T_j, satisfying Eqn(2.3.26), we have that there exists an eigenvalue λ_ℓ of A such that the following inequality holds.

$$|\lambda_\ell - \mu_i^j| \leq 2.5 \left[|\beta_{j+1} u_i^j(j)| (1 + 2\varepsilon_0) + 2\varepsilon_1\sigma j^{1/2} \right] \tag{2.3.29}$$

In particular Lemma 2.3.5 tells us that whenever the last component of the eigenvector corresponding to an eigenvalue of some Lanczos matrix is very very small, then that eigenvalue is a good approximation to an eigenvalue of the original matrix A. On the other hand Eqn(2.1.4) (and Paige's modification of it which includes the effects of finite precision arithmetic) tells us that if the off-diagonal entries β_{j+1} are not small, then the only way convergence can occur is if some eigenvectors of the Lanczos matrices have very very small last components. This argument is only valid for isolated eigenvalues. However, using a different argument it can be demonstrated that if there is an eigenvalue cluster in some T_j, then there is an eigenvalue of A in the interval defined by that cluster. Lemmas 2.3.3 and Lemma 2.3.5 combine to give us the following lemma which tells us that the significant losses in the orthogonality of the Lanczos vectors go hand in hand with the convergence of one or more of the eigenvalues of the Lanczos matrices to eigenvalues of the original matrix A.

LEMMA 2.3.6 Paige [1971,1980] Under the hypotheses of Theorem 2.3.1, significant losses in the orthogonality of the Lanczos vectors can occur only if one or more of the eigenvalues of the Lanczos matrices has converged to an eigenvalue of the original matrix A. Furthermore, such losses in orthogonality occur along the directions of the Ritz vectors corresponding to such eigenvalues.

To see this one needs only to replace the Ritz vector y_i^j in Eqn(2.3.23) by $V_j u_i^j$, and invoke Lemmas 2.3.3 and 2.3.5. In particular, it is clear that Eqn(2.3.23) implies that the Lanczos vectors are essentially orthogonal, unless the denominator of this equation becomes very small. This can happen only if β_{j+1} is very small for some j or if the last component of the corresponding eigenvector of the Lanczos matrix is pathologically small. In practice, it is not typical for the β_{j+1} to be pathologically small, although one can construct examples where this happens. Instead in practice one observes the convergence of one or more of the eigenvalues of the Lanczos matrices by the appearance of pathologically small components of the corresponding Lanczos matrix eigenvectors, or equivalently by very small $u_i^j(j)$. By Lemma 2.3.5 however this corresponds to the convergence of eigenvalues of the Lanczos matrices to eigenvalues of A. Lemma 2.3.6 can be used as an argument for the limited type of reorthogonalization which is done in several Lanczos procedures discussed in the literature.

Another very important result in Paige [1980] is the following lemma which states that the eigenvalues of any Lanczos matrix T_j must be contained in an interval which is only a very small perturbation of the interval defined by the eigenvalues of the given matrix A.

LEMMA 2.3.7 Paige [1980] Under the hypotheses of Theorem 2.3.1 for every j, the eigenvalues μ_i^j, $i = 1,...,j$ of the corresponding Lanczos matrix T_j satisfy

$$\left[\lambda_{\min}(A) - j^{5/2}\sigma\varepsilon_2\right] \le \mu_i^j \le \left[\lambda_{\max}(A) + j^{5/2}\sigma\varepsilon_2\right] \qquad (2.3.30)$$

where ε_2 is given in Eqn.(2.3.24).

The basic idea in the proof of Lemma 2.3.7 is to show that the eigenvalues μ_i^j of each of the Lanczos matrices are essentially Rayleigh quotients of the original matrix A corresponding to the corresponding Ritz vectors y_i^j. In particular from Theorem 2.3.1 we have that for any Ritz vector y_i^j corresponding to μ_i^j,

$$y_i^T A y_i - \mu_i^j(y_i^T y_i) = -\varepsilon_{ii} + y_i^T F_j u_i.$$

To simplify the notation the j superscripts were omitted. Here ε_{ii}^j is defined in Eqn(2.3.22). Invoking Eqns(2.3.25) and (2.3.27) we get that the Rayleigh quotient and computed eigenvalue differ by less than $\delta j\sigma\varepsilon_2$ where $\delta \ge |\beta_{j+1}u_i(j)|$. Thus with no reorthogonalization and in general even with the total loss of the linear independence of the associated Lanczos vectors, the convex hulls of the spectrums of the Lanczos matrices are contained in a small perturbation of the convex hull of the spectrum of A. We state one more Lemma which is important for the

Lanczos procedures described in Chapter 4.

LEMMA 2.3.8 Paige [1980] Under the hypotheses of Theorem 2.3.1 for some j let an eigenvalue μ_i^j and a corresponding normalized eigenvector u_i^j of the Lanczos matrix T_j be computed. That is, $T_j u_i^j = \mu_i^j u_i^j$. Then for any J>j there exists an eigenvalue μ_ℓ^J of T_J such that

$$| \mu_\ell^J - \mu_i^j | \leq | \beta_{j+1} u_i^j(j) |.$$

(2.3.31)

The proof of Lemma 2.3.8 is straight-forward. This lemma is important numerically. Specifically, it says that if a particular size Lanczos matrix is appropriate for a given eigenvalue in the sense that that eigenvalue has converged at that size as measured by the size of the last component of the corresponding eigenvector of the Lanczos matrix, then any larger size Lanczos matrix is just as suitable for the computation of that eigenvalue. This is very important from a numerical standpoint.

Paige [1980] also demonstrates that until one or more of the eigenvalues of the Lanczos matrices converge or as Paige calls it, stabilize, the basic Lanczos procedure with no reorthogonalization behaves essentially like a Lanczos procedure which incorporates total reorthogonalization of every Lanczos vector w.r.t. every other Lanczos vector. This is really just a statement that any significant losses in the orthogonality of the Lanczos vectors can occur only if the Lanczos recursion has been carried out far enough so that one or more eigenvalues of the A-matrix are being approximated well by eigenvalues of the corresponding Lanczos matrices. Thus, the losses in orthogonality are not just simple accumulations of roundoff errors. They are due to the roundoff error, for if there were none then there would not be any losses in orthogonality. However, the orthogonality problems are caused by the interaction of the roundoff errors with the convergence.

In Chapter 4 we will use the Lemmas and Theorems included in this Section in our development of our Lanczos procedures which do not use any reorthogonalization. In Section 2.5 we survey briefly the recent research on Lanczos procedures for solving real symmetric eigenelement problems or problems which are equivalent to real symmetric problems. First however in the next section we attempt to identify those basic properties which make one Lanczos procedure different from another. Lewis [1977] and Scott [1979] discuss a block Lanczos analog of the fundamental loss of orthogonality result given in Lemma 2.3.3.

SECTION 2.4   TYPES OF PRACTICAL LANCZOS PROCEDURES

Any practical Lanczos procedure must address the problems created by the losses in the orthogonality of the Lanczos vectors when the basic Lanczos recursion is implemented on a computer. These problems are two-fold. First, simple eigenvalues of A may appear as numerically-multiple eigenvalues of some T_j for large j. Second spurious eigenvalues may appear among the eigenvalues of some T_j. These are eigenvalues which are due to the losses in orthogonality of the Lanczos vectors and which should not be accepted as approximations to eigenvalues of A. We will have more to say about these problems in Chapter 4. There are two extreme approaches for dealing with these problems. One incorporates the total reorthogonalization of every Lanczos vector with respect to every previously-generated Lanczos vector. The other approach simply accepts the losses in orthogonality and then tries to deal directly with the problems introduced by these losses.

Total reorthogonalization is one way of maintaining the orthogonality, however it destroys the basic simplicity of the Lanczos recursion with respect to both the storage requirements and the number of arithmetic operations required. It requires that all of the previously-generated Lanczos vectors be available for the reorthogonalization of each Lanczos vector as it is generated, and therefore, the number of eigenvalues which can be computed is limited by the amount of easily accessible computer storage available. On the other hand a Lanczos procedure which does not use any reorthogonalization needs only the two most recently-generated Lanczos vectors at each stage, so has minimal computer storage requirements.

The literature contains many examples of proposed Lanczos procedures which fall somewhere in between total reorthogonalization and none at all. In Section 2.5 we will briefly survey many of the existing Lanczos procedures for eigenelement computations. Trying to distinguish between these various Lanczos procedures can lead to an enormous amount of confusion on the part of someone who is not familiar with Lanczos techniques. In this section we first attempt to clarify this situation.

We restrict our discussion to real symmetric matrices. However, the real symmetric Lanczos recursion can be generalized to handle other types of matrices. For example in Chapter 5 we provide an extension to singular value/vector computations for real, rectangular matrices. In Chapter 6 we provide an extension to diagonalizable complex symmetric matrices. In Chapter 4, Section 4.9, we show that the Lanczos recursion can be modified to use on Hermitian and on certain real symmetric generalized eigenvalue problems.

There are at least 5 different choices which a designer of a Lanczos procedure must make. Below we briefly specify what each of these is. We will use these 5 choices in our discussion in Section 2.5.

CHOICE 1. TYPE OF RECURSION The first question which must be dealt with is the selection of the type of basic Lanczos recursion to be used. Do we use a single-vector Lanczos recursion as A(2,7) defined by Eqns(2.3.2) - (2.3.5) together with Eqns(2.3.7) - (2.3.8), or a block analog of this recursion as defined in Eqns(2.4.1) - (2.4.2) below? In fact one could also use a hybrid of these two types of recursions as we will do in Chapter 7.

BASIC BLOCK LANCZOS RECURSION

Let A be a real symmetric nxn matrix. Specify q and define an nxq starting block of vectors Q_1, typically generated randomly such that $Q_1^T Q_1 = I_q$. Set $B_1 \equiv 0$, and $Q_0 \equiv 0$. Then for $j=1,2,...,s$ define Lanczos blocks Q_{j+1} and corresponding Lanczos matrix blocks A_j, and B_{j+1} by

$$Q_{j+1} B_{j+1} = P_j \equiv AQ_j - Q_j A_j - Q_{j-1} B_j^T \qquad (2.4.1)$$

where

$$A_j \equiv Q_j^T (AQ_j - Q_{j-1} B_j^T) \text{ and } Q_{j+1} B_{j+1} = P_j \qquad (2.4.2)$$

with B_{j+1} an upper triangular matrix obtained by either a QR factorization or the Gram-Schmidt orthogonalization of the unnormalized block P_j.

We use the notation Q_j instead of V_j to denote the jth Lanczos blocks because in the block procedures which we discuss in Chapter 7 we will modify the basic procedure by incorporating one type of limited reorthogonalization and thereby maintain a certain level of orthogonality between the blocks. Thus, for the block case, at least for our block procedures, the Lanczos blocks will be nearly orthogonal. We note that this is not necessarily true for other block procedures. See for example, Lewis [1977] which we will briefly describe in the next section.

In the block case the Lanczos matrices obtained are not tridiagonal. They are block tridiagonal of the form

$$T_s \equiv \begin{bmatrix} A_1 & B_2^T & & & \\ B_2 & A_2 & B_3^T & & \\ & & \ddots & & \\ & & & & B_s^T \\ & & & B_s & A_s \end{bmatrix}. \qquad (2.4.3)$$

The computational advantages of tridiagonal matrices do not extend to this case. However if the blocks are narrow, then one could treat the Lanczos matrices as band matrices and reasonably large Lanczos matrices could be handled. In Chapter 7 however, where we discuss our iterative block Lanczos procedures we will be working with relatively small Lanczos matrices, $sq \ll n$. q is the number of vectors in the first block and s is the number of blocks generated on an iteration. On each iteration the latest and best approximations to the desired eigenspace will be used as the starting block. Other block procedures however are not iterative and can involve the generation of large block tridiagonal matrices, see for example Lewis [1977] and Scott [1979].

The type of recursion which one chooses will depend primarily upon what one is trying to compute. If only a few extreme eigenvalues, extreme meaning either the algebraically-smallest or the algebraically-largest, a basis for the corresponding eigenspace is required, and the user expects one or more of the desired eigenvalues to be multiple, then a block Lanczos procedure may be preferable to a single-vector procedure. Our own experience has been that in most cases the single-vector Lanczos procedure with no reorthogonalization is preferable because typically it requires less storage, less computation, and produces more accurate results.

CHOICE 2.   AMOUNT OF REORTHOGONALIZATION USED   Perhaps the main question which must be addressed other than the choice of recursion which is to be used, is the question of how much reorthogonalization is to be used, and reorthogonalization w.r.t. what?   The amount of reorthogonalization allowed or specified in the Lanczos algorithms in the literature varies from none at all to complete reorthogonalization of each and every Lanczos vector generated w.r.t. all of the Lanczos vectors which have been generated previously. Reorthogonalization requires the use of Lanczos vectors either to use in a direct reorthogonalization of Lanczos vectors versus Lanczos vectors or to use to generate particular Ritz vectors which will then be used in the reorthogonalizations. This means keeping the Lanczos vectors on some readily accessible computer storage throughout the entire Lanczos computations or else regenerating them when they are needed which is impractical if the reorthgonalizations are done repeatedly.

Thus, reorthogonalization (other than that which only involves nearest-neighbors) effectively limits the number of eigenvalues which can be computed accurately. The primary focus of this book is on single-vector Lanczos procedures which use no reorthogonalization and therefore which do not suffer from this limitation. The single-vector Lanczos procedures presented in Chapters 4, 5, and 6 are all Lanczos procedures with no reorthogonalization. In Chapter 7 however, we deal with an iterative block Lanczos procedure and use a limited form of reorthogonalization. The single-vector Lanczos procedures with no reorthogonalization allow the computation of many and in some cases even all of the eigenvalues of a given matrix.

As we said earlier, initial proposals for fixing the Lanczos procedure incorporated the total reorthogonalization of each Lanczos vector w.r.t to all previously-generated ones. Some of the more recent research has focussed on limited reorthogonalization. Some of this, which makes ample sense according to Paige's results [1971], incorporates reorthogonalizations w.r.t converged Ritz vectors rather than w.r.t Lanczos vectors. Others use limited reorthogonalization of the Lanczos vectors themselves, such as one step or two-step reorthogonalizations of the Lanczos vectors. Our survey of the literature in the next section will be organized according to the type of reorthogonalization (if any) that was used in the procedure being discussed.

CHOICE 3.   SHORT OR LONG CHAINS OF VECTORS   In some sense this choice is not independent of Choices 1 and 2. Those choices determine the type of Lanczos matrix which will be generated, which in turn implicitly sets limits on the lengths of the chains of Lanczos vectors or equivalently sets limits on the sizes of the Lanczos matrices which can be handled reasonably. In the block case the Lanczos matrices are banded but not tridiagonal. The width of the band depends upon the number of vectors in each block. If reorthogonalization is used, depending upon how it is used, the Lanczos matrices may not even be banded. This may also be the case when the single-vector Lanczos procedure is used with reorthogonalization. In these situations there is a limit on the number of eigenvalues which can be computed reasonably.

For any Ritz vector approach such as Parlett and Scott [1979], where the Lanczos procedure 'periodically' computes Ritz vectors, long chains of Lanczos vectors cannot be used easily. However, for a single-vector Lanczos procedure which uses no reorthogonalization, long chains of Lanczos vectors or equivalently very large Lanczos matrices can be handled easily. No Ritz vector computations are required in the eigenvalue portion of such a procedure. Summarizing, any type of reorthogonalization limits the lengths of the Lanczos chains which are feasible, as does the incorporation of repetitive Ritz vector computations. Both therefore limit the amount of spectral information which a user can obtain about a given matrix.

CHOICE 4. ITERATIVE OR NONITERATIVE Whether or not a given Lanczos procedure is iterative or not is somewhat in the eye of the beholder. Most Lanczos procedures are 'iterative' in some sense. In an attempt to be more precise we make the following definition. However, many Lanczos procedures will fall somewhere between the extremes of iterative and noniterative as specified by this definition. We will call such procedures mixed.

DEFINITION 2.4.1 A Lanczos procedure is said to be iterative if within the Lanczos procedure a sequence of Lanczos matrices is generated, each of which corresponds to a different starting vector (or in the case of the block Lanczos procedures to a different starting block.) A Lanczos procedure will be called noniterative if a Lanczos matrix of size M is generated, eigenvalue/eigenvector computations on that matrix are performed, and then the procedure is terminated.

In many Lanczos procedures a starting vector or block is used to generate a nested family of Lanczos matrices. Eigenvalues (and/or eigenvectors) computations are then performed on certain members of this family as the size of these members is increased in some fashion. If the number of such members for which these computations are performed is large, then the procedure is really an iterative one, although not iterative in the sense of Definition 2.4.1. However, in some cases the numbers of members of this family for which eigenelements computations are performed is quite small, and in that case the procedure is not really iterative in any ordinary sense.

CHOICE 5. COMPUTE EIGENVALUES, RITZ VECTORS SIMULTANEOUSLY? From Lanczos procedure to Lanczos procedure there can be a wide variation in the amount of work done at each stage of the procedure. In some cases the eigenvalue and Ritz vector computations are done in separate stages. This is true of our single-vector procedures discussed in Chapters 4, 5 and 6. In other cases eigenvalues and eigenvectors (Ritz vectors) are computed simultaneously. This is true of our iterative block procedure discussed in Chapter 7. Thus at each internal Lanczos matrix eigenelement computation, there is the question of whether or not both the eigenvalues of the Lanczos matrices are computed along with the corresponding Ritz vectors of A. A Ritz vector computation requires an eigenvector of the Lanczos matrix along with all of the previously-generated Lanczos vectors. Because of the latter, a Lanczos procedure which requires the repeated computation of Ritz vectors will be expensive in terms of both the computer storage required and in terms of the number of arithmetic operations required. Table 2.4.1 provides a summary of the preceding discussion.

TABLE 2.4.1   Specification of a Lanczos Procedure

Type of Choice	Possible Choices
Reorthogonalization	Total: w.r.t. All Lanczos Vectors
	Limited: w.r.t. Local Lanczos Vectors
	Limited: w.r.t. Converged Ritz Vectors
	None: No Reorthogonalization
Iterative	New Starting Vector[a] on Each Iteration
Not Iterative	Only One Lanczos Matrix Considered
Mixed	Only One Starting Vector[a]. However Several Lanczos Matrices Considered
Type of Recursion	Single-Vector
	Block
	Hybrid
Ritz Vector Computations	Every 'Iteration'
	'Periodically'
	None
Lengths of Chains of Vectors	Limited in Length
	Short Lengths Only
	No Real Limit on Length

[a]   For block Lanczos recursions replace 'vector' by 'block'.

SECTION 2.5   RECENT RESEARCH ON LANCZOS PROCEDURES

Using the terminology in Section 2.4 we will attempt to describe briefly much of the recent research on Lanczos procedures for eigenelement computations for real symmetric problems. The discussion will proceed according to the amount of reorthogonalization used in the procedure being discussed. We first consider the so-called classical Lanczos procedures, those which require the total reorthogonalization of each Lanczos vector as it is generated w.r.t.

every Lanczos vector previously-generated. This is done to enforce the classical requirement of the global orthogonality of the entire set of Lanczos vectors generated. Included in this class are Newman and Pipano [1977], Ericsson and Ruhe [1980,1982], Ericsson [1983], and Golub and Underwood [1975,1977]. The first two papers propose single-vector Lanczos procedures. The last one proposes a block Lanczos procedure.

We will then discuss those Lanczos schemes which use some limited form of reorthogonalization. In this case the reorthogonalization may be w.r.t Lanczos vectors or it may be w.r.t. converged Ritz vectors. Papers included in this class which will be discussed are those by Cullum and Donath [1974], Lewis [1977], Parlett and Scott [1979], Scott [1979,1981] and Parlett [1982]. The Cullum and Donath block procedure is discussed in detail in Chapter 7, Section 7.4. Last but most assuredly not the least, we discuss briefly those Lanczos schemes which do not use any reorthogonalization at all. Papers included in this class are those by van Kats and van der Vorst [1976,1977], Edwards, Licciardello and Thouless [1979], Cullum and Willoughby [1979,1980,1981], and Parlett and Reid [1981]. The Cullum and Willoughby procedures are discussed in detail in Chapter 4. In this Chapter we will not make any attempt to provide rigorous justifications for any of these procedures. We simply state what these procedures are, what types of eigenelement problems they are designed to solve, and make some comments about their advantages and disadvantages. The reader is referred to the individual references for more details.

LANCZOS PROCEDURES, TOTAL REORTHOGONALIZATION

The Newman and Pipano reference is Section 6 of Chapter 10 in the NASTRAN Users Manual [1977]. The Lanczos procedure which is described there is designed for computing a few of the eigensolutions of a real symmetric generalized eigenvalue problem

$$Kx \equiv \lambda Mx. \tag{2.5.1}$$

In particular they consider large scale structural vibration and buckling problems. Structural vibration problems require a prespecified number of eigenvalues closest to a specified shift λ_0 along with the corresponding eigenmodes. For such problems the stiffness matrix K and the mass matrix M are both real symmetric and semidefinite matrices. The eigenvalues are all positive. Buckling problems require a prespecified number of the eigenvalues which are smallest in magnitude. In this case, the stiffness matrix K is real symmetric, positive definite, and the differential stiffness matrix M is a real symmetric indefinite matrix. The eigenvalues may be positive or negative.

In both cases, the original problem is converted to an equivalent shifted inverse form. For example for structural vibration problems this equivalent form is given in Eqns(2.5.2).

$$M\overline{K}^{-1}Mx = \mu Mx, \quad \overline{K} = (K - \lambda_0 M) = LDL^T, \quad \mu = 1/(\lambda - \lambda_0). \tag{2.5.2}$$

If the shift $\lambda_0 = 0$, then the following form is used.

$$L^{-1}ML^{-T}y = \mu y \quad \text{where} \quad y = L^T x, \quad \overline{K} = LL^T, \quad \mu = 1/\lambda. \tag{2.5.3}$$

In buckling problems, a small $\alpha > 0$ is chosen so that the resulting matrix $\overline{K} \equiv K + \alpha M$ is positive definite. In this case we have the same shifted inverse problem as that defined by Eqn(2.5.3) except that $\mu = 1/(\lambda + \alpha)$. In Eqns(2.5.2) and (2.5.3), L is a (block) lower triangular matrix and D is a (block) diagonal matrix.

The basic single-vector Lanczos recursion given by Eqns(2.1.1) - (2.1.2) is used along with total reorthogonalization of all of the Lanczos vectors generated. At each iteration the newly-generated Lanczos vector is reorthogonalized w.r.t all Lanczos vectors which have been generated up to that point in time. This reorthogonalization is actually done iteratively until the desired orthogonalization is achieved because simple reorthogonalization may not always yield vectors which are orthogonal to the desired accuracy. In each case the Lanczos procedure is being asked to compute a few of the eigenvalues of a transformed problem which are largest in magnitude.

This NASTRAN procedure is not iterative but it allows for iteration via restarting. If this procedure is restarted, then the new starting vector is chosen randomly and made orthogonal to any known eigenvectors. This orthogonality is maintained throughout the computations by explicitly reorthogonalizing each Lanczos vector w.r.t each such converged eigenvector. The authors state that the Lanczos matrices generated are never larger than $2\overline{q} + 10$ where $\overline{q}$ is the number of eigenvalues requested but not yet computed accurately.

The NASTRAN procedure consists of the following steps. First, the program determines a shift parameter by using various heuristics. This shift can be zero. Second, excessively small elements in the M-matrix are zeroed out. A tentative size for an appropriate Lanczos matrix is determined. The corresponding shifted K-matrix $\overline{K} \equiv K - \lambda_0 M$ is factored. The basic single-vector Lanczos recursion with total reorthogonalization is applied to a shifted-inverse version of the original problem, see Eqns(2.5.2) and (2.5.3). Eigenelements of the resulting real symmetric tridiagonal Lanczos matrix are computed using a QR algorithm. Error estimates are obtained, and if they are sufficiently small, the computed eigenelements are accepted. If the desired convergence has not yet been achieved, the Lanczos process is repeated with a different starting vector generated randomly and made orthogonal to any eigenvectors which have been computed sufficiently accurately. Subsequently, in each reorthogonalization the new Lanczos vector is also reorthogonalized w.r.t such previously-computed eigenvectors. Once acceptable eigenelements have been obtained, the computed quantities are transformed back to the relevant physical quantities.

Obviously this method is expensive both in terms of the amount of computer storage and the number of arithmetic operations required. Additional storage requirements caused by the reorthogonalizations, limit this approach to the computation of a few eigenelements. At each stage all of the previously-generated Lanczos vectors must be readily available for the reorthogonalizations.

The Ericsson and Ruhe [1980,1982] Lanczos algorithm is also designed for computing a specified number of eigenvalues of a generalized real symmetric eigenvalue problem in any part of the spectrum. It also applies the single-vector Lanczos recursion to a shifted and inverted problem, using factorizations of matrices. There are, however, several key differences between the NASTRAN program and the Ericsson and Ruhe approach. They start with an equation of the form Eqn(2.5.1), where K and M are both assumed to be real and symmetric, M is assumed to be positive semidefinite, and the null spaces of K and M are disjoint.

In contrast to the Newman-Pipano procedure, the Ericsson-Ruhe procedure is designed to be iterative. An iteration may involve a change in the shift, a change in the starting vector, or changes in both of these. They also assume that for any given shift μ that they have a factorization of the generally indefinite matrix $K - \mu M = LDL^T$ where L is a (block) lower triangular matrix and D is a (block) diagonal matrix, and a subroutine which solves $(K - \mu M)x = y$ using this factorization. This factorization is also used to keep running track of the number of eigenvalues less than the given shift μ. In the simplest case when D is diagonal, this is done by counting the negative entries in D. They also assume that a factorization $M = WW^T$ of the matrix M is available. The factor W may be rectangular but must have full rank.

The user specifies an interval of interest [a,b] and the Ericsson-Ruhe procedure iteratively generates and considers a sequence of shifts starting from the lower end of the specified interval. The eigenvalue counts obtained from the factorizations of the shifted K-matrices are used to monitor whether or not all of the eigenvalues in each given subinterval have been computed. For a given shift μ_s the indefinite matrix $K - \mu_s M$ is factored. A starting vector is determined randomly. If there are any known converged eigenvectors then the starting vector is orthogonalized w.r.t them. The Lanczos recursion with total reorthogonalization of each Lanczos vector is then applied to the inverted and shifted problem

$$W^T(K - \mu M)^{-1}Wz = \theta z \text{ where } z = W^Tx, \ \theta \equiv 1/(\lambda - \mu), \tag{2.5.4}$$

λ denotes an eigenvalue of the original problem, and x denotes an eigenvector of the original problem. The inverse of the matrix $K - \mu M$ is obtained by invoking the factorization and of course is not explicitly computed. Each application of the Lanczos procedure yields a certain number of converged eigenvalues. The procedure keeps track of which eigenelements have converged and then determines an appropriate new shift μ_{s+1} and/or a new starting vector for the next iteration.

The new matrix $(K - \mu_{s+1}M)$ is then factored and the Lanczos procedure (with total reorthogonalization) is applied to the corresponding shifted inverse problem defined as in Eqn(2.5.4), yielding additional converged eigenvalues. Iterations continue with the repeated choosing of a shift, (a starting vector), a factorization, an application of the Lanczos procedure, checking the number of converged eigenvalues versus the number in the interval until all of the desired eigenvalues (and eigenvectors) are obtained.

Ericsson and Ruhe [1980] discuss several ways of choosing the successive shifts. They also address the questions of how many eigenvalues can one expect to get for a given shift, and how many steps of the Lanczos algorithm to use on each iteration. Note that each choice of shift requires a new matrix factorization of the associated matrix $K - \mu M$. The authors discuss the possible use of selective reorthogonalization (see Parlett and Scott [1979]) in their discussion of the number of arithmetic operation counts required. However in their actual program, they use total reorthogonalization of the Lanczos vectors. They also state that in practice the length of the Lanczos chain used on each iteration, that is the size of the Lanczos matrix considered, is determined primarily by storage considerations.

By using a sequence of shifts, factorizations, and related checks that all of the eigenvalues in a given interval have actually been computed, the Ericsson and Ruhe [1980] procedure can sweep across an interval picking up eigenelements as it goes. This ability to sweep across an interval picking up a few eigenvalues on each sweep, somewhat mollifies the limitations of Lanczos procedures which require the full reorthogonalization of the Lanczos vectors.

We note that the Pipano-Newman procedure could also be used in a similar iterative mode simply by restarting on each iteration with a different shift and/or starting vector. This would require the incorporation of a strategy for choosing appropriate shifts. They also factor a matrix of the form $\overline{K} \equiv K - \mu M$ so that they also have available a count on the number of eigenvalues $< \mu$. Their formulation of the shifted inverse problem is somewhat different from the Ruhe and Ericsson formulation, compare Eqns(2.5.2), (2.5.3), and (2.5.4), because they do not factor the matrix M. However, the effect upon the eigenvalue spectrum is the same for both types of transformations.

In contrast to the Newman-Pipano procedure the Ericsson-Ruhe procedure iterates over a sequence of shifts which are chosen by the program, uses the counts of the number of eigenvalues in various subintervals obtained from the intermediate factorizations to avoid missing any eigenvalues, incorporates a clever correction to each eigenvector computed, and because a factorization of the M matrix is also used, solves a different shifted-inverse version of the original problem. However, both procedures rest upon the idea of computing eigenvalues near some specified point in the spectrum of the original problem by computing those eigenvalues of a related shifted and inverted problem which are largest in magnitude. Moreover, both procedures use factorizations of shifted matrices of the form $K - \mu M$, both use randomly-generated starting vectors, and both use a Lanczos procedure with total reorthogonalization of the Lanczos vectors to generate relatively small Lanczos matrices, i.e. short chains of Lanczos vectors.

The preceding two procedures used the single-vector form of the basic Lanczos recursion. We now consider a block Lanczos procedure which uses total reorthogonalization of the blocks, see Golub and Underwood [1977]. Section 2.4 contains the basic block Lanczos recursion, see Eqns(2.4.1) - (2.4.2), and the definition of a basic iterative Lanczos procedure. The Golub-Underwood block procedure is iterative, uses short chains of blocks on each iteration (i.e. small Lanczos matrices), computes Ritz vectors on each iteration, and reorthogonalizes each Lanczos block as it is generated w.r.t all previously-generated Lanczos blocks.

The user specifies a block size q and the number of blocks s to be generated on each iteration. The block size must be greater than or equal to the number of eigenvalues (and eigenvectors) desired. It is assumed that the size of the Lanczos matrices qs $<<$ n, the order of the original matrix. A starting block for the first iteration is determined randomly. If some eigenvectors are known a priori then this block is orthogonalized w.r.t such vectors, as are all succeeding Lanczos blocks. This prevents the recomputation of these known eigenvectors. On each iteration the basic block recursion is applied for s-steps to obtain a sequence of orthonormal blocks $\overline{Q}_s \equiv \{Q_1,...,Q_s\}$ which spans the corresponding Krylov space

$$\mathcal{K}^s(A;Q_1) \equiv sp\{Q_1,AQ_1,...,A^{s-1}Q_1\}$$

corresponding to the first block. (The iteration superscript is omitted in order to simplify the notation).

At each iteration the corresponding Lanczos matrix is defined by Eqn(2.4.2). As convergence occurs columns in the P_1 block, see Eqn(2.4.1), can become 'zero'. The Golub-Underwood procedure checks for this and if and when it occurs, any such column is replaced by a randomly-generated vector which is orthogonalized against all of Q_1 and any preceding portions of the corresponding block Q_2. Thus they keep the sizes of the sub-blocks fixed as the iterations proceed.

The q algebraically-largest eigenvalues, $\{\mu_1,...,\mu_q\}$, and corresponding eigenvectors of the Lanczos matrix, $U_1 \equiv \{u_1,...,u_q\}$, are computed. The eigenvalues are taken as approximations to the desired eigenvalues of A. The corresponding Ritz vectors $\widetilde{Q}_1 \equiv \overline{Q}_s U_1$ are computed and used as approximations to eigenvectors of A. Convergence is checked by computing the 'new' second block $\widetilde{P}_1 \equiv A\widetilde{Q}_1 - \widetilde{Q}_1 A_1$. If convergence has not yet occurred, the iteration proceeds using $\widetilde{Q}_1$ as the starting block, and the rest of the procedure is repeated using the resulting new Lanczos matrix. Iterations continue until the desired convergence is obtained.

On each iteration, each block Q_j is reorthogonalized w.r.t all preceding Lanczos blocks obtained on that iteration. Therefore all of the Lanczos blocks generated within an iteration must be kept in readily available storage since they are used repeatedly throughout the block generation. The resulting reorthogonalization terms which nominally belong in the Lanczos matrix are not explictly incorporated into these matrices.

In practice the number of blocks which can be generated within a given iteration is determined by the amount of computer memory available. One source of difficulties with any iterative block Lanczos procedure is that its convergence rate depends upon there being a decent gap between the smallest eigenvalue being approximated by the procedure and the next smallest eigenvalue of A, and being able to generate a sufficient number of sub-blocks on each iteration. Thus, the choice of block size is critical. For a given amount of computer storage, an increase in block size decreases the number of sub-blocks which can be generated on any given iteration. This can have a very negative effect upon the resulting convergence. This problem is addressed in Chapter 7, Section 7.5 where we present a hybrid Lanczos procedure (with limited reorthogonalization) which combines features from both the block and the single-vector procedures. In the discussions of block procedures we will assume that the user wants the algebraically-largest eigenvalues of A. The algebraically-smallest ones can be obtained by simply using -A rather than A.

LANCZOS PROCEDURES WITH LIMITED REORTHOGONALIZATION

Thus, we are led to consider procedures which require less orthogonalization or even none. Paige [1971] provided an explanation for the huge losses in orthogonality which typically occur when the basic Lanczos recursion is implemented, showing that they were caused by the combination of the convergence of various eigenvalues with the effects of roundoff and cancellation errors. He showed that these losses were correctable by incorporating reorthogonalization w.r.t any corresponding such converged Ritz vectors, because the losses in orthogonality occurred in the directions of the Ritz vectors. This result of Paige did not become well-known until the late 70's. In the intervening time period Cullum and Donath [1974] incorporated such a correction but for different reasons.

This block procedure differs from the Golub-Underwood approach in several ways. First, there is no reorthogonalization until some computed Ritz vector has converged. Convergence is determined in the process of generating the second block within a given iteration. If convergence is observed, then one or more of the vectors of the unnormalized second block P_1 will be very small, essentially 'zero' from a numerical standpoint. When this occurs the size of the second and succeeding blocks is decreased accordingly, often resulting in a corresponding increase in the number of blocks which can be generated on a given iteration and thereby accelerating the convergence of the remaining eigenvalues to be computed. In order for this decrease in block size to be legitimate, that is for subsequent Lanczos vectors to be orthogonal to those converged eigenvectors which are not being allowed to generate descendants in the recursion, it is necessary to explicitly reorthogonalize each new Lanczos block w.r.t each such converged eigenvector. Thus we have this limited form of reorthogonalization w.r.t converged Ritz vectors. All of the vectors required for these reorthogonalizations reside in the first Lanczos sub-blocks.

The iterative block Lanczos procedures have the property that the approximate eigenvectors are the first block in the chain of Lanczos vectors generated on each iteration. The fact that any relevant converged eigenvectors are retained as columns in the first block on subsequent iterations means that succeeding eigenvalues and corresponding Ritz vectors can be computed with the same degree of accuracy as the ones already computed. We called this an implicit deflation. Note that this is not true if converged vectors are deflated explicitly. In that case succeeding eigenvectors are computed with decreasing accuracy.

Thus this procedure is iterative, uses short chains or equivalently small Lanczos matrices, computes Ritz vectors on each iteration, but uses a very limited form of reorthogonalization, namely the reorthogonalization of the Lanczos vectors w.r.t any computed converged Ritz vectors. We note here that other Ritz vectors which we are not computing could also have converged. However, we do not attempt to identify such vectors and do not reorthogonalize w.r.t them. Observe that with this type of reorthogonalization it is only necessary to have enough storage for 3 blocks of vectors rather than for all of the Lanczos blocks. The other Lanczos vectors which are needed for the Ritz vector computations could be saved on auxiliary storage and called back just for the Ritz vector computations. However, in the actual implementation of this procedure, we kept all of the Lanczos blocks generated within each iteration in main storage. If this is done then the storage requirements for the Golub and Underwood and the Cullum and Donath procedures are essentially the same. See Chapter 7, Section 7.4 for more details.

Lewis [1977] also uses the basic block Lanczos recursion but as a noniterative procedure. He is dealing with a specific application where his matrix is a finite difference discretization of Laplace's equation on a region with a highly irregular boundary. The eigenelements of this matrix are to be used in the study of the tides of oceans in various places in the world. Lewis [1977] contains a brief explanation of the particular application. One particular matrix described is sparse, of order $n=1919$ and has only pairs of eigenvalues (not necessarily distinct) with the exception of a zero eigenvalue of unknown odd multiplicity. The extreme eigenvalues are not the ones of interest. What is required are the eigenvalues in the interior interval of the spectrum, $[.000025,.0245]$. The overall spectrum is defined by the interval $[0,2.92]$.

Lewis applies the basic block Lanczos recursion to this matrix, using a starting block with two vectors and generating a long chain of Lanczos blocks. Due to the fact that he wants many interior eigenvalues, the Lanczos matrices may be larger than the given matrix. Lewis reorthogonalizes each Lanczos block Q_{j+1} as it is generated w.r.t its two closest neighbor blocks Q_{j-1} and Q_j. Therefore, the amount of storage required is not excessive since there are only two vectors in each block and only three blocks need to be in storage at any given point in the recursion. We note that the global orthogonality of the resulting blocks is not preserved. Lewis points out that one difference between a block Lanczos procedure and a single-vector procedure is that losses of orthogonality in the vectors can occur which are not caused by convergence. Thus, convergence yields a loss of orthogonality, however the implication does not go in the other direction.

Each of the resulting Lanczos matrices is treated as a band matrix. The Schwarz procedure for reducing a band matrix to tridiagonal form is used along with Sturm sequencing packages from EISPACK to compute the required eigenvalues of the resulting Lanczos matrix. Required eigenvectors of the Lanczos matrices are obtained by inverse iteration.

Lewis reports that for the matrices he considered, local reorthogonalization was essential. One should note however that he did not use the block analog of the modified Gram-Schmidt version of the Lanczos recursions given in Eqns(2.3.2) - (2.3.5) and (2.3.7) - (2.3.8) and that may have had some effect upon his numerical results. He monitored the orthogonality of the Lanczos blocks by computing the singular values of the off-diagonal matrices B_{j+1}. Lewis notes the difficulty in knowing at what point in the recursions to make eigenvalue, eigenvector computations on the resulting Lanczos matrix.

Since global orthogonality of the Lanczos blocks is not preserved, he has the problem of determining which eigenvalues of the Lanczos matrix are the appropriate approximations to the desired eigenvalues and which ones are extraneous. His tests for deciding which eigenvalues are candidates for 'good' approximations to eigenvalues of A incorporate the following heuristics: (1) Eigenvalues of the Lanczos matrices which persist as the size of these matrices is increased are probably eigenvalues of A. (2) Numerically-multiple eigenvalues of the Lanczos matrices are eigenvalues of A. (3) The last components of a given eigenvector of a Lanczos matrix corresponding to an eigenvalue μ of that matrix are a good indicator of the accuracy of that eigenvalue as an approximation to an eigenvalue of A. Lewis states however that the only true test on the accuracy which he accepts is a check on the residual norms obtained using the corresponding Ritz vectors.

Thus, the Lewis [1977] procedure uses the basic block Lanczos recursion with the block size equal to the maximum multiplicity of the desired eigenvalues. For the particular application he was considering, Lewis knew that this multiplicity was two. A very limited form of reorthogonalization is used, simple nearest and next-nearest neighbors, so that relatively long chains of vectors can be generated. The procedure is not iterative, and it is designed to compute many eigenvalues, even interior ones. Ritz vectors are not computed repeatedly, except to verify convergence when convergence is indicated by various heuristics used.

Parlett and Scott [1979] combines the implicit deflation idea in the Cullum and Donath procedure together with an economical method in Kahan and Parlett [1976] for monitoring the orthogonality of the Lanczos vectors to obtain a single-vector Lanczos procedure with limited reorthogonalization. They call the resulting procedure, Lanczos with selective reorthogonalization. The basic idea is to maintain a level of independence among the Lanczos vectors V_j as measured by the quantity

$$\kappa_j \equiv \| I - V_j V_j^T \|. \qquad (2.5.5)$$

Observe that full orthogonality of the Lanczos vectors would require that $\kappa_j \approx \varepsilon$ where ε denotes the machine epsilon. The selective reorthogonalization procedure maintains $\kappa_j \approx \sqrt{\varepsilon}$. Two mechanisms for monitoring the size of κ_j as j is increased are discussed in Parlett [1982] and will not be repeated here. Research on methods for monitoring losses in orthogonality continues. Much of the following discussion was taken from Parlett [1980, Chapter 13] and Scott [1981].

Using such a monitor to indicate when reorthogonalization is required, the Parlett-Scott procedure reorthogonalizes the Lanczos vectors only on those steps of the recursion where it is deemed necessary. Specifically, the basic Lanczos recursion is applied until at some recursion step j the upper bound specified for the orthogonality measure κ_j is violated. At such a point in the recursion, certain Ritz vectors are computed and the Lanczos vectors v_j and v_{j+1} are both orthogonalized against such Ritz vectors. Theoretically, this means that later Lanczos vectors will be orthogonal w.r.t these Ritz vectors.

Thus, 'periodically' the generation of the Lanczos matrix is stopped, eigenvalue/eigenvector computations on the current Lanczos matrix are made, certain Ritz vectors are computed, two Lanczos vectors are reorthogonalized w.r.t these 'converged' Ritz vectors, and then the procedure picks up where it left off continuing to generate a larger Lanczos matrix. When all of the desired eigenvalues and corresponding Ritz vectors have been generated to the desired accuracy, then the procedure terminates.

At pauses those eigenvalues of the corresponding Lanczos matrix whose eigenvectors have the property that their last components are less than $\sqrt{\varepsilon} \| A \|$ are selected as 'converged' and their corresponding Ritz vectors are used in the reorthogonalizations. The success of this procedure for a given matrix obviously depends upon the number of Ritz vectors which must be considered at each pause and the length of the chain of Lanczos vectors required to compute the desired eigenvalues. Since all of the Lanczos vectors are required for each required Ritz vector computation, the Lanczos vectors must either be kept on readily available storage or else be regenerated every time the Ritz vectors have to be computed. Note that the reorthogonalization terms are not included in the Lanczos matrices. The basic argument for not including them is that perturbations of order $\sqrt{\varepsilon}$ in any Lanczos matrix result in only $O(\varepsilon)$ perturbations in the eigenvalues of such a matrix and can be easily tolerated.

Selective reorthogonalization eliminates the copies of Ritz eigenvalues and vectors which occur if no reorthogonalization at all is incorporated. The number of Lanczos steps is kept to a minimum. Note however that this does not guarantee that the amount of work is kept to a minimum. Parlett and Scott also say that multiple eigenvalues can be found, copy by copy, and that this method can be used as a black box with no delicate parameters to set other than the desired accuracy and the amount of fast storage available. One would have to argue with the last statement. If in fact at each stage, selective reorthogonalization w.r.t all of the converged Ritz vectors is allowed, then one does not have any a priori measure of the amount of storage or of the number of Ritz vectors which will be required at each stage of the recursion. This will obviously depend upon the particular eigenvalue distribution and for some problems could easily exceed the storage available. For example this could easily be the case if the small end of the spectrum is desired and there is a significant difference in the magnitudes of the eigenvalues at the small and the large ends. The program would have to be able to handle Ritz vector computations and reorthogonalizations w.r.t the large end of the spectrum before the desired small end had converged sufficiently.

Scott [1979] proposes a block Lanczos algorithm with selective reorthogonalization. He argues that a block Lanczos version solves two of the main problems associated with the single-vector Lanczos procedures. First eigenvalues are missed only when the corresponding eigenvectors are orthogonal to the entire subspace spanned by the block of starting vectors so one decreases the chances of missing an eigenvalue. Second, multiple eigenvalues and a basis for the corresponding invariant subspace can be found up to the sizes of the blocks. Third, the given matrix is multiplied by an entire block of vectors each time. This could be beneficial if the matrix being used is not stored in main memory. He explores the possibility of reorthogonalizing only w.r.t those converged Ritz vectors which are wanted. This is of course what is done in the Cullum and Donath procedure. There the other end of the spectrum was ignored and allowed to replicate as much as it wanted. Scott [1981] notes that this invalidates his a priori estimates obtained for selective reorthogonalization. However, one can still judge convergence by looking at residual norms of Ritz vectors.

He considers two types of problems: (1) Find all of the eigenvalues at one end of the spectrum; (2) Find all of the eigenvalues outside of some interval. This procedure mimics the corresponding single-vector procedure with selective reorthogonalization. It is mixed not iterative in the sense of using different starting vectors, although Scott allows for restarting. The lengths of the chains are limited by storage and computation costs. Scott requires that all of the 'converged' Ritz vectors be kept in main memory.

Scott incorporates not only selective reorthogonalization but also local reorthogonalization like Lewis [1977]. He states that for each step j in the recursion only a very few eigenvalues of the corresponding Lanczos matrix are computed. This is permissible because of the observed way in which basic Lanczos procedures converge. See for example, Tables 4.7.5 - 4.7.10 in

Chapter 4. In particular the eigenvalues which are most interior in the spectrum and with the smallest gaps converge last. If the storage limitations have been reached and the desired convergence has not yet been achieved, then the program computes and stores all 'converged' eigenvectors. Unconverged eigenvectors are also computed and the new starting block is formed by taking linear combinations of these vectors. This block is orthogonalized against all converged Ritz vectors, and the orthogonality of the resulting Lanczos blocks with respect to these converged vectors is monitored as the recursion proceeds. The Lanczos blocks are reorthogonalized w.r.t these converged vectors as needed.

Thus, Scott [1979] uses a block procedure with limited reorthogonalization. This reorthogonalization is of two types, selective and local. The chains of vectors cannot therefore be too long. Otherwise the storage demands due to the converged Ritz vectors can become excessive. Ritz vectors are computed 'periodically'. The procedure is not iterative in the ordinary sense but is in a mixed sense as we indicated above for the single-vector procedure with selective reorthogonalization.

Limiting the amount of reorthogonalization allowed corrects many of the 'problems' encountered in using a Lanczos procedure with total reorthogonalization. The resulting computer storage requirements are nominally significantly less than those required for procedures with total reorthogonalization. These 'limited' procedures are however more complicated to program, and in the end the amount of information which one can obtain about the spectrum of a very large real symmetric matrix is still severely limited by the amount of storage available. For very large problems one is typically able to compute relatively few eigenvalues (and corresponding eigenvectors). Moreover, if a shifted-inverse version of the given problem cannot be used, then one is also limited to computing the extreme eigenvalues of the original problem. The exception to these statements is the procedure of Lewis [1977] which used only local reorthogonalization so that if the block size is very small like 2, is not seriously limited by storage considerations. The Lewis [1977] procedure however did not include a simple test for determining which of the computed eigenvalues were genuine and which were spurious.

LANCZOS PROCEDURES WITH NO REORTHOGONALIZATION

In the remainder of this paper we consider Lanczos procedures similar to Lewis [1977], except that each is a single-vector Lanczos procedure and none of these procedures uses any type of reorthogonalization of the Lanczos vectors. Thus, of the three classes of Lanczos procedures the members of this class have the smallest storage requirements. Three of these procedures, van Kats-van der Vorst [1976,1977], Edwards et al [1979], and Parlett-Reid [1981] differ primarily from the Cullum-Willoughby procedure w.r.t the mechanism used to separate the 'good' Ritz values of the Lanczos matrices from the 'spurious' or extraneous ones caused by the losses of orthogonality of the Lanczos vectors. The basic idea in the identification tests used in the three procedures by van Kats-van der Vorst, Edwards et al, and Parlett-

Reid is the same whereas the key idea used in the Cullum-Willoughby identification test is quite different as we explain briefly below and in detail in Section 4.5 of Chapter 4. All four of these procedures use the modified, basic single-vector Lanczos recursion given by Eqns(2.3.2) - (2.3.5) and (2.3.7) - (2.3.8) without any reorthogonalization of any Lanczos vectors.

The discussion of the van Kats-van der Vorst procedure and that of the Edwards et al procedure is confined purely to the differences between the identification tests used. The Parlett-Reid procedure differs in other respects so more comments will be made about it. All three rest upon the basic premise that if for a given $\delta > 0$ and size m one can show that for all $k \geq 0$, there is an eigenvalue of the Lanczos matrix T_{m+k} in the interval $[\mu - \delta, \mu + \delta]$ for all $k \geq 0$, then the original matrix A has an eigenvalue in this interval. Thus, these procedures try to identify which eigenvalues of the Lanczos matrices are converging as eigenvalues of these matrices as we let the size of the Lanczos matrices increase.

First consider the van Kats-van der Vorst procedure [1976,1977]. Their identification test is based upon the interlacing theorem, see for example Jennings [1977], relating eigenvalues of any Lanczos matrix T_j to the eigenvalues of any Lanczos matrix T_J with $J > j$. In particular if we denote the eigenvalues of T_j by $\mu_1^j \geq \mu_2^j \geq ... \geq \mu_j^j$, then the interlacing theorem states that

$$\mu_i^j \geq \mu_i^{j-1} \geq \mu_{i+1}^j \tag{2.5.6}$$

That is, in any given interval between two eigenvalues of the Lanczos matrix T_j we must have an eigenvalue of the smaller matrix T_{j-1}. The details of the test are spelled out in van Kats and van der Vorst.

Thus, their identification test labels those eigenvalues which have converged in the sense that for some j and some user-specified tolerance they are eigenvalues of both T_j and T_{j-1}. The tolerance is allowed to change as the computations proceed. This can lead to some fuzziness in the identification process, as indicated by Example 4.1 in their 1977 paper where for certain size Lanczos matrices more than 40 eigenvalues were identified as 'good' although the order of the given matrix was n=40. This was due to the presence of what we will call 'spurious' eigenvalues close to the 'good' ones.

The van Kats-van der Vorst work clearly illustrated that the Lanczos procedure with no reorthogonalization could be used to compute reliably large numbers of eigenvalues of very large matrices. However, it raised the question of whether or not a sharper identification test could be devised which would eliminate the problems caused by the spurious eigenvalues which were close to good eigenvalues. In practice the eigenvalues were computed for example every 10 steps of the Lanczos procedure and those obtained for the Lanczos matrix $T_{10 \cdot k}$ were compared with those obtained for the Lanczos matrix $T_{10 \cdot (k-1)}$. Van Kats and van der Vorst provide an interesting example of a matrix with clusters of eigenvalues at each of the extremes of the spectrum and well-separated eigenvalues in the center of the spectrum for which the middle eigenvalue converges first when their Lanczos procedure is applied.

This Lanczos procedure with no reorthogonalization (as does any such procedure which does not require reorthogonalization of the Lanczos vectors) allows one to contemplate computing many or even all of the eigenvalues of very large matrices in subintervals or even all of the spectrum. The storage requirements are minimal. Storage for only two Lanczos vectors is required, along with storage for the Lanczos tridiagonal matrices generated and that required to generate the matrix-vector multiplies Ax. Thus long chains of vectors, or equivalently huge Lanczos matrices, can be generated and many eigenvalues of the original matrix can be computed.

Edwards, Licciardello and Thouless [1979] in their studies of electrons in disordered systems need to determine many of the eigenvalues of very large matrices. Their identification test can be viewed as a refinement of the van Kats-van der Vorst approach. They do not have to compute two sets of eigenvalues, only those of T_m are necessary. The identification test is done using Sturm sequencing. They claim to get better results than with the identification test of Paige [1971]. Paige proposed that for a given Lanczos matrix, error estimates be computed for all of the eigenvalues computed using the last components of the corresponding eigenvectors of the Lanczos matrix and that these estimates be used to determine which eigenvalues are 'good'. This approach runs into difficulties whenever one or more of the eigenvalues of interest is nearly multiple as an eigenvalue of the Lanczos matrix or is one of a cluster of eigenvalues in the original matrix.

The Edwards et al approach requires the user to specify a precision δ to which the eigenvalues of the given matrix are to be computed. Two eigenvalues of a Lanczos matrix T_m which differ by less than δ are assumed to correspond to the same eigenvalue of A. Any such numerically-multiple eigenvalue of T_m is accepted as an eigenvalue of A. Furthermore, as in van Kats and van der Vorst [1977], it is assumed that if for all $k \geq 0$, T_{m+k} has an eigenvalue in the interval $[\mu - \delta, \mu + \delta]$, then A has an eigenvalue in that interval.

The determinants of the shifted, real symmetric tridiagonal Lanczos matrices $(T_m - \mu I)$ satisfy the recursion

$$d_{j+1} = (\alpha_{j+1} - \mu)d_j - \beta_{j+1}^2 d_{j-1} \tag{2.5.7}$$

with $d_0 = 1$ and $d_{-1} = 0$. The number of negative ratios of determinants $r_k(\mu) \equiv d_{k-1}/d_k$ for $k = 1, \ldots, j$ gives the number of eigenvalues of T_j which are $< \mu$. These ratios satisfy the following recursion which can be used to compute them.

$$r_{j+1}(\mu) = (\alpha_{j+1} - \mu - \beta_{j+1}^2 r_j)^{-1}. \tag{2.5.8}$$

This is just a restatement of the Sturm sequencing property.

Edwards et al first compute the eigenvalues μ_k^m, k = 1,2,...,m, of a user-specified Lanczos matrix T_m. Then for each k, Sturm sequence counts are run from j=1,2,..., $J(\mu_k^j)$ using $\mu_k^j + \delta$ and $\mu_k^j - \delta$ for which either

$$r_J(\mu_k^j + \delta) \geq r_J(\mu_k^j - \delta) \text{ and } p_J(\mu_k^j + \delta) = p_J(\mu_k^j - \delta) + 1, \text{ or}$$
$$p_J(\mu_k^j + \delta) > p_J(\mu_k^j - \delta) + 1. \tag{2.5.9}$$

In Eqns(2.5.9), $p_J(t)$ denotes the number of values of $j \leq J$ for which the ratios $r_j(t) < 0$. These tests yield the smallest size Lanczos matrix for which the given eigenvalue of T_m is an eigenvalue of some Lanczos matrix to within the user-specified tolerance δ. If $J(\mu_k^j) = m$, then the eigenvalue μ_k^j is not accepted an approximation to an eigenvalue of the original matrix A. Thus, as in the van Kats-van der Vorst procedure, this procedure accepts a given eigenvalue as an approximation to an eigenvalue of A only after it has converged to within a specified tolerance, which in this case is δ.

The authors report results obtained using single, double and extended precision arithmetic. The higher the precision, the smaller the Lanczos matrix needed to compute the desired eigenvalues. For a test matrix of size n=512, they report that using single precision all of the eigenvalues of this matrix were obtained to the desired accuracy using a Lanczos matrix of size m = 1100. Corresponding results in double precision were obtained by m = 800. In extended precision m = 700 was required.

The van Kats-van der Vorst and Edwards et al procedures (as well as the Cullum-Willoughby) do not give computable rules for determining how large the Lanczos matrix must be in order to compute the desired eigenvalues. Of course the required size varies with the particular matrix being considered. Parlett-Reid [1981] state that their primary objective is to rectify this situation. They present a procedure which automatically determines the appropriate size Lanczos matrix.

The procedure described in Parlett and Reid [1981] does not use any reorthogonalization. The underlying assumption in their identification test is the same as that in van Kats and van der Vorst [1977] and in Edwards et al. Thus, at any given point in their procedure they accept an eigenvalue only after they have guaranteed its accuracy. Convergence of eigenvalue approximations is monitored by tracking the eigenvalues of the Lanczos matrices T_j as j is increased. Eigenvalues of T_{j+1} are compared with eigenvalues of T_j and convergence is observed essentially when a given eigenvalue appears as an eigenvalue of both of these matrices or more correctly as an eigenvalue of all T_{j+k} for $k \geq 0$.

The obvious way to track eigenvalues is to use a bisection procedure which keeps track of all of the eigenvalues of each Lanczos matrix and their changes as j is increased. Parlett and Reid found however that this would involve too much computation and storage. Thus, they developed various heuristic approximating procedures to track approximately the progress of all of the Ritz values of the sequence of Lanczos matrices as j is increased, where the accuracy to which the Ritz values are determined is limited. This accuracy is comparable to the changes in these values since the previous step so that only those Ritz values which have 'converged' as eigenvalues of the Lanczos matrices are known accurately at any particular stage of the computations.

At each stage they store the two end points of each subinterval of the spectrum which contains one or more eigenvalues of the corresponding Lanczos matrix, along with counts on the number of eigenvalues of the Lanczos matrix less than each of these end points. These counts are determined by counting sign changes of determinant ratios. These determinants are defined in Eqn(2.5.7). The values of the corresponding ratios are stored at each stage. The strategy for putting down more end points of intervals is based upon some complicated and somewhat tricky heuristics which are used to identify end points which are eigenvalues of both T_j and of T_{j+1} for some j. Rational interpolation functions of low order are employed along with checks on the accuracy of various end points being saved. Accuracy is checked by computing eigenvectors of the Lanczos matrices and looking at the sizes of the last components of these computed eigenvectors. The objective of these computations is to compute all of the eigenvalues in a user-specified interval to a user-specified accuracy.

This procedure requires some mechanism for rejecting certain subintervals determined by the computations. Another set of heuristics is used to make these identifications. The procedure requires storage for the generation of Lanczos matrices, a vector to store the computed eigenvalues, arrays to store the end points of the intervals determined, to store the determinant ratios, and to store the counts of the eigenvalues in each subinterval. The authors state that the total storage required, other than the amount required to generate the matrix-vector multiplies, is $(2n + 2j + 8.5c)$ words where c is the number of eigenvalues to be computed. This estimate is based upon the assumption that the lengths of the arrays needed to keep track of the end points of the intervals, the determinant ratios and the number of eigenvalues in each subinterval is of length not more than 3c. This procedure is available in the Harwell Library [1981] as subroutine EA14AD. Comparisons of the results obtained for this procedure with those obtained by Cullum and Willoughby [1979] are given and shown to be very similar.

Summarizing the Parlett-Reid procedure, we see that it is a single-vector Lanczos procedure with no reorthogonalization designed for computing many eigenvalues in user-specified subintervals of the spectrum of a real symmetric matrix. The procedure is in some sense iterative in that running approximate values of the eigenvalues of the intermediate Lanczos matrices are kept at each stage in the recursion. However, it is not iterative in the sense that different starting vectors are incorporated. A long or short chain of Lanczos vectors can be generated depending upon what is to be computed. Ritz vectors are not computed during the eigenvalue computations. One drawback is that the amount of storage which will be required cannot be determined a priori because one does not know a priori how many end points will be generated by the procedure.

The final Lanczos procedure which we consider is the single-vector Lanczos procedure Cullum and Willoughby [1979,1980,1981]. It is discussed in detail in Chapter 4. This procedure uses no reorthogonalization of the Lanczos vectors and is designed to compute either a very few or very many (sometimes even all) of the eigenvalues of large, sparse matrices. It is discussed in detail in Chapter 4. It differs from the preceding procedures of van Kats-van der Vorst, Edwards et al, and Parlett-Reid in the mechanism used for identifying which eigenvalues of the Lanczos matrices are 'good' and which of them are 'spurious' and should therefore be disregarded. The identification tests in the three preceding procedures are based upon the convergence of eigenvalues of the Lanczos matrices as the size is increased. Thus, at each stage within one of those three procedures an eigenvalue is checked to see if it is good, after it has appeared as an eigenvalue of two or more Lanczos matrices of increasing size. Other eigenvalues are not accepted as representative of the spectrum of the original matrix A, so that at each stage the user only sees those eigenvalues which have been identified as accurate approximations to eigenvalues of A.

The strategy in the Cullum-Willoughby procedures is quite different. The identification test does not rest on how the eigenvalues of the Lanczos matrices are changing as the size of the Lanczos matrices is increased. Instead it is a test which for any size Lanczos matrix identifies not the 'converged' eigenvalue approximations but rather those eigenvalues of the Lanczos matrix in question which are due to the losses in orthogonality. The test picks out the 'spurious' eigenvalues directly. The remaining eigenvalues of the Lanczos matrix in question are accepted as 'good' regardless of their actual accuracy. Thus, at each stage the user gets an overall picture of the spectrum of the given original matrix.

The identification test which we use does not require the user to supply tolerances for the various computations. The identification is clear-cut and not fuzzy. This test has its basis in the relationship between the Lanczos tridiagonalization and the conjugate gradient method for solving systems of linear equations. This relationship is discussed in detail in Section 4.3 of Chapter 4. Instead of using comparisons between various size Lanczos matrices, for example, $T_{10 \cdot k}$ and $T_{10 \cdot (k+1)}$ as in the van Kats and van der Vorst procedure it uses the given Lanczos matrix and the submatrix of the Lanczos matrix obtained by deleting the first row and column of that matrix. This test is explained in detail in Section 4.5 of Chapter 4. Like the Edwards et al procedure only one set of eigenvalues must be computed. The cost of our identification test is a small addition to the cost of the computation of these eigenvalues. Our single-vector Lanczos procedures are simple to program and do not require any complicated heuristics. The eigenvalue-eigenvector computations are split into two steps. First eigenvalues are computed using a program with minimal storage requirements. Then after the desired eigenvalues are computed sufficiently accurately, eigenvectors for some user-specified subset of these eigenvalues are computed by a separate program. Suitable programs are provided in Volume 2 of this book.

Thus, the Cullum-Willoughby procedure uses the single-vector Lanczos recursion with no reorthogonalization. The basic eigenvalue procedure does not compute any Ritz vectors. Short or long chains of vectors or equivalently small or large Lanczos matrices are generated depending upon what is to be computed. This procedure is not iterative in the sense of changing starting vectors. Typically, subsets of the eigenvalues of two or three Lanczos matrices would be computed in order to obtain the desired eigenvalues accurately. These eigenvalue computations need only be done in those subintervals where convergence has not already been achieved.

This eigenvalue procedure uses minimal storage. The identification test allows the determination of the desired eigenvalues to very high accuracy. No fuzzy tolerances are required. For many matrices this procedure maximizes the number of eigenvalues which can be computed. Typically by $m=3n$, the user will have successfully computed most of the eigenvalues of the given matrix. This may not be true for very stiff matrices because of the possible heavy replication for such matrices of the dominant eigenvalues. Thus for a given amount of storage, of the procedures discussed this procedure provides the maximum amount of information about the given matrix. In addition the ideas generalize to other problems such as the computation of eigenelements of Hermitian matrices and the computation of singular values and vectors of real, rectangular matrices. See Chapters 4 and 5, respectively. Moreover, a procedure for nondefective complex symmetric matrices has also been devised using these ideas. See Chapter 6. Volume 2 contains the associated FORTRAN codes.

CHAPTER 3

TRIDIAGONAL MATRICES

SECTION 3.1   INTRODUCTION

We have said that one of the primary advantages of the basic single-vector Lanczos eigenvalue procedure is that it replaces a general real symmetric matrix A by a family of real symmetric tridiagonal matrices T_k, k = 1,2,.... However, standard procedures for computing eigenvalues and eigenvectors of small and medium size real symmetric matrices (see EISPACK [1976,1977]) also transform the given matrix into an equivalent real symmetric tridiagonal matrix before the eigenvalues and eigenvectors are computed. The difference is that in those procedures the resulting tridiagonal matrix is obtained by applying a sequence of orthogonal transformations to the original matrix.   Orthogonal transformations typically cause significant fill-in; and therefore, those procedures require that the user provide enough storage for an entire matrix of the size being considered. If the given matrix is very large, a very large amount of storage is required.   For example if the matrix is of order 1000 and each floating point number occupies 8 bytes of storage, then 4 million bytes of storage will be required.   On the other hand, the basic single-vector Lanczos procedure with no reorthogonalization defined in Chapter 2 does not suffer from this problem because the given matrix is not explicitly modified during the Lanczos tridiagonal reduction process.   Before proceeding we present an extreme example of the fill-in which can occur when orthogonal transformations are used to explicitly reduce a general real symmetric matrix to real symmetric tridiagonal form.

EXAMPLE 3.1.1   (Possible Fill-in)   Define a real symmetric matrix

$$A \equiv \begin{bmatrix} \alpha & g^T \\ g & B \end{bmatrix}.$$

Define a Householder transformation (Stewart [1973]) $U = I - 2uu^T$ where $u^T \equiv (0, v^T)$ with $\| v \| = 1$, and v is chosen to reduce the first row and column of A. If we apply this transformation to A, setting $B_* = B - 2[vw^T + wv^T]$, $w = Bv - \rho v$, $\rho = v^T Bv$, and $e_1^T = (1,0,...,0)$, we reduce A to

$$A_+ \equiv \begin{bmatrix} \alpha & \beta e_1^T \\ \beta e_1 & B_* \end{bmatrix}$$

For the purposes of illustration we trace the fill-in which occurs in each of the B_* which are generated on the first three reduction steps of a matrix A with the following pattern of nonzeros. In this discussion x denotes a nonzero entry, all other entries are 0, and the vector v

76

defines the Householder transformation used at that step.

$$A_1 \equiv \begin{bmatrix} x & x & & & & & & x \\ x & x & x & & & & & \\ & x & x & x & & & & \\ & & x & x & x & & & \\ & & & x & x & x & & \\ & & & & x & x & x & \\ & & & & & x & x & x \\ x & & & & & & x & x \end{bmatrix} \qquad v = \begin{bmatrix} x \\ 0 \\ 0 \\ 0 \\ 0 \\ 0 \\ 0 \\ x \end{bmatrix} \qquad w = \begin{bmatrix} x \\ x \\ 0 \\ 0 \\ 0 \\ 0 \\ x \\ x \end{bmatrix}$$

After one step in the reduction we obtain the following matrix, note the fill-in.

$$A_2 \equiv \begin{bmatrix} x & x & & & & & & \\ x & x & x & & & & x & x \\ & x & x & x & & & & x \\ & & x & x & x & & & \\ & & & x & x & x & & \\ & & & & x & x & x & \\ & x & & & & x & x & x \\ & x & x & & & & x & x \end{bmatrix} \qquad v = \begin{bmatrix} x \\ 0 \\ 0 \\ 0 \\ x \\ x \end{bmatrix} \qquad w = \begin{bmatrix} x \\ x \\ 0 \\ x \\ x \\ x \end{bmatrix}$$

After two steps we have the following matrix.

$$A_3 \equiv \begin{bmatrix} x & x & & & & & & \\ x & x & x & & & & & \\ & x & x & x & & x & x & x \\ & & x & x & x & & x & x \\ & & & x & x & x & & \\ & & x & & x & x & x & x \\ & & x & x & & x & x & x \\ & & x & x & & x & x & x \end{bmatrix} \qquad v = \begin{bmatrix} x \\ 0 \\ x \\ x \\ x \end{bmatrix} \qquad w = \begin{bmatrix} x \\ x \\ x \\ x \\ x \end{bmatrix}$$

By the end of the third step the remaining lower portion of the matrix has filled in.

$$A_4 \equiv \begin{bmatrix} x & x & & & & & & \\ x & x & x & & & & & \\ & x & x & x & & & & \\ & & x & x & x & x & x & x \\ & & & x & x & x & x & x \\ & & & x & x & x & x & x \\ & & & x & x & x & x & x \\ & & & x & x & x & x & x \end{bmatrix}$$

This very simple example clearly illustrates the possible problems with fill-in. The original matrix was tridiagonal except for the addition of an extra entry in each of the codiagonal corners. Thus, when subjected to orthogonal transformations sparse matrices may become dense matrices. We now proceed with the primary discussion. We will see that except for diagonal matrices, real symmetric tridiagonal matrices, in terms of both the storage requirements and the arithmetic operation counts required to perform certain operations on them, are computationally optimal.

DEFINITION 3.1.1 An mxm matrix $T_m \equiv (t_{ij})$, $1 \leq i,j \leq m$, is tridiagonal if and only if $t_{ij} = 0$ for $|i-j| > 1$ $1 \leq i,j \leq m$.

If a tridiagonal matrix T is symmetric, then it is completely defined by 2m-1 values. The amount of computer storage required to specify any mxm tridiagonal matrix is always $\leq 3m - 2$ words. Below we discuss the special properties of tridiagonal matrices which are relevant for our discussions of the single-vector Lanczos eigenelement procedures. We are particularly interested in symmetric tridiagonal matrices. However, many of the lemmas which are included are valid for general complex tridiagonal matrices. The proofs of these lemmas are no more difficult for the general case than for the symmetric case.

Some of what we discuss is only valid for real symmetric tridiagonal matrices. In particular we will see that the eigenvalue computations for real symmetric tridiagonal matrices are particularly simple and cost effective. See Section 3.5. For such matrices, if only the eigenvalues in some small portion of the spectrum are desired, these can be computed without computing all of the eigenvalues of the matrix. The storage requirements for such eigenvalue computations are minimal, depending linearly upon the size of the tridiagonal matrix. Even if we want to compute all of the eigenvalues of a real symmetric mxm tridiagonal matrix, this can be accomplished in $O(m^2)$ operations. We need a few definitions.

DEFINITION 3.1.2 Let m be fixed and for any $1 \leq i \leq j \leq m$ define

$$T_{i,j} \equiv \begin{bmatrix} \alpha_i & \beta_{i+1} & & \\ \gamma_{i+1} & \alpha_{i+1} & & \\ & & \ddots & \beta_j \\ & & \gamma_j & \alpha_j \end{bmatrix} \tag{3.1.1}$$

where $\beta_1 \equiv \gamma_1 \equiv 0$ and for $i \leq k \leq j$, α_k, β_k, γ_k are specified complex numbers. Define

$$\delta_k \equiv \gamma_k \beta_k, \quad 2 \leq k \leq m. \tag{3.1.2}$$

Unless otherwise stated, we always assume that $\delta_k \neq 0$ for $2 \leq k \leq m$. Moreover, we use the following notation.

$$T_k \equiv T_{1,k} \text{ and } \hat{T}_k \equiv T_{k,m}, \text{ for } 1 \le k \le m. \tag{3.1.3}$$

For the Lanczos eigenvalue applications we will see (empirically) that it is not unreasonable to assume that the $\delta_k \ne 0$. Insight into the behavior of the single-vector Lanczos procedures can be obtained by examining expressions which involve characteristic polynomials of the corresponding symmetric tridiagonal Lanczos matrices T_j and $\hat{T}_j$ $j = 1,2,...,m$, generated by these procedures. Therefore, it is convenient to make the following additional definitions.

DEFINITION 3.1.3 For any tridiagonal matrix T_m and any $k = 1,2,...,m$ we denote the characteristic polynomials of the submatrices T_k and of $\hat{T}_k$ defined in Eqn(3.1.3) by

$$a_k(\mu) \equiv \det(\mu I - T_k) \text{ and } \hat{a}_k(\mu) \equiv \det(\mu I - \hat{T}_k). \tag{3.1.4}$$

The following lemma is well-known. Eqns(3.1.5) - (3.1.7) follow directly from the Laplace expansion theorem for computing determinants of matrices. See for example, Wilkinson [1965, pp.423-6]. Eqns(3.1.5) and (3.1.6) play a key role in effectively computing the determinant ratios which are needed in the Sturm sequencing computations in our single-vector Lanczos eigenvalue procedures. See Section 3.5.

LEMMA 3.1.1 The determinants of the principal submatrices T_k and $\hat{T}_k$ of any mxm tridiagonal matrix T_m satisfy the following recursions where $\delta_j \equiv \gamma_j \beta_j$ for $2 \le j \le m$.

$$a_k(\mu) = (\mu - \alpha_k) a_{k-1}(\mu) - \delta_k a_{k-2}(\mu) \quad \text{with } a_{-1}(\mu) \equiv 0, \ a_0(\mu) \equiv 1 \tag{3.1.5}$$

$$\hat{a}_k(\mu) = (\mu - \alpha_k) \hat{a}_{k+1}(\mu) - \delta_{k+1} \hat{a}_{k+2}(\mu) \quad \text{with } \hat{a}_{m+2}(\mu) \equiv 0, \ \hat{a}_{m+1}(\mu) \equiv 1 \tag{3.1.6}$$

$$a_m(\mu) = (\mu - \alpha_k) a_{k-1}(\mu) \hat{a}_{k+1}(\mu) - \delta_k a_{k-2}(\mu) \hat{a}_{k+1}(\mu) - \delta_{k+1} a_{k-1}(\mu) \hat{a}_{k+2}(\mu). \tag{3.1.7}$$

Proof. Eqn(3.1.5) can be obtained by expanding $\det(\mu I - T_k)$ around the bottom row of $(\mu I - T_k)$. Eqn(3.1.6) can be obtained by expanding $\det(\mu I - \hat{T}_k)$ around the top row of $(\mu I - \hat{T}_k)$. Eqn(3.1.7) corresponds to an expansion of $\det(\mu I - T_m)$ around the kth row of $(\mu I - T_m)$. □

SECTION 3.2 ADJOINT AND EIGENVECTOR FORMULAS

DEFINITION 3.2.1. The adjoint of any given matrix $(\mu I - T_m)$, which we will denote by adj(μI $-$ T$_m$), is characterized by the following relationships.

$$(\mu I - T_m)\text{adj}(\mu I - T_m) \equiv \text{adj}(\mu I - T_m)(\mu I - T_m) = a_m(\mu)I. \qquad (3.2.1)$$

In particular, if μ is not an eigenvalue of T_m, then the inverse $(\mu I - T_m)^{-1} =$ adj(μI $-$ T$_m$)/a$_m$(μ).

We first derive formulas for the elements of adj(μI $-$ T$_m$) in terms of the characteristic polynomials of the submatrices T_k for $1 \le k \le m - 1$ and $\hat{T}_k$ for $2 \le k \le m$. We will then, in the case that μ is a simple eigenvalue of T_m, express the adjoint in terms of the left and right eigenvectors of T_m. This will give us expressions for the eigenvectors of T_m in terms of these characteristic polynomials.

LEMMA 3.2.1 Let μ be a free parameter. Let $S \equiv (s_{ij}) = \text{adj}(\mu I - T_m)$. Then the entries of adj(μI $-$ T$_m$) are given by the following equations.

$$s_{ij} = \begin{cases} a_{i-1}(\mu)\hat{a}_{j+1}(\mu)\left[\displaystyle\prod_{k=i+1}^{j} \beta_k\right] & \text{for } i<j \\[2ex] a_{i-1}(\mu)\hat{a}_{i+1}(\mu) & \text{for } i=j \\[2ex] a_{j-1}(\mu)\hat{a}_{i+1}(\mu)\left[\displaystyle\prod_{k=j+1}^{i} \gamma_k\right] & \text{for } i>j. \end{cases} \qquad (3.2.2)$$

From Eqn(3.2.2) we have that

$$\text{trace}[\text{adj}(\mu I - T_m)] = \sum_{k=1}^{m} a_{k-1}(\mu)\hat{a}_{k+1}(\mu). \qquad (3.2.3)$$

Proof. Choose i,j such that $1 \le i \le j \le m$. Let $R'(\mu)$ be the $(m-1) \times (m-1)$ submatrix obtained from $(\mu I - T_m)$ by deleting row j and column i. If $i<j$ and $\ell = (j-i)$, then

$$R'(\mu) \equiv \begin{bmatrix} (\mu I - T_{i-1}) & 0 & 0 \\ -\gamma_i e_1 e_{i-1}^T & R_*(\mu) & 0 \\ 0 & -\gamma_{j+1} e_1 e_\ell^T & (\mu I - \hat{T}_{j+1}) \end{bmatrix}$$

where $R_*(\mu)$ is the $\ell \times \ell$ lower triangular matrix

From Eqn(3.2.1) we have that $s_{ij} = (-1)^{i+j}\det(R'(\mu))$. But the formulas given in Eqn(3.2.2) follow directly from this relationship. Eqn(3.2.3) is an immediate consequence of Eqn(3.2.2).

□

LEMMA 3.2.2 Let μ and ν be two free parameters then

$$\frac{a_m(\mu) - a_m(\nu)}{\mu - \nu} = \sum_{k=1}^{m} a_{k-1}(\mu)\hat{a}_{k+1}(\nu). \tag{3.2.4}$$

Proof. We establish Eqn(3.2.4) by induction on m. Since by definition $a_0(\mu) \equiv 1$ and $\hat{a}_{m+1}(\mu) \equiv 1$, Eqn(3.2.4) is true when m=1. Therefore, proceed with the induction by assuming that it is true for all $j \leq m-1$. To simplify the notation define $\Delta_\ell \equiv (a_\ell(\mu) - a_\ell(\nu)) / (\mu - \nu)$. By direct substitution we obtain the following recursion

$$\Delta_m = a_{m-1}(\mu) + (\nu - \alpha_m)\Delta_{m-1} - \delta_m \Delta_{m-2}.$$

Then using the induction assumption, we have that

$$\Delta_m = a_{m-1}(\mu) + (\nu - \alpha_m) \sum_{k=1}^{m-1} a_{k-1}(\mu)\hat{a}_{k+1,m-1}(\nu) - \delta_m \sum_{k=1}^{m-2} a_{k-1}(\mu)a_{k+1,m-2}(\nu) .$$

If we rearrange the terms in these summations, using the facts that $\hat{a}_{m+1}(\nu) = 1$, $\hat{a}_{m,m-1}(\nu) = 1$, and $\hat{a}_m(\nu) = (\nu - \alpha_m)$, we obtain the following equality.

$$\Delta_m = a_{m-1}(\mu)\hat{a}_{m+1}(\nu) + a_{m-2}(\mu)\hat{a}_m(\nu) + \sum_{k=1}^{m-2} a_{k-1}(\mu)[(\nu - \alpha_m)a_{k+1,m-1}(\nu) - \delta_m a_{k+1,m-2}(\nu)].$$

Now if we apply Eqn(3.1.5) to the quantities inside the summation, we obtain Eqn(3.2.4). □

The following limiting form of Eqn(3.2.4) follows immediately from Eqn(3.2.4) since the characteristic polynomials are continuous functions.

COROLLARY 3.2.2

$$a'_m(\mu) = \frac{\partial}{\partial\mu}[a_m(\mu)] = \sum_{k=1}^{m} a_{k-1}(\mu)\hat{a}_{k+1}(\mu). \tag{3.2.5}$$

Therefore for any μ

$$\text{trace}[\text{adj}(\mu I - T_m)] = \sum_{k=1}^{m} a_{k-1}(\mu)\hat{a}_{k+1}(\mu) = a'_m(\mu). \tag{3.2.6}$$

Now using Lemma 3.2.1 and Corollary 3.2.2, we easily obtain determinant formulas for the eigenvectors of any mxm tridiagonal matrix T_m in terms of the characteristic polynomials of the submatrices T_j, $1 \leq j \leq m$. These formulas will not be used in the actual implementations of our Lanczos procedures; however, they will be useful in the arguments which we give in Chapter 4 supporting the use of the single-vector Lanczos recursion without any reorthogonalization of the Lanczos vectors.

LEMMA 3.2.3 Let T_m be a mxm tridiagonal matrix. Let μ be a simple eigenvalue of T_m. Choose left and right eigenvectors of T_m such that

$$
\begin{aligned}
T_m u &= \mu u, \ u \neq 0 \\
v^T T_m &= \mu v^T, \ v \neq 0.
\end{aligned}
\tag{3.2.6}
$$

Then we have that $v^T u \neq 0$ and for u and v such that $v^T u = 1$,

$$
\mathrm{adj}(\mu I - T_m) = a'_m(\mu) \, uv^T.
\tag{3.2.7}
$$

Proof. Define $S(\mu) \equiv \mathrm{adj}(\mu I - T_m)$. Since μ is simple, the left and right null spaces of $(\mu I - T_m)$ are one-dimensional and $a'_m(\mu) \neq 0$. Note that by Eqn(3.2.6), u spans the right null space and v spans the left null space. From Eqn(3.2.1) we have that

$$
S(\mu)(\mu I - T_m) = (\mu I - T_m)S(\mu) = 0.
$$

Therefore, the rows (columns) of $S(\mu)$ must be in the left (right) null space of $(\mu I - T_m)$. Therefore, for some constant c

$$
S(\mu) = c u v^T.
$$

But then we have that $\mathrm{trace}(S) = c[v^T u] = a'_m(\mu) \neq 0$, so that $c = a'_m(\mu)$. $\square$

To simplify some of the notation in the following discussions we introduce the following definition.

DEFINITION 3.2.2 For $k = 2, \ldots, m$ and $2 \leq j \leq k + 1$, define

$$
\pi_{j,k}(\delta) \equiv
\begin{cases}
1 & \text{for } j - k = 1 \\
\prod_{\ell=j}^{k} \delta_\ell & \text{for } j \leq k
\end{cases}
\tag{3.2.8}
$$

and let $\pi_k \equiv \pi_{2,k}(\delta)$, $\hat{\pi}_k \equiv \pi_{k,m}(\delta)$.

From Lemma 3.2.3, we obtain the following determinant formulas for the left and the right eigenvectors corresponding to any simple eigenvalue of any tridiagonal matrix.

COROLLARY 3.2.3 For any simple eigenvalue μ of a mxm tridiagonal matrix T_m, we have the following relationships between the components of the left and right eigenvectors v and u of T_m corresponding to μ. (See Eqn(3.2.6).) We let u(k) and v(k) denote respectively, the kth components of the vectors u and v. For k=2,3,...,m,

$$u(k) = [a_{k-1}(\mu)u(1)]/\pi_k(\beta) \quad \text{and} \quad u(k) = [\hat{a}_{k+1}(\mu)u(m)]/\hat{\pi}_{k+1}(\gamma), \quad (3.2.9)$$

$$v(k) = [a_{k-1}(\mu)v(1)]/\pi_k(\gamma) \quad \text{and} \quad v(k) = [\hat{a}_{k+1}(\mu)v(m)]/\hat{\pi}_{k+1}(\beta), \quad (3.2.10)$$

$$u(k)v(k) = [a_{k-1}(\mu)\hat{a}_{k+1}(\mu)] / a'_m(\mu), \quad (3.2.11)$$

$$u(1)v(m) = \pi_m(\beta) / a'_m(\mu) \quad \text{and} \quad u(m)v(1) = \pi_m(\gamma) / a'_m(\mu). \quad (3.2.12)$$

Eqns(3.2.9) - (3.2.12) are straight-forward consequences of Lemmas 3.2.1 and 3.2.3. If we use the determinant recursions given in Eqns(3.1.5) - (3.1.6), then these formulas for the eigenvectors become very easy to implement. Unfortunately, however, this method for calculating eigenvectors is not numerically stable (see Wilkinson [1965, pp.316-321]) and should not be used. In our discussions we make use of these relationships only to demonstrate theoretically various properties of single-vector Lanczos procedures. With slightly more work than using the determinant recursions, the right and left eigenvectors u and v can be computed in a stable way using inverse iteration. This method is described very briefly in Section 3.4.

The following Lemma provides us with one of the fundamental identities which we use in our arguments to substantiate the use of the single-vector Lanczos recursion without any reorthogonalization.

LEMMA 3.2.4 Let T_m be a mxm tridiagonal matrix. Let μ be a free parameter. Then for any $1 \leq k \leq m$, we have the following determinant relations.

$$\hat{\pi}_{k+1}(\delta)a_{k-1}(\mu) = a_{m-1}(\mu)\hat{a}_{k+1}(\mu) - a_m(\mu)a_{k+1,m-1}(\mu) \quad (3.2.13)$$

$$\pi_k(\delta)\hat{a}_{k+1}(\mu) = \hat{a}_2(\mu)a_{k-1}(\mu) - a_m(\mu)a_{2,k-1}(\mu). \quad (3.2.14)$$

Proof. We use induction to establish Eqn(3.2.13). The proof for Eqn(3.2.14) is similar. Since $\hat{\pi}_{m+1} = 1$, $a_{m+1,m-1}(\mu) = 0$ and $a_{m+1,m}(\mu) = 1$, Eqn(3.2.13) is valid for k = m. Therefore, consider the case k = m-1. We have that $\hat{\pi}_m = \delta_m$, $\hat{a}_m = (\mu - \alpha_m)$ and $a_{m,m-1} = 1$. Therefore from Eqn(3.1.4) we have that

$$\delta_m a_{m-2}(\mu) = a_{m-1}(\mu)(\mu - \alpha_m) - a_m(\mu)1.$$

Thus, Eqn(3.2.13) is valid for k=m-1. Now assume that it is valid down to some k where

$2 \leq k \leq m-1$. Then by Eqn(3.1.5) for m and for m-1,

$$\hat{a}_k(\mu) = (\mu-\alpha_k)\hat{a}_{k+1}(\mu) - \delta_{k+1}\hat{a}_{k+2}(\mu)$$
$$a_{k,m-1}(\mu) = (\mu-\alpha_k)a_{k+1,m-1}(\mu) - \delta_{k+1}a_{k+2,m-1}(\mu)$$

Using these expressions, rearranging and then using the induction on k we obtain

$$\hat{a}_k(\mu)a_{m-1}(\mu) - a_{k,m-1}(\mu)a_m(\mu) = \hat{\pi}_{k+1}[(\mu-\alpha_k)a_{k-1}(\mu) - a_k(\mu)].$$

But by Eqn(3.1.5) the right-hand side of this equation is just $\hat{\pi}_k a_{k-2}(\mu)$. Therefore, Eqn(3.2.14) holds for any k such that $1 \leq k \leq m$. ☐

In the Lanczos discussions for a given tridiagonal Lanczos matrix T_m we will be primarily interested in values of these determinants at values of μ which are eigenvalues of T_m. For any such μ, we have the following identity which we will use in Chapter 4 to justify the 'identification' test which we use in our Lanczos procedures. It is a direct consequence of Lemma 3.2.4.

COROLLARY 3.2.4 Under the hypotheses of Lemma 3.2.4, assume that μ is an eigenvalue of T_m. That is, that $a_m(\mu) = 0$. Then

$$\hat{a}_2(\mu)a_{m-1}(\mu) = \prod_{k=2}^{m} \delta_k. \tag{3.2.15}$$

SECTION 3.3   COMPLEX SYMMETRIC OR HERMITIAN TRIDIAGONAL

DEFINITION 3.3.1 Let $A \equiv (a_{ij})$, $1 \leq i,j \leq n$, be a complex matrix. A is complex symmetric if and only if $A^T \equiv A$. It is Hermitian if and only if $A^H \equiv A$.

If a matrix $A \equiv B + iC$ is complex symmetric, then $B^T = B$ and $C^T = C$. That is, both the real and the imaginary parts of the matrix are real symmetric matrices. If a matrix is Hermitian then $B^T = B$ and $C^T = -C$. That is, the real part is real symmetric and the imaginary part is skew symmetric.

The determinant, adjoint and eigenvector formulas and recursions which were derived in the preceding two sections are applicable to any general mxm complex tridiagonal matrix. The single-vector Lanczos procedures for real symmetric problems which are discussed in Chapter 4 all generate real symmetric tridiagonal matrices, and in Section 3.5 we will specialize the discussions to such matrices. Real symmetric tridiagonal matrices possess many properties not shared by general tridiagonal matrices. In Chapter 6 we consider a single-vector Lanczos procedure for nondefective complex symmetric matrices. In this case the Lanczos matrices are themselves complex symmetric and tridiagonal. More recent research Cullum and Willoughby

[1984] has looked at the the question of devising a Lanczos procedure for general nondefective nonsymmetric matrices. This research has used a version (see Eqns(6.3.2) - (6.3.3)) of the generalized Lanczos recursion which generates Lanczos matrices which are again complex symmetric and tridiagonal. In that context the generality of the class of complex symmetric tridiagonal matrices is important. The following lemma tells us that in fact this class of matrices encompasses all irreducible tridiagonal matrices, in the sense that any irreducible tridiagonal matrix is diagonally similar to a complex symmetric tridiagonal matrix.

LEMMA 3.3.1 Let T_m be any irreducible, tridiagonal mxm matrix. That is, $\delta_k \neq 0$, for $2 \leq k \leq m$. Then the nonsingular diagonal matrix $D \equiv (d_{ij})$ defined by the recursion

$$d_{jj} \equiv d_j \equiv d_{j-1}\sqrt{\gamma_j/\beta_j} \text{ for } j>1 \text{ and } d_1 \equiv 1, \qquad (3.3.1)$$

transforms T_m into the matrix $\tilde{T}_m \equiv D^{-1}T_mD$, which is complex symmetric. Moreover if we define $x \equiv D^{-1}z$, then

$$T_m z = \mu z \text{ if and only if } \tilde{T}_m x = \mu x. \qquad (3.3.2)$$

Proof. It is a straightforward exercise to demonstrate that

$$\tilde{T}_m = \begin{bmatrix} \alpha_1 & \sqrt{\gamma_2\beta_2} & & & \\ \sqrt{\gamma_2\beta_2} & \alpha_2 & & & \\ & \sqrt{\gamma_3\beta_3} & & & \\ & & & \alpha_{m-1} & \sqrt{\gamma_m\beta_m} \\ & & & \sqrt{\gamma_m\beta_m} & \alpha_m \end{bmatrix} \qquad (3.3.3)$$

which is obviously symmetric. □

COROLLARY 3.3.1 Let T_m be an irreducible, Hermitian, $m \times m$ tridiagonal matrix, then the diagonal matrix transformation D defined in Lemma 3.3.1 is unitary, and the corresponding matrix $\tilde{T}_m \equiv D^{-1}T_mD$ is real symmetric with positive off-diagonal entries.

Proof. Clearly, the diagonal entries α_k must be real. Let $T_m(k,k+1) \equiv r_{k+1}e^{-i\theta_{k+1}}$, where $r_{k+1} > 0$. Then since T_m is Hermitian, $T_m(k+1,k) \equiv r_{k+1}e^{i\theta_{k+1}}$. But this implies that $d_k = e^{i\theta_k}d_{k-1}$, for $2 \leq k \leq m$. Obviously, D is unitary. Furthermore, $\tilde{T}_m(k,k) = \alpha_k$ and $\tilde{T}_m(k,k+1) = \tilde{T}_m(k+1,k) = r_{k+1}$. □

COROLLARY 3.3.2 Let T be an irreducible, real symmetric, $m \times m$ tridiagonal matrix. Then the matrix $\tilde{T}$ obtained from T by changing the signs of one or more of the off-diagonal entries β_{k+1} has the same eigenvalues as T.

Proof. Let $D \equiv \text{diag}\{d_1, d_2, ..., d_m\}$ where $d_1 = 1$. For $k = 1, ..., m-1$ set

$$d_{k+1} \equiv d_k[\ \pm 1]$$

where for each k, -1 is chosen if the sign of β_{k+1} is to be changed, and $+1$ is chosen if the sign of β_{k+1} is not changed. (Essentially we are considering all those matrices which can be obtained from T by changing one or more of the signs of off-diagonal entries in T.) Then the tridiagonal matrix $\tilde{T} \equiv D^{-1}TD$ has the same eigenvalues as T. But, clearly

$$\tilde{T}(k,k) = T(k,k) = \alpha_k \text{ and}$$
$$\tilde{T}(k,k+1) = (d_k/d_{k+1})\beta_{k+1}, \quad \tilde{T}(k+1,k) = (d_{k+1}/d_k)\beta_{k+1}$$

so that $\tilde{T}$ is the matrix obtained from T by changing the signs of certain off-diagonal entries of T as specified by the choices of the d_k. $\qquad\qquad\square$

Thus, from Corollary 3.3.1 we have that the eigenvalues of a tridiagonal Hermitian matrix T can be computed by computing the eigenvalues of the corresponding real symmetric tridiagonal matrix $\tilde{T}$ obtained from T by replacing the off-diagonal entries by their magnitudes. Eigenvectors u of T are obtained by computing the eigenvectors $\tilde{u}$ of $\tilde{T}$ and setting $u = D^{-1}\tilde{u}$. From Corollary 3.3.2 we have that the eigenvalues of a real symmetric tridiagonal matrix are independent of the signs of the off-diagonal entries. Lemma 3.3.1 tells us that the eigenvalues of complex symmetric tridiagonal matrices do not possess any special properties.

SECTION 3.4   EIGENVECTORS, USING INVERSE ITERATION

In the single-vector Lanczos procedures which we describe in Chapters 4, 5, and 6 of this book we use inverse iteration on tridiagonal matrices in two key computations. These two computations are done on the tridiagonal Lanczos matrices. First it is used to obtain error estimates for the computed eigenvalue approximations. Second it is used to compute suitable eigenvectors of the Lanczos matrices for use in computing the corresponding approximate eigenvectors of A. More will be said about each of these subjects in Sections 4.7 and 4.8 of Chapter 4. In these Lanczos procedures, the matrices which we will be dealing with are either real symmetric or nondefective, complex symmetric tridiagonal matrices.

The basic idea behind inverse iteration is the following. Let A be a nondefective mxm matrix with a real orthonormal set of eigenvectors X, so that $A = X\Lambda X^T$. If we have an accurate approximation μ to some eigenvalue of A, then we solve the following system of equations

$$(A - \mu I)w = z \qquad\qquad\qquad (3.4.1)$$

where z is a randomly-generated vector to obtain the vector

$$w = \sum_{j=1}^{n} [(x_j^T z)/(\lambda_j - \mu)]x_j \qquad (3.4.2)$$

But, observe that if μ is a good approximation to some eigenvalue λ_J of A, then the term or terms in this summation involving λ_J will be dominant. If in fact the eigenvalue under consideration is simple, is well separated from the other eigenvalues, and the starting vector z has a nontrivial projection on the eigenvector corresponding to that eigenvalue, then in practice w will be a good approximation to the desired eigenvector. For more details on this procedure see Peters and Wilkinson [1979]. In this section we restrict our considerations to tridiagonal matrices.

As indicated above the key to the successful use of inverse iteration is the sufficient amplification of the desired eigenvector through solving Eqn(3.4.1). In practice we need a scale invariant means of measuring the growth in the norm of the solution w w.r.t. the norm of the original starting vector z. This is accomplished by scaling the initial vector using an approximation for the norm of the given matrix. The vector z is reduced in size by this estimate prior to any computations. In particular for the tridiagonal mxm Lanczos matrices T_m, it is scaled so that

$$\sum_{k=1}^{m} |z(k)| = \varepsilon_{tol} \quad \text{where} \qquad (3.4.3)$$

$$\varepsilon_{tol} = m * \varepsilon_{mach} * (\sum_{k=1}^{m} |\alpha_k| + \sum_{k=2}^{m} \beta_k). \qquad (3.4.4)$$

If μ is a close approximation to a true isolated eigenvalue of A, then typically only one step of inverse iteration is necessary to obtain the desired eigenvector u such that $T_m u = \mu u$.

In our single-vector Lanczos procedures we use inverse iteration only on the Lanczos matrices. Therefore, we are solving systems of equations

$$T^{\mu} u \equiv (T_m - \mu I)u = z \qquad (3.4.5)$$

where T^{μ} is tridiagonal and it is irreducible. We consider the solution of Eqn(3.4.5) in detail. For this discussion there is not any reason why T^{μ} has to be symmetric. Therefore, we let

$$T^{\mu}(k,k) \equiv \alpha_k - \mu, \text{ and}$$
$$T^{\mu}(k,k+1) \equiv \gamma_k, \quad T^{\mu}(k+1,k) \equiv \beta_{k+1}.$$

From the irreducibility we have that $\delta_k \equiv \beta_k \gamma_k \neq 0$ for $2 \leq k \leq m$. If T^{μ} were not irreducible we would solve each irreducible subsection separately.

We solve Eqn(3.4.5) by Gaussian elimination, see for example Stewart [1973], progressively reducing Eqn(3.4.5) to a triangular system

$$Ru = \tilde{z} \qquad (3.4.6)$$

which is easily solved using back substitution. Because T is tridiagonal, R will have at most three nonzero entries in each row and these will be in the kth, $(k+1)$st and $(k+2)$nd columns. In the following discussion we will denote these entries as follows.

$$r_k \equiv R_{k,k}, \ r'_k \equiv R_{k,k+1} \text{ and } \bar{r}_k \equiv R_{k,k+2}.$$

Because T_m is tridiagonal we can successively eliminate variable u_{k-1} from the kth equation with only one elimination step.

Denote the result of the $(k-1)$st elimination step by

$$au_{k-1} + bu_k = \tilde{z}_{k-1} \qquad (3.4.7)$$

where a and b are coefficients that are computed and thus vary as we proceed through the matrix. Note that if k=2 then $a = \alpha_1$, and $b = \beta_2$. Before we perform the next elimination step we compare Eqn(3.4.7) with the next equation in the sequence of equations defined by the matrix,

$$\gamma_k u_{k-1} + \alpha_k u_k + \beta_{k+1} u_{k+1} = z_k. \qquad (3.4.8)$$

Note that if $k = m$, then the $\beta_{k+1} u_{k+1}$ term does not appear in the above equation.

Pivoting for size is critical for maintaining the numerical stability of the elimination process. Pivoting corresponds to a simple reordering of the sequence of equations specified by the given matrix T_m. In a tridiagonal matrix such pivoting is easily implemented. Pivoting procedures, see for example, Wilkinson [1965, pp.212] typically consider the size of the elements along the kth row (or column) of the current reduced matrix and select the maximal element for the next pivot. In fact on the kth step they are constrained to consider only the maximal elements in the first k rows. For a tridiagonal matrix this type of pivoting reduces to a simple comparison of the magnitude of the coefficients of the next variable to be eliminated in two adjoining equations. Thus, there are only two possible situations. In Case 1 no interchange of rows occurs; it requires 1 divide, 2 multiplications and 2 additions. Case 2 requires an interchange of successive rows; it requires 1 divide, 3 multiplications and 2 additions. To be specific consider the following.

CASE 1 $|a| \geq |\gamma_k| > 0$. If this is true then no interchange of rows is necessary. We let $d \equiv (\gamma_k/a)$, then $r_{k-1} \equiv a$, $r'_{k-1} \equiv b$ and $\bar{r}_{k-1} \equiv 0$. The modified equation resulting from this elimination step is specified by Eqn(3.4.7) with the coefficients a and b defined by $a = \alpha_k - db$, $b = \beta_{k+1}$, and the modified kth component of the right-hand side of the equation is given by $\tilde{z}_k = z_k - d\tilde{z}_{k-1}$.

CASE 2 $|a| < |\gamma_k|$. In this case we must interchange the two equations (rows) before we perform the elimination step. To interchange the rows we set $r_{k-1} \equiv \gamma_k$, $r'_{k-1} \equiv \alpha_k$, and $\bar{r}_{k-1} \equiv \beta_{k+1}$. We then set TEMP $= \tilde{z}_{k-1}$ and set the (k-1)st component of the right-hand side $\tilde{z}_{k-1} = z_k$. The modified kth equation is obtained by setting d $= a/\gamma_k$ and the new coefficients a $= b - d\alpha_k$, b $= -d\beta_{k+1}$. The corresponding modified component of the right-hand side is obtained by setting $z_k = $ TEMP $- d\tilde{z}_{k-1}$.

After completing the elimination steps we obtain the triangular system given in Eqn(3.4.6) which is easily solved using the following back substitution formula for k = m, m-1, ..., 1.

$$u_k = (z_k - r'_k u_{k+1} - \bar{r}_k u_{k+2}) / r_k. \tag{3.4.9}$$

The system Eqn(3.4.5) is very ill-conditioned because μ is an eigenvalue of T_m. The final pivot which we compute, r_m, should be essentially zero. In this situation, as was done in EISPACK [1976,1977], we set $r_m \equiv [\varepsilon_{mach} \Sigma_m]$ where Σ_m denotes the sum of the absolute values of all of the elements of T_m and ε_{mach} denotes the machine epsilon of the computer arithmetic being used.

SECTION 3.5   EIGENVALUES, USING STURM SEQUENCING

For any real symmetric matrix we have the Sturm sequencing property. See for example, Jennings [1977, Chapter 9].

DEFINITION 3.5.1 (Sturm Sequencing Property) Given any real symmetric $n \times n$ matrix A and any real number μ, the number $c(\mu)$ of disagreements in sign between the consecutive values of the sequence of the determinants $a_j(\mu)$, $0 \le j \le n$, of the principal minors $A_j(\mu)$ of the matrix $(\mu I - A)$ equals the number of eigenvalues of A which are larger than μ where $a_0 \equiv 1$.

Therefore, given any real symmetric matrix A and any two scalars μ and ν, the number of eigenvalues of A in the interval $(\mu, \nu]$ is easily determined if the corresponding counts $c(\mu)$ and $c(\nu)$ for the Sturm sequences corresponding to μ and to ν are available. The number of eigenvalues in the interval is simply the difference $c(\mu) - c(\nu)$.

For a general real symmetric matrix these counts can be obtained reliably whenever it is possible to factor the given matrix as follows.

$$A = WDW^T \tag{3.5.1}$$

where D is a diagonal matrix and W is nonsingular, see for example Parlett [1980, Chapter 1]. The numbers of negative, of positive and of zero eigenvalues of A are equal to respectively, the number of negative, of positive, and of zero entries on the diagonal of D. This relationship is based upon Lemma 3.5.1 and Definition 3.5.1. A proof of Lemma 3.5.1 can be found for example in Parlett [1980, p.46]. In this section we focus on real symmetric tridiagonal matrices.

LEMMA 3.5.1 The number of positive, negative or zero eigenvalues of any Hermitian matrix is preserved under congruent transformations.

DEFINITION 3.5.2 For any matrix A and any nonsingular matrix W, the matrix WAW^T is called a congruence transformation of A.

When the matrix in question is tridiagonal, the determinants of the principal minors are generated by the simple recursion given in Eqn(3.1.5). However, if we attempt to use this recursion directly, we will encounter numerical problems. As we proceed down the diagonal of a given matrix, the principal minors can vary radically in magnitude, numerically exceeding the allowable range of numbers representable by the computer on which the recursion is being executed. However, a numerically-stable implementation of the Sturm sequencing can be obtained by using ratios of the determinants of successive principal minors. For a mxm tridiagonal matrix these corresponding ratios satisfy the following recursion. For $i = 2,...,m$ and with $r_1 \equiv (\mu - \alpha_1)$, we obtain

$$r_i(\mu) = (\mu - \alpha_i) - \beta_i^2/r_{i-1}(\mu) \quad \text{with } r_i \equiv a_i(\mu)/a_{i-1}(\mu). \qquad (3.5.2)$$

The count $c(\mu)$ of the number of eigenvalues $>\mu$ is determined by the number of negative ratios. See Wilkinson [1965, pp.426-429] for a discussion of the numerical stability of determinant calculations for Hessenberg matrices. The tridiagonal matrix is a very special case of a Hessenberg matrix.

To demonstrate how one uses the Sturm sequence property to compute eigenvalues of a given real symmetric tridiagonal matrix T_m consider the following. We wish to compute the algebraically smallest eigenvalue μ_m^m of T_m on the interval $a < \mu \leq b$. As we said earlier, if X is a test value for μ, then a Sturm sequence calculation provides $c(X)$, the number of eigenvalues of T_m which are greater than X. If $c(a) = c(b)$, then there is no μ in the given interval. Otherwise we set $XL = a$, $XU = b$ and $X = (XL + XU)/2$, the midpoint of the interval. If $c(X) < c(XL)$ then set $XL = XL$ and $XU = X$. Otherwise $c(X) = c(XL)$ and we set $XL = X$ and $XU = XU$. The successive subintervals (XL,XU), each enclosing at least one eigenvalue of A, are bisected until the endpoints of the resulting subinterval are closer than a given tolerance. The midpoint of this small interval is then accepted as an approximation to an eigenvalue of A. This type of procedure is called bisection.

A bisection program, BISECT, is included in the EISPACK library [1977, pp.157-165]. It uses the convergence tolerance

$$\varepsilon_{conv} \equiv 2*\varepsilon_{mach}*[\,|XL| + |XU|\,] + \varepsilon_1$$

where ε_1 is an absolute error tolerance and typically $\varepsilon_1 \equiv \varepsilon_{mach} * \| A \|$. Our bisection subroutine BISEC (see Volume 2, Chapter 2) which we use in our 'real' symmetric single-vector eigenvalue procedures, is a modified form of the EISPACK program BISECT. We use BISEC to compute the eigenvalues of the real symmetric tridiagonal matrices generated by the single vector Lanczos recursions. BISEC uses less storage than BISECT and includes the modifications necessary to implement our single-vector Lanczos procedures.

The key problem in the implementation of a single vector Lanczos procedure with no reorthogonalization is the identification of the extra or spurious eigenvalues of the Lanczos tridiagonal matrices T_m which appear because of the losses in orthogonality between the Lanczos vectors generated. The identification test which we use to sort the 'bad' eigenvalues from the 'good' ones rests upon certain relationships between the eigenvalues of the Lanczos tridiagonal matrices T_m, $j = 1, 2, \ldots, m$ and the eigenvalues of the associated $\hat{T}_2$ matrices. See Eqn(3.1.3). This test is carried out directly in our BISEC subroutine, using Sturm sequencing and with essentially no cost over that which is required to compute the eigenvalues of T_m.

Bisection procedures are numerically stable, see Barth, Martin and Wilkinson [1967], in the sense that the computed eigenvalues are close in magnitude to those of the given matrix and in the sense that the computed eigenvalues are the exact eigenvalues of a matrix that is close to the original one. It is theoretically possible that the determinant of some minor may vanish, see Ortega [1960]. Suppose that for some k, $a_{k-1}(\mu) = 0$. Then from Eqn(3.1.5) $a_k(\mu) = -\beta_k^2 a_{k-2}(\mu)$ so that $a_k(\mu)$ must have the opposite sign of $a_{k-2}(\mu)$. Furthermore we have that $a_{k+1}(\mu)/a_k(\mu) = (\mu - \alpha_{k+1})$. This says that we should increment the Sturm sequencing count by 1 and then restart the r sequence at $k+1$. Our BISEC program mimics the EISPACK bisection subroutine. In both programs if some ratio $r_{k-1}(\mu) = 0$, then in Eqn(3.5.2) the quantity $\beta_k^2/r_{k-1}(\mu)$ is replaced by $\beta_k/\varepsilon_{mach}$. Note that since we have by construction that $\beta_k > 0$ for all k,

$$r_k(\mu) \cong -\beta_k/\varepsilon_{mach} < 0 \quad \text{and} \quad r_{k+1} \cong (\mu - \alpha_{k+1}).$$

A bisection subroutine is particularly useful when only the eigenvalues in some subinterval of the spectrum are required. It can be used to directly compute the eigenvalues in any portion of the spectrum without having to compute any unwanted eigenvalues. In the next Chapter we focus on the main topic of this book, Lanczos procedures for the computation of eigenvalues and eigenvectors of real symmetric problems.

CHAPTER 4

LANCZOS PROCEDURES WITH NO REORTHOGONALIZATION
FOR REAL SYMMETRIC PROBLEMS

SECTION 4.1   INTRODUCTION

We now address the main topic of this book, the construction and implementation of practical Lanczos procedures with no reorthogonalization. In this chapter we concentrate on real symmetric problems. In particular we consider real symmetric matrices, Hermitian matrices, and certain real symmetric generalized eigenvalue problems. However, the basic ideas introduced in this chapter are not limited to these 'real' symmetric problems. Chapter 5 extends these ideas to the computation of singular values and singular vectors of large, real rectangular matrices. Chapter 6 addresses the computation of eigenvalues and eigenvectors of nondefective complex symmetric matrices. Thus in the next three chapters, we present a family of Lanczos procedures which do not use any reorthogonalization. Codes for each of these procedures are contained in Volume 2 of this book.

As discussed in Section 2.3, the basic Lanczos procedure does not possess the nice theoretical properties given in Section 2.2 when it is implemented on a computer with finite precision arithmetic. In practice the Lanczos vectors will typically not be orthogonal or even linearly independent. Thus, the theoretical estimates given in Section 2.2 are not applicable to straight-forward implementations of the basic recursion, and therefore cannot be used to justify the use of the basic Lanczos procedure for computing eigenvalues of a given matrix. The basic recursion must be modified to obtain a practical Lanczos algorithm.

In Section 2.4 we listed several possible modifications. The so-called classical approaches attempt to maintain the orthogonality of the Lanczos vectors. However, orthogonality can be maintained only by incorporating some sort of explicit reorthogonalization of the Lanczos vectors. In the literature this reorthogonalization has taken the form of reorthogonalizations w.r.t. either other Lanczos vectors or w.r.t. one or more of the associated Ritz vectors. Either type of reorthogonalization requires that the user keep all of the Lanczos vectors on some readily available storage for recall either for direct use in the reorthogonalizations or for the computation of some subset of the corresponding Ritz vectors which will then be used in the reorthogonalizations. Therefore, in either case extra storage is needed and this effectively limits the number of eigenvalues which can be computed. If the given matrix is large, storing its Lanczos vectors quickly consumes huge amounts of auxiliary storage.

In this book we focus primarily on the non-classical (some say unorthodox) approach of not using any reorthogonalization and then devising a scheme for unraveling the effects of the losses in orthogonality. In this chapter and in Chapters 5 and 6 we define practical Lanczos procedures with no reorthogonalization. The only exception will be Chapter 7 where we will define iterative block Lanczos procedures which use some reorthogonalization w.r.t. certain Ritz vectors. In order to justify the use of the Lanczos recursion without any reorthogonalization we must first demonstrate that even though the orthogonality or even the linear independence of the Lanczos vectors is lost, the eigenvalues which we want will still appear as eigenvalues of the Lanczos matrices if we make these tridiagonal matrices large enough. At first glance there is no obvious reason why this should be true. The classical error estimates (see Section 2.2) which require the orthogonality of the Lanczos vectors have been extended to the case of simply linearly independent Lanczos vectors, see for example Parlett [1980]. However these arguments cannot be extended to the case of linearly dependent Lanczos vectors. Furthermore, as we will see in the discussion of the numerical results, the losses in orthogonality may make it necessary to carry the recursion out beyond the order of the matrix in order to get the required information. The classical arguments clearly do not provide any justification for using Lanczos matrices which are larger than the original matrix.

The second problem which must be addressed in order to obtain a practical Lanczos procedure with no reorthogonalization is devising an identification test for picking out that subset of those eigenvalues of any Lanczos matrix which are legitimate approximations to eigenvalues of the original matrix. We choose to reword this as the problem of devising a test for identifying those computed eigenvalues of the Lanczos matrices which are 'spurious'; that is, they are the result of the losses in orthogonality and should not be accepted as approximations to eigenvalues of the original matrix. All other computed eigenvalues will be accepted as such approximations regardless of how accurate these approximations are. This change in viewpoint provides several advantages.

As we pointed out in Section 2.5 several Lanczos procedures which do not use any reorthogonalization have appeared in the literature. Each of these contains some mechanism for selecting an appropriate subset of the eigenvalues of the Lanczos matrices generated. However all of these procedures, with the exception of ours, are based upon identifying those eigenvalue approximations which have converged. For such procedures any 'good' eigenvalues which have not yet converged, for example any eigenvalues of the Lanczos matrices which approximate eigenvalues of the original matrix to only two or three digits, are not accepted as approximations to eigenvalues of the original matrix. In these methods the user must provide a tolerance which is used to identify which eigenvalues have converged. This mode of identification generally reduces the resolution achievable and can sometimes lead to mistakes in the identification process, see Cullum, Willoughby and Lake [1983]. Moreover, this mode of identification prevents the user from getting an overall picture of the spectrum of the matrix being analyzed until total convergence has occurred. At each stage in such a Lanczos procedure the user would see only that portion of the spectrum of the original matrix whose eigenvalue approximations have converged.

The Lanczos procedures which we have devised and will discuss are simple. There are no complicated heuristics. They provide high accuracy and resolution and have minimal storage requirements. These procedures operate in two phases. (This is typical of Lanczos procedures with no reorthogonalization.) The eigenvalue and eigenvector (Ritz vector) computations are separated. Eigenvalues are computed first, using very little computer storage. Then for some subset of those eigenvalue approximations which are accurate, corresponding Ritz vectors are computed. Either a few or many eigenvalues can be computed, depending upon what the user wants. Moreover, the cost of computing corresponding Ritz vectors can be estimated cheaply before any expensive Ritz computations are performed. We will elaborate upon each of these points as we proceed through this chapter. Our procedures, like any other single-vector Lanczos procedures, cannot directly determine the multiplicities of the eigenvalues of the original matrix.

Our single-vector procedures are not iterative in the sense that many Lanczos matrices are used in the eigenvalue computations. However, they are iterative in the sense that the size of the Lanczos matrix required for the user-specified computation depends upon the eigenvalue distribution in the matrix being considered and upon what portion of the spectrum is being computed. In Section 4.7 we will illustrate the type of convergence one can expect and make some comments about rates of convergence. The observed behavior is very interesting. In some cases extreme eigenvalues of the original matrix can be computed readily, but in other cases it is almost as much work to compute a few extreme eigenvalues as it is to compute a lot or even most of the eigenvalues of the original matrix. What we will call the gap stiffness plays a key role in the effective rate of convergence of these Lanczos procedures with no reorthogonalization.

The arguments which we give to justify the use of a Lanczos procedure with no reorthogonalization and which we use to construct a test to identify which of the eigenvalues of the Lanczos matrices are only artifacts of the losses in orthogonality of the Lanczos vectors, rest upon the fundamental equivalence of Lanczos tridiagonalization to the method of conjugate gradients for solving linear systems of equations Ax=b when A is a positive definite matrix. Verification of this well-known equivalence is given in Section 4.2, and a generalization of this equivalence to the practical case of finite precision arithmetic is given in Section 4.3. In Section 4.4 we use this equivalence to obtain a plausibility argument that we should expect to be able to compute the eigenvalues which we want by simply applying the Lanczos recursion with no reorthogonalization and then computing eigenvalues of the associated Lanczos matrices, if we choose a large enough Lanczos matrix. In Section 4.5 we use this equivalence to discover a simple mechanism for identifying that subset of the eigenvalues (of any Lanczos matrix which we are considering) which are 'spurious' and should not be kept as approximations to eigenvalues of the original matrix. Any assumptions which we make will be labelled clearly as we proceed through this chapter.

In Section 4.6 we examine the 'spurious' eigenvalues more closely by providing an example which illustrates the two types of spuriousness which can be encountered. Section 4.7 contains a discussion of our Lanczos procedure for computing eigenvalues of real symmetric matrices. Section 4.8 contains a discussion of our Lanczos procedure for computing corresponding eigenvectors. Both sections contain typical numerical results. Section 4.9 contains extensions of these procedures to Hermitian matrices and to certain real symmetric generalized problems.

SECTION 4.2   AN EQUIVALENCE, EXACT ARITHMETIC

The arguments which we use for justifying the use of a Lanczos procedure with no reorthogonalization are based upon a generalization of the well-known equivalence between the Lanczos tridiagonalization procedure for solving a system of linear equations

$$Ax = b \qquad (4.2.1)$$

and the conjugate gradient optimization procedure for solving the same system of equations, Hestenes and Stiefel [1952] and Hestenes [1956]. The arguments demonstrating this equivalence require that A be positive definite. In order to use such an argument in the practical case it will be necessary to generalize this equivalence to the case considered by Paige, namely Lanczos tridiagonalization in finite precision arithmetic. However, first in this section we consider this equivalence, assuming infinite precision arithmetic. Then in Section 4.3 we show how these arguments can be localized so that the effects of finite precision arithmetic can be incorporated. Householder [1964] discusses this basic equivalence assuming infinite precision arithmetic, and many references to this equivalence can be found in the literature, see for example Paige and Saunders [1975] and Reid [1971].

LANCZOS TRIDIAGONALIZATION   Let A be a real symmetric nxn matrix. Assume that A has n distinct eigenvalues. Lanczos tridiagonalization has been defined previously in Eqns(2.1.1) - (2.1.2). We can use it to solve Eqn(4.2.1) by first generating the Lanczos matrix T_n, and then solving the resulting real symmetric tridiagonal system of equations

$$T_n y \ = \ V_n^T b \ \text{ and setting } x \ = \ V_n y. \qquad (4.2.2)$$

If A has fewer than n distinct eigenvalues, then less than n steps of the recursion will suffice. In Eqn(4.2.2), $V_n \equiv \{v_1,...,v_n\}$.

CONJUGATE GRADIENTS   The conjugate gradient procedure uses recursions which appear to be very different from those used in Lanczos tridiagonalization For any positive definite matrix we define the 'cost' function

$$f(x) \equiv \ [r(x)^T A^{-1} r(x)]/2 \ \text{ where } r(x) \equiv \ - Ax + b. \qquad (4.2.3)$$

Equivalently, we can define $f(x) \equiv \ [e(x)^T A \ e(x)]/2$ where $e(x) \equiv \ - x + A^{-1}b$.

The vector $r(x)$ is both the residual of Eqn(4.2.1) at x, and the negative of the gradient of $f(x)$ at x. The conjugate gradient procedure, when it is applied to Eqn(4.2.1), is an optimization procedure which determines the vector x^* for which the error function $f(x)$ is minimized. This vector is the solution of Eqn(4.2.1). This is called the conjugate gradient method because it takes a sequence of gradients of the function $f(x)$ (residuals, $r_i = r(x_i)$) and conjugates them with respect to the matrix A, thereby obtaining directions which are A-conjugate. We denote these directions by p_i, $i = 1,2,...$. In the procedure the function $f(x)$ is sequentially minimized along each of these directions which are generated as the optimization proceeds, generating iterates x_{i+1}. In exact arithmetic, after at most n such line minimizations we reach the minimizing point of $f(x)$ over all of E^n.

The quantities needed in the conjugate gradient procedure are generated using the following recursions

$$Ap_i = \omega_i(r_i - r_{i+1}) \tag{4.2.4}$$

$$p_{i+1} = r_{i+1} + \gamma_i p_i \tag{4.2.5}$$

where the scalars ω_i and γ_i satisfy

$$\omega_i \equiv p_i^T A p_i / p_i^T r_i \tag{4.2.6}$$

$$\gamma_i \equiv -p_i^T A r_{i+1} / p_i^T A p_i. \tag{4.2.7}$$

It is not difficult to demonstrate that

$$\gamma_i \equiv \sigma_i^2 \text{ where } \sigma_i \equiv \rho_{i+1}/\rho_i, \ \rho_1 \equiv 1 \text{ and } \rho_i \equiv \|r_i\|. \tag{4.2.8}$$

Since A is a nonsingular matrix we have from Eqn(4.2.4) that the corresponding iterates x_i satisfy the recursion $x_{i+1} = x_i + p_i/\omega_i$. Using the value of ω_i given by Eqn(4.2.6), we have that x_{i+1} corresponds to minimizing $f(x)$ along the line $x = x_i + \mu p_i$, $\mu \neq 0$. The definition of γ_i makes the direction p_{i+1} A-conjugate to the previous direction p_i. That is, for $1 \leq i \leq n - 1$ we have that $p_i^T A p_{i+1} = 0$. Theoretically, it is easy to demonstrate that the directions $P_n \equiv \{p_1,...,p_n\}$ are A-conjugate. That is, $P_n^T A P_n$ is a diagonal matrix. Hestenes and Stiefel [1952] demonstrated that for any positive definite matrix A, the corresponding function $f(x)$ defined in Eqn(4.2.3) can be minimized globally by sequentially minimizing it along n conjugate directions.

We demonstrate that if all of the computations required are performed in infinite precision arithmetic, then these two procedures are 'equivalent'.

DEFINITION 4.2.1 Two procedures are equivalent if and only if the quantities generated by one of these procedures determines all of the quantities which would be generated by the other procedure. Specifically, the Lanczos tridiagonalization procedure and the conjugate gradient optimization procedure applied to Eqn(4.2.1) are equivalent if and only if given the scalars α_i, β_{i+1} and the Lanczos vectors v_i defined by the Lanczos tridiagonalization, then these quantities determine corresponding quantities ω_i, γ_i, r_i, and p_i which satisfy the conjugate gradient relationships and vice-versa.

This discussion is primarily an academic exercise since in practice we have only finite precision arithmetic and the effects of the finite precision must be included in our analyses of our Lanczos procedures with no reorthogonalization if these analyses are to have any validity. We include the exact case here because it gives the reader a simple introduction to the basic relationships between these two procedures without the added complications of terms due to the finite precision arithmetic.

ASSUMPTION 4.2.1 For the rest of this section we assume that A is a positive definite, real symmetric nxn matrix with n distinct eigenvalues.

Whether or not A has n distinct eigenvalues is not a critical factor in the following discussions. The arguments could be modified to handle the general case. Select a unit starting vector v_1 for the Lanczos recursion which has a nonzero projection on every eigenspace of A. Since the terms generated by the conjugate gradient procedure depend only upon the initial residual r_1 and not upon the starting vector x_1, and since the convergence in n steps is independent of the starting vector as long as the corresponding residual has a nonzero projection on every eigenspace, we can always set $r_1 = v_1$, or vice-versa. We will denote any sets of 'residuals', 'Lanczos vectors', or 'conjugate directions' by respectively,

$$R_m \equiv \{r_1,...,r_m\}, \quad V_m \equiv \{v_1,...,v_m\}, \quad P_m \equiv \{p_1,...,p_m\}. \tag{4.2.9}$$

DEFINITION 4.2.2 Let L_σ denote the bidiagonal matrix with 1's on the diagonal and entries $-\sigma_i$ on the subdiagonal, and D_δ denote any diagonal matrix with diagonal entries δ_i, $1 \le i \le n$.

Theorem 2.2.1 states that Lanczos tridiagonalization generates orthonormal sets of vectors V_j, $1 \le j \le n$. Therefore $v_{n+1} = 0$ and from Eqn(2.1.4) we have that

$$AV_n = V_n T_n. \tag{4.2.10}$$

Similarly, for conjugate gradients we have that theoretically the sets P_j, $1 \le j \le n$, are A-conjugate, the sets R_j, $1 \le j \le n$, are orthonormal, and $r_{n+1} = 0$. For a proof of these facts see for example Hestenes [1980]. Therefore from Eqns(4.2.4) and (4.2.5) respectively, we get that

$$AP_n = R_n L_1 D_\omega \tag{4.2.11}$$

$$R_n = P_n L_\gamma^T. \tag{4.2.12}$$

The proof of the equivalence of these two procedures rests upon defining the proper relationship between the scalar coefficients α_i and β_{i+1} generated by the Lanczos recursion and the scalar coefficients ω_i and γ_i generated by the conjugate gradient recursions. It is perhaps easiest to see what this relationship should be by starting with Lanczos tridiagonalization. Theorem 2.2.1 in Chapter 2 tells us that each Lanczos matrix is positive definite since A is. Furthermore, Corollary 3.3.2 in Chapter 3 tells us that the eigenvalues of these Lanczos matrices are independent of the signs of the β_{i+1}. Furthermore, changing the signs of the β_{i+1} simply multiplies the resulting Lanczos vectors by ± 1. In the discussions in Sections 4.2 and 4.3 we choose to assume that each $\beta_{i+1} \leq 0$. However in the actual implementations discussed in Sections 4.7 and 4.8 we select $\beta_{i+1} \geq 0$.

Because each T_m is positive definite, it can be factored without any pivoting to obtain

$$T_m = L_\sigma D_\omega L_\sigma^T. \tag{4.2.13}$$

where L_σ is the bidiagonal matrix with 1's on the diagonal and the entries $-\sigma_i$ in the (i+1,i)-th location. D_ω is a diagonal matrix with diagonal entries ω_i. Since T_m is positive definite each pivot $\omega_i > 0$. Eqn(4.2.13) is simply the matrix version of the following set of nonlinear recursions connecting the two sets of scalars α_i, β_{i+1} and ω_i, γ_i.

$$\begin{aligned} \alpha_{i+1} &= \omega_{i+1} + \omega_i \gamma_i \\ \beta_{i+1} &= -\sigma_i \omega_i. \end{aligned} \tag{4.2.14}$$

We will use Eqn(4.2.14) to define the correspondence between the scalar parameters generated by Lanczos tridiagonalization and the scalar parameters generated by the conjugate gradient optimization procedure. We also need the following two statements which follow easily from Eqns(4.2.7) - (4.2.8).

$$L_1 = D_\rho^{-1} L_\sigma D_\rho \tag{4.2.15}$$

$$L_\gamma^T = D_\rho^{-1} L_\sigma^T D_\rho. \tag{4.2.16}$$

LANCZOS TRIDIAGONALIZATION YIELDS CONJUGATE GRADIENTS We first show that given the quantities defined by applying the Lanczos recursion, we can construct a complementary set of quantities which satisfy the conjugate gradient relationships.

LEMMA 4.2.1 Given a positive definite, real symmetric nxn matrix A with n distinct eigenvalues, select a unit starting vector v_1 which has a nonzero projection on every eigenvector of A. Apply the basic Lanczos recursion to A to generate corresponding Lanczos matrices T_m, m = 1,2,...,n. we then use the scalars ω_i defined by Eqn(4.2.13) are all positive and the scalars σ_i defined by Eqn(4.2.13) are all nonnegative. Therefore we can use Eqn(4.2.8) to define corresponding positive scalars ρ_i and γ_i, i=1,...,n starting with $\rho_1 = 1$.

Proof. Each Lanczos matrix is positive definite. Therefore the factorization in Eqn(4.2.13) is well-defined and each of the pivots $\omega_i > 0$. Furthermore, since by construction all of the $\beta_{i+1} \leq 0$, we have from Eqn(4.2.14) that each $\sigma_i \geq 0$. $\qquad \square$

LEMMA 4.2.2   Under the hypotheses of Lemma 4.2.1 define vectors

$$R_n \equiv V_n D_\rho \text{ and } r_{n+1} = 0, \qquad\qquad (4.2.17)$$

and P_n using Eqn(4.2.11). Then $R_n = P_n L_\gamma^T$. That is Eqn(4.2.5) is valid for i=1,2,...,n.

Proof.   Transform the matrix $AV_n D_\rho$ into the matrix $AP_n L_\gamma^T$ by sequentially using the following relationships:

$$AV_n = V_n T_n, \text{ Eqn(4.2.10)}$$
$$V_n = R_n D_\rho^{-1}, \text{ Eqn(4.2.17)}$$
$$T_n = L_\sigma D_\omega L_\sigma^T, \text{ Eqn(4.2.13)}$$
$$R_n = AP_n D_\omega^{-1} L_1^{-1}, \text{ Eqn(4.2.11)}$$

together with Eqns(4.2.15) and (4.2.16) for L_1^{-1} and L_γ^T.   □

LEMMA 4.2.3   Under the hypotheses of Lemma 4.2.2 the direction vectors P_n are A-conjugate. In particular, $P_n^T A P_n = D_\rho^2 D_\omega$.

Proof.   Transform the matrix $L_\gamma P_n^T A P_n$ into the matrix $L_\gamma D_\rho^2 D_\omega$ by successively using the relationships

$$AP_n \equiv R_n L_1 D_\omega, \text{ Eqn(4.2.11)}$$
$$R_n^T = L_\gamma P_n^T, \text{ Eqn(4.2.12)}$$
$$R_n^T R_n = D_\rho^2, \text{ Eqn(4.2.17)}$$

together with the formulas for L_1 and for L_γ given in Eqns(4.2.15) and (4.2.16).   □

Finally we must prove that the scalars defined by Eqn(4.2.13) satisfy the conjugate gradient equations, Eqns(4.2.6) and (4.2.7). We use the conjugacy of the directions p_i and the orthogonality of the residuals r_i which was just established in Lemmas 4.2.2 and 4.2.3.

LEMMA 4.2.4   Under the hypotheses of Lemma 4.2.2 the scalars ω_i and γ_i defined by Equations (4.2.13) satisfy the conjugate gradient relations Eqn(4.2.6) and (4.2.7).

Proof.   From Eqn(4.2.5) (Lemma 4.2.2) and Lemma 4.2.3 we have that for each i

$$0 = p_i^T A p_{i+1} = p_i^T A r_{i+1} + \gamma_i p_i^T A p_i.$$

Therefore, the desired formula for the γ_i is a simple consequence of the conjugacy of the vectors p_i.

From Lemma 4.2.2 we have that $P_n = R_n L_\gamma^{-T}$ Thus, since L_γ^{-T} is upper triangular each direction p_i is some linear combination of the residuals r_k, for k≤i. Since $R_n \equiv V_n D_\rho$, and the

V_n are orthogonal, we have that the R_n are also orthogonal. Therefore, we have that

$$r_j^T p_i = 0, \text{ for } 1 \leq i \leq n, \, i < j \leq n + 1. \tag{4.2.18}$$

That is the residuals are orthogonal to the directions of search. Using this orthogonality and Eqn(4.2.11) we have for each i that

$$p_i^T A p_i = \omega_i p_i^T r_i,$$

but this is just the conjugate gradient formula, Eqn(4.2.6) for the ω_i. $\qquad\square$

Together Lemmas 4.2.1 - 4.2.4 demonstrate that in exact arithmetic the quantities generated by Lanczos tridiagonalization define corresponding quantities which satisfy all of the conjugate gradient relationships. Now consider the converse relationship.

CONJUGATE GRADIENTS YIELDS LANCZOS TRIDIAGONALIZATION Let a starting residual r_1 with unit norm be specified. (This defines a starting point x_1.) Then use the conjugate gradient procedure to generate directions p_i, residuals r_i, and scalars ω_i and γ_i. Note that the corresponding quantities ρ_i and σ_i are all well-defined. Define matrices L_σ, D_ω and T_m using Eqn(4.2.13). Clearly this is equivalent to defining the scalars α_i and β_{i+1} in the associated Lanczos procedure by using Eqn(4.2.14).

LEMMA 4.2.5 Let A be a positive definite, real symmetric nxn matrix with n distinct eigenvalues. Use the conjugate gradient relationships to generate vectors p_i, r_i, and scalars ω_i, γ_i, σ_i, ρ_i. Use Eqns(4.2.13) to define scalars α_i and β_{i+1} and Eqn(4.2.17) to define corresponding vectors V_n. Then $v_{n+1} = 0$,

$$V_n^T V_n = I_n \text{ and } A V_n = V_n T_n.$$

Furthermore, the scalars α_i and β_{i+1} satisfy the Lanczos formulas (in exact arithmetic).

$$\alpha_i = v_i^T A v_i \text{ and } \beta_{i+1} = v_{i+1}^T A v_i.$$

Proof. The orthogonality is obvious since the residuals are orthogonal. We obtain Eqn(4.2.10), $AV_n = V_n T_n$, by transforming $AV_n D_\rho$ into $V_n T_n D_\rho$ by applying the following relationships sequentially:

$$R_n = P_n L_\gamma^T, \text{ Eqn(4.2.12)}$$
$$AP_n = R_n L_1 D_\omega, \text{ Eqn(4.2.11)}$$
$$R_n = V_n D_\rho, \text{ Eqn(4.2.17)}$$
$$L_\gamma^T = D_\rho^{-1} L_\sigma^T D_\rho, \text{ Eqn(4.2.16)}$$
$$L_1 = D_\rho^{-1} L_\sigma D_\rho, \text{ Eqn(4.2.15)}$$
$$T_n = L_\sigma D_\omega L_\sigma^T, \text{ Eqn(4.2.13)}$$

Thus, we have obtained the basic Lanczos recursion. It is now trivial to use this recursion and the orthogonality of the Lanczos vectors to obtain the required Lanczos formulas for the α_i and β_{i+1}. □

We have demonstrated the equivalence of these two procedures, assuming infinite precision arithmetic. Therefore, in that case, any properties of the conjugate gradient procedure can be immediately translated into corresponding properties for the conjugate gradient procedure. In practice, however, we have finite precision arithmetic and the arguments given here are not valid because the global orthogonality (conjugacy) is no longer valid. We are interested however in the practical case because we are constructing Lanczos procedures with no reorthogonalization. What, if any kind of 'equivalence' of these two procedures can one hope to establish in the practical case?

From Chapter 2, Theorem 2.3.1 we know that one-step, nearest-neighbor orthogonality of the Lanczos vectors is maintained as long as the off-diagonal entries of the Lanczos matrices do not get too small. In the next section we demonstrate that this is sufficient to obtain a generalization of the equivalence given in this section. For our discussions of Lanczos procedures we will only need an equivalence which allows us to construct associated conjugate gradient quantities whenever we are given quantities which have been generated using the basic Lanczos recursion. We can do this using only nearest-neighbor orthogonality of the Lanczos vectors. The obvious question is why do we care about such equivalences? This will be made clear in the next 3 sections. In Section 4.4 we use this equivalence to develop a plausibility argument for why we can expect the Lanczos recursion with no reorthogonalization to remain well-behaved and to actually generate the quantities we want to compute, namely approximations to the eigenvalues and eigenvectors of the original matrix A. In Section 4.5 we exploit this relationship between these two procedures to obtain the identification test which we use to transform the basic Lanczos recursion into a practical Lanczos procedure.

SECTION 4.3   AN EQUIVALENCE, FINITE PRECISION ARITHMETIC

We consider only a one-sided equivalence. The arguments rigorous. We use the error analysis of Paige [1976] to incorporate the effects of the finite precision arithmetic. First in Lemmas 4.3.1 - 4.3.6 and Theorems 4.3.2 - 4.3.5 we demonstrate that given the quantities defined by applying the Lanczos recursion to a positive definite real symmetric matrix using Eqns(4.3.1) - (4.3.2), the associated quantities defined using Eqns(4.2.14) and (4.3.3) - (4.3.5) approximately satisfy the corresponding conjugate gradient relationships defined in Eqns(4.2.4) - (4.2.7). This argument uses the nearest-neighbor near-orthogonality of the Lanczos vectors and Assumptions 4.3.1 - 4.3.5.

We then demonstrate in Theorem 4.3.6 that the norms of the associated conjugate gradient residuals r_k defined through this correspondence must decrease (not necessarily monontonically) as the number of steps in the Lanczos recursion is increased. Furthermore, the limiting value must be 0, although because of the finite precision arithmetic we cannot take a limit in the normal sense. This argument uses the fact that the conjugate gradient optimization procedure, when it is used for solving Eqn(4.2.1), is very robust, in the sense that this procedure will converge even if the directions generated are not more than one-step, nearly-A-conjugate and the line minimizations on each iteration are very crude.

These results tell us that we can expect a Lanczos procedure with no reorthogonalization to behave rationally as long as the off-diagonal entries β_{i+1} do not become 'too small'. In Section 4.4 we use these results to argue that the Lanczos Phenomenon is real. (That argument is not totally rigorous. We will point out the difficulty with it.) In Section 4.5 we use these results to identify a mechanism for sorting the 'spurious' eigenvalues of the Lanczos matrices generated using the Lanczos recursions in Eqns(4.3.1) - (4.3.2) from the 'good' eigenvalues.

Let A be a real, symmetric matrix of order n. We restate the version of the Lanczos recursion used by Paige [1976]. For i=1,2,...,m define scalars α_i and β_{i+1} and Lanczos vectors v_i by the following recursion.

$$\beta_{i+1}\, v_{i+1} = Av_i - \beta_i v_{i-1} - \alpha_i v_i \quad \text{where} \tag{4.3.1}$$

$$\alpha_i = v_i^T(Av_i - \beta_i v_{i-1})$$

$$|\beta_{i+1}| = \|Av_i - \beta_i v_{i-1} - \alpha_i v_i\| \tag{4.3.2}$$

$$v_1 \equiv \text{random vector}, \quad \|v_1\| \equiv 1, \quad v_0 \equiv 0, \quad \beta_1 \equiv 0.$$

The Lanczos matrices T_m have diagonal entries α_i, $1 \le i \le m$, and superdiagonal (and subdiagonal) entries β_{i+1}, $1 \le i \le m-1$.

DEFINITION 4.3.1 The Lanczos Phenomenon Let A be a real symmetric nxn matrix. Let v_1 be a starting vector for the Lanczos recursions in Eqns(4.3.1) - (4.3.2) and use these recursions to generate Lanczos matrices T_m, m = 1,2,... . Then for large enough M, every distinct eigenvalue of A will appear as an eigenvalue of the Lanczos matrix T_m for all $m \ge M$.

Paige [1971], [1976], [1980] analyzes the convergence of eigenvalues of the Lanczos matrices to eigenvalues of A. Paige [1971,1980] demonstrated that the initial loss of global orthogonality in the Lanczos vectors is caused by the convergence of some eigenvalue of a Lanczos matrix. See Lemma 2.3.3 in Chapter 2. He then used this result to demonstrate that some eigenvalue of the family of Lanczos matrices T_m, m = 1,2,... must have converged by the time m = n. He then showed that converged eigenvalues of the Lanczos matrices, that is eigenvalues which persisted as the order of the Lanczos matrix was increased, were eigenvalues of A. In Paige [1971] he stated that this proof could be extended to prove the convergence of more eigenvalues. However, the arguments which he used do not lend themselves to an extension which would explain the Lanczos phenomenon. In Paige [1980] he again considered the question of the convergence of the eigenvalues of the Lanczos matrices as m is increased but this analysis is restricted to $m \leq M$ where M is the smallest value of m by which some eigenvalue of a Lanczos matrix has converged. Thus it only applies until global orthogonality is lost so it cannot be used to discuss the Lanczos Phenomenon. Godunov and Prokopov [1970] and Paige [1972] give examples of the Lanczos phenomenon for very small test matrices.

To be specific we list the four main assumptions which we make below. There are two additional assumptions which we will list as we need them. However they are technical assumptions having to do with the sizes of the machine epsilon relative to the size of the norm of A, the size of the Lanczos matrix m, and the number of nonzero elements in any row of A. The Lanczos procedure does not require that the original matrix A be positive definite. However, we need this assumption in discussing conjugate gradient constructions. Clearly this is not a restriction theoretically since for some shift σ the matrix $A + \sigma I$ is positive definite and the eigenvalues of A are easily obtained from those of this shifted matrix.

ASSUMPTION 4.3.1 A is a real symmetric, positive definite matrix of order n with q distinct eigenvalues $\lambda_1 > \lambda_2 > ... > \lambda_q$, and the starting vector v_1 used in the Lanczos recursion has a non-zero projection on each eigenspace of A.

ASSUMPTION 4.3.2 For all i which we consider, we assume that there exists a $\beta^* > 0$, such that $|\beta_{i+1}| > \beta^*$. In practice the β_{i+1} do not become small as long as there are eigenvalues which have not yet converged. Table 4.7.2 gives some examples of the size β_i encountered in some of our tests.

ASSUMPTION 4.3.3 Each Lanczos matrix T_m which we consider is positive definite. See Lemma 2.3.7 for Paige's justification of this assumption.

We use the correspondence between Lanczos tridiagonalization and the conjugate gradient procedure which was defined in Section 4.2. We repeat the definitions here before stating Assumption 4.3.4. Let Lanczos vectors v_i and scalars α_i, $1 \leq i \leq m$, and β_{i+1}, $1 \leq i \leq m - 1$, be generated using Eqns(4.3.1) - (4.3.2). Then for each i, we define scalars ω_i and γ_i by the factorization of the Lanczos matrix T_m given in Eqn(4.2.13) or equivalently as the solutions of the set of nonlinear recursions given in Eqn(4.2.14). No assumption is made concerning the relationship between the order m of the Lanczos matrix and the order n of the given matrix A.

Using these scalars we then define 'residuals' r_i, and 'directions' p_i by the following relationships. For $i = 1,2,...,m$

$$\rho_{i+1} \equiv \sigma_i \rho_i \text{ with } \rho_1 \equiv 1, \tag{4.3.3}$$

$$Au_i = \beta_{i+1} v_{i+1} + \omega_i v_i, \tag{4.3.4}$$

$$p_i = \rho_i u_i \text{ and } r_i = \rho_i v_i. \tag{4.3.5}$$

We have used the optimization version of the conjugate gradient recursions instead of a related 3-term recurrence, see Rutishauser [1959], typically utilized in the numerical algebra literature. We do this because (as the reader will see) when the conjugate gradient procedure is viewed as an optimization procedure, it provides a direct explanation for how only local orthogonality of the Lanczos vectors can be sufficient for convergence of a Lanczos procedure with no reorthogonalization. It also provides a simple interpretation of the use of Lanczos matrices which are larger than the size of the original matrix.

ASSUMPTION 4.3.4 There exists a scalar R such that the norms ρ_i of the associated conjugate gradient residuals defined in Eqn(4.3.5) satisfy

$$\rho_j / \rho_i \leq R, \quad \text{for all } j > i. \tag{4.3.6}$$

Theorem 4.3.2 demonstrates that typically this assumption is satisfied.

In tests on several examples where we first computed the Lanczos quantities using Eqns(4.3.1) - (4.3.2) and then used Eqn(4.3.3) to compute the associated ρ_i, we observed that the norms ρ_i of the residuals r_i varied nonmonotonically as i was increased, but the variation was controlled in the sense of Eqn(4.3.6) where R was relatively small. For example, for a test Poisson matrix of order 528 with zero boundary conditions, with m = 4n any R > 2.52 satisfies Eqn(4.3.6). For a test Poisson matrix of the same order but with zero normal derivative boundary conditions, with m=4n any R > 7.25 satisfies (4.3.6). For details on the construction of these matrices see Section 4.6. The ratios ρ_j / ρ_i for $j \geq i$ appear repeatedly in the arguments used to show that the conjugate gradient relationships are approximately satisfied.

Theorem 4.3.2 provides the justification for Assumption 4.3.4. First however, in Theorem 4.3.1 we look at the ratios in Eqn(4.3.6) assuming that all of the computations are done in infinite precision arithmetic. Then in Theorem 4.3.2 we consider these ratios using finite precision arithmetic. The proofs of these two theorems are quite different. Theorem 4.3.2 assures us that Assumption 4.3.4 is quite reasonable.

THEOREM 4.3.1 (Exact Arithmetic) Let A be a positive definite, real symmetric matrix. Given a starting vector v_1 use Eqns(4.3.1) and (4.3.2) to generate Lanczos vectors and matrices, and then construct corresponding conjugate gradient quantities using the correspondence defined by Eqns(4.2.13) and (4.3.3) - (4.3.5). Then any $R \geq \sqrt{2}[\lambda_{max}(A)/\lambda_{min}(A)]$ satisifes Eqn(4.3.6), $\rho_j / \rho_i \leq R$ for any $j > i$.

Proof. We have by Eqn(4.2.18) that for each j, $v_{j+1}^T u_j = 0$. Furthermore, we have that $u_j = v_j + \sigma_{j-1} u_{j-1}$. Therefore, for each j

$$\| u_j \|^2 = 1 + \sum_{k=0}^{j-2} [\sigma_{j-1}^2 \cdots \sigma_{j-1-k}^2].$$

Therefore, we have that for any $j \geq i$,

$$(\rho_j/\rho_i) = \prod_{\ell=i}^{j-1} \sigma_\ell \leq \|u_j\|.$$

From Eqn(4.3.4) we have that

$$Au_j = (\beta_{j+1}v_{j+1} + \omega_j v_j).$$

Therefore, using the fact that for each j, $v_{j+1}^T v_j = 0$, the fact that each 2x2 principal submatrix of T_m is positive definite, the fact that from Eqns(4.2.14) that each $\omega_i \leq \alpha_i$, and that each α_i is a Rayleigh quotient of A we get the following upper bound on $\|u_j\|$.

$$\|u_j\| \leq [\alpha_j^2 + \alpha_j \alpha_{j+1}]^{1/2} / \lambda_{min}(A) \leq \sqrt{2}[\lambda_{max}(A)/\lambda_{min}(A)].$$

$\square$

Theorem 4.3.1 implies that R in Eqn(4.3.6) is related to the stiffness of the matrix. However this theoretical bound is much larger than what we observed in practice. Of course if we were to use infinite precision then there can be at most n nonzero ρ_j, so obviously there would be a global upper bound for all the ratios in that case. However, the interesting point about Theorem 4.3.1 is that the arguments only used local relationships, local orthogonality and exact line searches. In Theorem 4.3.2 we show that such a bound is also valid when finite precision arithmetic is used. We need the following Lemma.

LEMMA 4.3.1. Let L_σ be the bidiagonal mxm matrix with $L(i,i) \equiv 1$, $1 \leq i \leq m$, and $L(i+1,i) \equiv -\sigma_i$, $1 \leq i \leq (m-1)$. Then if for each i, $\sigma_i \equiv \rho_{i+1}/\rho_i$, we have that for any $i < j \leq m$,

$$L_\sigma^{-1}(j,i) = \prod_{k=i}^{j-1} \sigma_k = \rho_j/\rho_i. \tag{4.3.7}$$

Proof. We need to compute the (j,i)-th element of the inverse of L_σ. We can obtain the ith column of L_σ^{-1} by solving the equation

$$L_\sigma x = e_i.$$

But, L_σ is lower triangular so that if $k < i$ then $x(k) = 0$. If $k = i$ then $x(k) = 1$. Therefore, for any $k > i$ we get that

$$x(k) = \prod_{\ell=i}^{k-1} \sigma_\ell = \rho_j/\rho_i.$$

But, $x(j) = L_\sigma^{-1}(j,i)$. $\square$

Paige [1976,1980] makes the following assumptions. We will be using his estimates repeatedly throughout our discussions. See Theorem 2.3.1 in Chapter 2.

ASSUMPTION 4.3.5. We have that $\|A\| \equiv \lambda_{max}(A)$. Define

$$\alpha \equiv \||A|\|/\lambda_{max}(A) \text{ and}$$

$$n_z = \text{maximum number of nonzeros per row} \tag{4.3.8}$$

The results from Paige [1976,19780] which we use require that the following be true where n is the order of A, m is the order of the Lanczos matrix, ε_0 and ε_1 are defined in Eqn(2.3.15), and ε_2 is defined in Eqn(2.3.24).

$$4m(3\varepsilon_0 + \varepsilon_1) << 1$$
$$2m[7 + n_z\alpha + 3(n+4)]\varepsilon << 1 \tag{4.3.9}$$
$$[\lambda_{min}(A) - \varepsilon_2 m^{5/2}\lambda_{max}(A)] > 0$$

where ε_0 and ε_1 are defined in Eqn(2.3.15) and ε_2 is defined in Eqn(2.3.24).

THEOREM 4.3.2 (Finite Precision Arithmetic)   Let A be a real symmetric, positive definite nxn matrix. Apply the Lanczos recursions (4.3.1) - (4.3.2) to A, obtaining Lanczos vectors and Lanczos matrices. Use Eqns(4.2.13) and (4.3.3) - (4.3.5) to construct corresponding conjugate gradient quantities. Under Assumptions 4.3.1, 4.3.3 and 4.3.5 we have that Assumption 4.3.4 is valid. In particular we have that for a given m,

$$\rho_j/\rho_i \leq \frac{2\lambda_{max}(A)[1 + \varepsilon(3n+4)/2]}{[\lambda_{min}(A) - \varepsilon_2\lambda_{max}(A)m^{5/2}]} \text{ for } i<j\leq m.$$

Proof. By Lemma 2.3.7 in Chapter 2 we have that T_m is positive definite. Therefore, the conjugate gradient quantities are well-defined and from Eqn(4.2.13) we have that

$$L_\sigma^{-1} = D_\omega L_\sigma^T T_m^{-1}.$$

But from Lemma 4.3.1 we have that for j>i,

$$\rho_j/\rho_i = L_\sigma^{-1}(j,i).$$

Therefore, we have that

$$\rho_j/\rho_i = \omega_j(e_j^T - \sigma_j e_{j+1}^T)T_m^{-1}e_i = \omega_j[T_m^{-1}(j,i) - \sigma_j T_m^{-1}(j+1,i)].$$

Define $S_m \equiv P T_m P^T$ where P is the permutation matrix $P \equiv (e_m, e_{m-1}, ..., e_1)$. Define a complementary set of pivots $\hat{\omega}_i$, $m\geq i\geq 1$ corresponding to S_m. Here we define $\hat{\omega}_m = \alpha_m$. It is a straight-forward exercise to show that these pivots satisfy the following recursion.

$$\hat{\omega}_k = \alpha_k - \beta_{k+1}^2/\hat{\omega}_{k+1}, \quad k = m,m-1,...,1.$$

Each pivot is positive since S_m is positive definite.

From Lemma 3.2.1 we have that for any k and i, k>i that

$$T_m^{-1}(k,i) = (\prod_{\ell=i+1}^{k} \beta_\ell)a_{i-1}\hat{a}_{k+1}/a_m.$$

Using this formula together with the definitions of the $\hat{\omega}_i$, we get that

$$\rho_j/\rho_i = T_m^{-1}(j,i) [\omega_j + \alpha_j - \hat{\omega}_j].$$

From this equation we see that we can obtain a bound on this ratio if we can bound the elements $|T_m^{-1}(j,i)|$. However, since T_m^{-1} is positive definite, the largest entry of this matrix must be on the diagonal. Suppose the contrary that for some $i<j$, $|T_m^{-1}(j,i)| > \max[T_m^{-1}(j,j), T_m^{-1}(i,i)]$. Then using the vector x whose ith component is 1 and whose jth component is $-\text{sgn}[T_m^{-1}(j,i)]$, we get that $x^T T_m^{-1} x < 0$. But this contradicts the positive definiteness.

We can obtain a bound on these elements as follows. By Lemma 3.2.1 we have that

$$T_m^{-1}(j,j) = a_{j-1}\hat{a}_{j+1}/a_m = \left(\prod_{\ell=1}^{j-1} \mu_\ell^{j-1} \right)\left(\prod_{\ell=j+1}^{m} \hat{\mu}_\ell^{j+1} \right) / \prod_{\ell=1}^{m} \mu_\ell^{m}.$$

Here μ_ℓ^{j-1} and $\hat{\mu}_\ell^{j+1}$ denote respectively the eigenvalues of the submatrices T_{j-1} and $\hat{T}_{j+1}$ of T_m. If we apply a permutation matrix to T_m which interchanges the jth and mth rows and columns of T_m, then we obtain a matrix with the same eigenvalues as T_m but which has a leading principal matrix of order m-1 composed of the two submatrices T_{j-1} and $\hat{T}_{j+1}$. But by the interlacing theorem we know that the eigenvalues of leading principal submatrices of T_m interlace the eigenvalues of T_m. Therefore, the totality of all of the eigenvalues of these two submatrices forms a set of values which interlace the eigenvalues of T_m. We denote the rearranged and renumbered set of eigenvalues by ν_j, $1 \leq j \leq m - 1$, where the ordering is chosen so that

$$0 < \mu_1^m \leq \nu_1 \leq \mu_2^m \leq \cdots \leq \nu_{m-1} \leq \mu_m^m.$$

But using these inequalities in the expression for $T_m^{-1}(j,j)$ and invoking Lemma 2.3.7, we obtain the following inequality

$$|T_m^{-1}(j,j)| \leq \frac{[\nu_1 \cdots \nu_{m-1}]}{[\mu_2^m \cdots \mu_m^m \mu_1^m]} \leq \frac{1}{\mu_1^m} \leq \frac{1}{[\lambda_{\min}(A) - \varepsilon_2\lambda_{\max}(A)m^{5/2}]}$$

Combining these inequalities we get the upper bound given in Eqn(4.3.9). ☐

LANCZOS TRIDIAGONALIZATION YIELDS CONJUGATE GRADIENTS We follow the same basic outline used in Cullum and Willoughby [1980]. Here however instead of making the assumption which was made there that all of the significant errors due to the finite precision arithmetic could be well-represented by losses in the local orthogonality of the Lanczos vectors, we use the results of Paige's error analysis of the Lanczos recursion, Paige [1976,1980], and directly introduce errors at each stage in the recursion and trace the effects of these errors as we recurse.

From Paige [1976] we have the following relationships for the quantities generated using the Lanczos recursion defined by Eqns(4.3.1) - (4.3.2). We will use these results repeatedly throughout the discussions in this chapter. See Chapter 2, Theorem 2.3.1 for a statement of the associated theorem. These estimates are to within the first order in ε, which denotes the machine epsilon.

$$AV_j = V_jT_j + \beta_{j+1}v_{j+1}e_j^T + \Delta W_j \qquad (4.3.10)$$

$$|v_i^Tv_i - 1| \le \varepsilon_0, \quad 1 \le i \le j \qquad (4.3.11)$$

$$\|\delta w_j\| \le \varepsilon_1 \lambda_{max}(A) \text{ where } \Delta W_j \equiv \{\delta w_1,...,\delta w_j\}. \qquad (4.3.12)$$

$$|\beta_{i+1}v_i^Tv_{i+1}| \le 2\varepsilon_0\lambda_{max}(A), \quad 1 \le i \le j-1 \qquad (4.3.13)$$

where we have used the notation

$$\varepsilon_0 \equiv (n+4)\varepsilon, \ \varepsilon_1 \equiv (7+n_z\alpha)\varepsilon \qquad (4.3.14)$$

Paige [1976] also shows that for $i=1,2,...,j$

$$|\alpha_i| \le \lambda_{max}(A)[1+\varepsilon(3n+4)/2]$$
$$|\beta_{i+1}| \le \lambda_{max}(A)[1+\varepsilon(2n+6)] \qquad (4.3.15)$$

Therefore, the entries in the Lanczos matrices are essentially bounded by the norm of the original matrix A.

In the remaining discussions we are working in finite precision arithmetic. The matrix A is real symmetric and positive definite. The Lanczos recursions defined in Eqns(4.3.1) - (4.3.2) are used to generate Lanczos matrices and vectors. Eqns(4.2.13) and (4.3.3) - (4.3.5) are used to construct corresponding conjugate gradient quantities. We first want to demonstrate that for any i the directions p_i and p_{i+1} are nearly A-conjugate. We use the normalized vectors $u_i = p_i/\rho_i$ and $v_i = r_i/\rho_i$ in the arguments. We introduce the following definitions because these expressions will occur frequently in the proofs which follow.

$$\varepsilon'_i \equiv v_i^Tv_i - 1$$
$$\bar{\varepsilon}_i \equiv \beta_{i+1}v_{i+1}^Tv_i \qquad (4.3.16)$$

$$F_i \equiv [\bar{\varepsilon}_i(\sigma_i-1/\sigma_i) + \beta_{i+1}(\varepsilon'_i-\varepsilon'_{i+1})] \qquad (4.3.17)$$

$$H_i \equiv 1 + (i-1)R^2 \qquad (4.3.18)$$

$$\varepsilon^* \equiv \max(\varepsilon_0,\varepsilon_1) = \varepsilon\max[(n+4), (7+n_z\alpha)] \qquad (4.3.19)$$

LEMMA 4.3.2. Under Assumptions 4.3.1 - 4.3.3 and 4.3.5 to first order in ε we have that

$$|F_i| \leq 2\varepsilon_0 \lambda_{max} (A)[1 + \lambda_{max} (A)/\beta^*] \tag{4.3.20}$$

Proof. Using the definitions of ω_i and σ_i given in Eqn(4.2.14), we get that $-\sigma_i\beta_{i+1} = \alpha_{i+1}-\omega_{i+1}$. Since $\omega_j>0$ for all j, it is easy to prove using Eqn(4.3.15) that

$$\sigma_i \leq \alpha_{i+1}/|\beta_{i+1}| \leq \lambda_{max} (A)[1 + \varepsilon(3n + 4)/2]/\beta^*. \tag{4.3.21}$$

Moreover, we have that

$$\sigma_i = -\beta_{i+1}/\omega_i \geq \beta^*/[\lambda_{max} (A)(1 + \varepsilon(3n + 4)/2)].$$

Therefore we have the same bound for σ_i and for $1/\sigma_i$. Furthermore,

$$|\sigma_i - 1/\sigma_i| \leq \lambda_{max} (A)[1 + \varepsilon(3n + 4)/2]/\beta^*.$$

Combining these comments with Eqn(4.3.15) for $|\beta_{i+1}|$ we obtain the upper bound given in Eqn(4.3.20). $\qquad\square$

We now state the one-step conjugacy result.

THEOREM 4.3.3. Under Assumptions 4.3.1 - 4.3.5, for each i the successive directions u_i and u_{i+1} defined from the Lanczos vectors using Eqn(4.3.4) are nearly-A-conjugate. Specifically, to first order terms in ε,

$$|u_i^T A u_{i+1}| \leq M_i \equiv 2\varepsilon^*\lambda_{max} (A)H_i \left[\frac{2\lambda_{max} (A)}{\lambda_{min} (A)} + 1 + \frac{\lambda_{max} (A)}{\beta^*}\right]. \tag{4.3.22}$$

The proof of Theorem 4.3.3 utilizes several intermediate results which we give separately as lemmas.

LEMMA 4.3.3. Under Assumptions 4.3.1 - 4.3.3 and 4.3.5, the directions u_i defined in Eqn(4.3.4) satisfy

$$\begin{aligned} Au_{i+1} &= Av_{i+1} + \sigma_i Au_i - \delta w_{i+1} \quad \text{or equivalently} \\ AV_m &= AU_m L_\sigma^T + \Delta W_m \end{aligned} \tag{4.3.23}$$

where ΔW_m is defined in Eqn(4.3.10).

Proof. From the definition in Eqn(4.3.4) together with one step of Eqn(4.3.10) we obtain

$$\begin{aligned} Au_{i+1} &= \beta_{i+2}v_{i+2} + \omega_{i+1}v_{i+1} = \\ Av_{i+1} &+ (\omega_{i+1} - \alpha_{i+1})v_{i+1} - \beta_{i+1}v_i - \delta w_{i+1} \end{aligned}$$

Using Definitions (4.2.14) twice followed by one application of Eqn(4.3.4) we obtain Eqn(4.3.23). $\qquad\square$

LEMMA 4.3.4. Under Assumptions 4.3.1 - 4.3.3 and 4.3.5, for each $i \geq 2$, the quantities σ_i and u_i defined in Eqns(4.2.14) and (4.3.4) from the Lanczos quantities satisfy

$$u_i^T A u_{i+1} \; - \; \sigma_i \sigma_{i-1} u_i^T A u_{i-1} \;\; = \;\; F_i - [\sigma_i \delta w_i + \delta w_{i+1}]^T u_i. \qquad (4.3.24)$$

Furthermore, we have that

$$\| u_i \| \; \leq \; \frac{\lambda_{max}(A)}{\lambda_{min}(A)}[2 + \varepsilon(3.5n + 8) + .5(|\varepsilon'_i| + |\varepsilon'_{i+1}|)] \; \leq \; \frac{\lambda_{max}(A)}{\lambda_{min}(A)}[2 + 4.5\varepsilon] \qquad (4.3.25)$$

Proof. From Lemma 4.3.3,

$$u_{i+1} - \sigma_i \sigma_{i-1} u_{i-1} \; = \; v_{i+1} + \sigma_i u_i - A^{-1} \delta w_{i+1} - \sigma_i \sigma_{i-1} u_{i-1}.$$

Rearranging this equation and applying Lemma 4.3.3 again we obtain

$$v_{i+1} + \sigma_i[v_i - A^{-1} \delta w_i] - A^{-1} \delta w_{i+1}.$$

Taking the inner product of this expression with Au_i, we get that

$$u_i^T A u_{i+1} - \sigma_i \sigma_{i-1} u_i^T A u_{i-1} =$$
$$u_i^T A v_{i+1} + \sigma_i u_i^T A v_i - u_i^T[\sigma_i \delta w_i + \delta w_{i+1}]$$

Now if we use definition Eqn(4.3.4) to replace Au_i in the above expression and then use definitions Eqns(4.3.16) and (4.2.14), the right-hand side of the above equation reduces to

$$\beta_{i+1}[\varepsilon'_{i+1} - \varepsilon'_i] + \bar{\varepsilon}_i[\sigma_i - 1/\sigma_i] - u_i^T[\sigma_i \delta w_i + \delta w_{i+1}]$$

which is just the right-hand side of Eqn(4.3.24).

The proof of Eqn(4.3.25) follows trivially from the definition of the u_i, and Eqns(4.3.15). Specifically, we have that

$$\| u_i \| \; \leq \; \| A^{-1} \| [| \beta_{i+1} | \, \| v_{i+1} \| + \omega_i \| v_i \|].$$

A direct application of Eqns(4.3.15) and (4.3.16) yields Eqn(4.3.25). □

LEMMA 4.3.5. Under Assumptions 4.3.1 - 4.3.3 and 4.3.5, to first order in ε, we obtain the following equality.

$$u_1^T A u_2 \;\; = \;\; F_1 - u_1^T(\sigma_1 \delta w_1 + \delta w_2). \qquad (4.3.26)$$

Proof. From Lemma 4.3.3 and the Lanczos recursion in Eqn(4.3.10) we get that

$$Au_2 = Av_2 + \sigma_1 Au_1 - \delta w_2$$
$$Au_1 = \beta_2 v_2 + \omega_1 v_1$$

Therefore,

$$u_1^T A u_2 = u_1^T A v_2 + \sigma_1 u_1^T A u_1 - u_1^T \delta w_2.$$

But, using the definition of u_1 in Eqn(4.3.4), $\omega_1 \equiv \alpha_1$, and Eqns(4.3.16) we obtain

$$u_1^T A v_2 = \beta_2 v_2^T v_2 + \omega_1 v_1^T v_2 = \beta_2 (1 + \varepsilon'_2) + (\bar{\varepsilon}_1 \omega_1)/\beta_2.$$

Using the definition of Au_1, the fact that $\omega_1 = \alpha_1$, the Lanczos recursion in Eqn(4.3.10), and Eqns(4.3.15) we obtain

$$\sigma_1 u_1^T A u_1 = \sigma_1 \bar{\varepsilon}_1 - \beta_2 (1 + \varepsilon'_1) - \sigma_1 u_1^T \delta w_1.$$

Therefore, we obtain Eqn(4.3.26). $\qquad\qquad\square$

We need the following simple bound.

LEMMA 4.3.6. For any i

$$|v_i^T A^{-1} \delta w_i| \leq \varepsilon_1 (1 + |\varepsilon'_i|)^{1/2} \lambda_{max}(A)/\lambda_{min}(A). \qquad (4.3.27)$$

Proof. Simply apply definition Eqn(4.3.16), use the fact that A is positive definite, and the bound in Eqn(4.3.12). $\qquad\qquad\square$

Now we can obtain a proof of Theorem 4.3.3.

Proof of Theorem 4.3.3 Define $y_k = u_k^T A u_{k+1}/\rho_k \rho_{k+1}$. Then from Lemma 4.3.4, for $k \geq 2$,

$$y_k - y_{k-1} = [F_k - u_k^T(\sigma_k \delta w_k + \delta w_{k+1})] / (\rho_k \rho_{k+1}).$$

Summing from $k=2$ to $k=i$, and using the fact from Lemma 4.3.4 that $y_1 = [F_1 - u_1^T(\sigma_1 \delta w_1 + \delta w_2)]/\rho_1 \rho_2$, we have that

$$u_i^T A u_{i+1} = \sum_{k=1}^{i} (F_k - t_k)(\rho_i \rho_{i+1}/\rho_k \rho_{k+1}).$$

where

$$t_k \equiv u_k^T[\sigma_k \delta w_k + \delta w_{k+1}].$$

By Assumption 4.3.4 we have

$$\sum_{k=1}^{i} (\rho_i \rho_{i+1}/\rho_k \rho_{k+1}) \leq H_i \quad \text{and} \quad \sum_{k=1}^{i} (\rho_i \rho_{i+1}/\rho_k \rho_k) \leq H_i.$$

From Eqn(4.3.20) we have for all k, $|F_k| \leq 2\varepsilon^* \lambda_{max}(A)[1 + \lambda_{max}(A)/\beta^*]$. Thus, to obtain the desired result we need only to determine an upper bound for the t_k. That is we have

$$\sum_{k=1}^{i} t_k \rho_i \rho_{i+1} / \rho_k \rho_{k+1} =$$

$$\sum_{k=1}^{i} u_k^T [\delta w_{k+1} \rho_i \rho_{i+1} / \rho_k \rho_{k+1} + \delta w_k \rho_i \rho_{i+1} / \rho_k \rho_k].$$

Therefore using the bound on $\|u_k\|$ given in Eqn(4.3.25), the bound on the $\|\delta w_k\|$ given in Eqn(4.3.12) and H_i we get the desired result. $\square$

In the remaining discussions in this section we are viewing the quantities defined in Eqns(4.2.13) and (4.3.3) - (4.3.5) from the Lanczos tridiagonalization quantities as though they were obtained by applying a minimization procedure to the function f(x) defined in Eqn(4.2.3). To be specific we define the starting iterate to be $x_1 \equiv 0$, and therefore in the definition of f(x) we set $b = r_1$. We accept the p_i as the directions of movement in this minimization procedure and since A is nonsingular we can treat each r_i as the negative of the gradient of f(x) at an iterate x_i defined by this residual. In particular we define the ith iterate as

$$x_i \equiv A^{-1}(r_1 - r_i). \tag{4.3.28}$$

If our constructions are consistent, then we should also be able to show that these iterates are obtained by the sequential movement along the directions p_i. To be specific we should be able to show that for each i,

$$x_{i+1} = x_i + p_i / \omega_i. \tag{4.3.29}$$

However this is easy to do. From the definitions given in Eqn(4.3.3) and Eqn(4.2.12) we have that

$$\rho_i A u_i = -\omega_i \rho_{i+1} v_{i+1} + \omega_i \rho_i v_i.$$

But if we now apply the definitions given in Eqn(4.3.5) followed by the definition of x_i given in Eqn(4.3.28) we get Eqn(4.3.29).

Thus, we have a complete construction; directions, iterates, gradients (residuals), and step sizes. Theorem 4.3.3 tells us that the direction of movement p_i is nearly A-conjugate to the previous direction of movement p_{i-1}. Then we must show that the line searches are accurate enough. That is, that the step size $1/\omega_i$ along each direction is sufficiently good. We do this by demonstrating in Theorem 4.3.4 that each residual r_{i+1} (which is just the negative of the gradient of f at the (i+1)st iterate, x_{i+1}) is sufficiently orthogonal to the corresponding direction of movement p_i. If $f(x_{i+1})$ were the minimum of f on the line $x = x_i + ap_i$, then the gradient at x_{i+1} would be orthogonal to the direction p_i.

We then demonstrate in Theorem 4.3.5 that under Assumptions 4.3.1 - 4.3.6 (See Eqns(4.3.31)) that the directions p_i and residuals r_i defined by Eqns(4.3.4) - (4.3.5), and the scalars γ_i and ω_i defined in Eqn(4.2.13) satisfy all of the conjugate gradient relationships approximately. Finally in Theorem 4.3.6 we prove that $|f(x_i)|$ decreases monotonically as the number of steps i in the Lanczos recursion is increased. Furthermore, the limit is 0. From this we get that the norms of the corresponding residuals $\|r_i\|$ 'converge' towards 0 as i is increased. Of course because of the finite precision arithmetic we will not be able to increase i forever. However, we can make it very large. In the next two sections we will use this 'convergence' to demonstrate certain properties of Lanczos procedures with no reorthogonalization.

THEOREM 4.3.4 Under Assumptions 4.3.1 - 4.3.5 the scalars σ_i and the vectors u_i defined by Eqn(4.2.14) and (4.3.3) - (4.3.5) from the Lanczos vectors and scalars satisfy

$$\sigma_i |v_{i+1}^T u_i| \leq N_i \equiv (i+1)R \left\{ \varepsilon^* \left[\frac{2\lambda_{max}(A)}{\beta^*} + \frac{\lambda_{max}(A)}{\lambda_{min}(A)} \right] + \frac{M_i}{\beta^*} \right\} \quad (4.3.30)$$

where M_i was defined in Eqn(4.3.22) in Theorem 4.3.3.

Proof. From Eqn(4.2.14), and from Eqn(4.3.4) for Au_{i+1} we have that

$$\sigma_i u_{i+1}^T Au_i / \omega_{i+1} = \sigma_i v_{i+1}^T u_i - \sigma_i \sigma_{i+1} v_{i+2}^T u_i .$$

From Lemma 4.3.3 we have that

$$\sigma_i \sigma_{i+1} v_{i+2}^T u_i = \sigma_{i+1} v_{i+2}^T u_{i+1} - \sigma_{i+1} \bar{\varepsilon}_{i+1}/\beta_{i+2} + \sigma_{i+1} v_{i+2}^T A^{-1} \delta w_{i+1}.$$

Combining and rearranging, we get

$$\sigma_{i+1} v_{i+2}^T u_{i+1} - \sigma_i v_{i+1}^T u_i =$$
$$-\sigma_i u_{i+1}^T Au_i/\omega_{i+1} + \sigma_{i+1} \bar{\varepsilon}_{i+1}/\beta_{i+2} - \sigma_{i+1} v_{i+2}^T A^{-1} \delta w_{i+1}.$$

Summing over i, we have

$$\sigma_{i+1} v_{i+2}^T u_{i+1} = \sigma_1 v_2^T u_1 + \sum_{k=1}^{i} -\sigma_k u_{k+1}^T Au_k/\omega_{k+1} +$$
$$\sum_{k=2}^{i+1} \sigma_k [\bar{\varepsilon}_k/\beta_{k+1} - v_{k+1}^T A^{-1} \delta w_k].$$

Using Lemmas 4.3.3 and 4.3.6 together with Eqn(4.3.13) on u_1, we get

$$|\sigma_{i+1} v_{i+2}^T u_{i+1}| \leq \sum_{k=1}^{i+1} \sigma_k [2\varepsilon_0 \lambda_{max}(A)/\beta^* + \varepsilon_1 \lambda_{max}(A)/\lambda_{min}(A)] +$$
$$\sum_{k=1}^{i} |u_{k+1}^T Au_k| (\sigma_k \sigma_{k+1}/|\beta_{k+2}|).$$

But, from Theorem 4.3.3,

$$|u_{k+1}^T Au_k| / |\beta_{k+2}| \leq M_k/\beta^*.$$

We have $H_k \leq H_i$ for $i \geq k$, and therefore by Assumption 4.3.4 we have that

$$\sum_{k=1}^{i} \sigma_k \sigma_{k+1} \leq iR, \quad \text{and} \quad \sum_{j=1}^{i+1} \sigma_j \leq (i+1)R,$$

from which Eqn(4.3.30) follows. □

ASSUMPTION 4.3.6. For each i, the $\bar{\varepsilon}_i$, ε'_{i+1} and N_i satisfy

$$|\bar{\varepsilon}_i/\beta_{i+1} - \sigma_i \varepsilon'_{i+1}| < \beta^*/[2(1 + \varepsilon(3n+4)/2)\lambda_{max}(A)]$$

$$|\bar{\varepsilon}_i/\beta_{i+1}| < \beta^*/\lambda_{max}(A)$$

$$[N_i/(1 - N_i - N_{i-1} - \varepsilon_0 - \varepsilon_1 \lambda_{max}(A)/\lambda_{min}(A))] < 1/4 \qquad (4.3.31)$$

$$[1 - \varepsilon_0 - N_{i-1} - \varepsilon_1 \lambda_{max}(A)/\lambda_{min}(A)] > 3/4$$

$$0 < \left\{ \frac{1 \pm N_i \pm N_{i-1} \pm \varepsilon[(n+4) + (7 + n_\chi \alpha)\lambda_{max}(A)/\lambda_{min}(A)]}{1 - [\varepsilon(3n+8)\lambda_{max}(A)/\lambda_{min}(A)]} \right\} < 2.$$

The bounds given in Eqns(4.3.31) are sufficient not optimal and could be weakened considerably. Using Theorems 4.3.3 and 4.3.4 we now show that the quantities defined in our correspondence approximately satisfy the remaining conjugate gradient relationships, Eqns(4.2.6) and (4.2.7). It is important to note that in the subsequent discussion the bound N_i defined in Theorem 4.3.4 need not be small. For example it is sufficient if it is $\leq 1/8$. See Eqns(4.3.31).

THEOREM 4.3.5 Under Assumptions 4.3.1 - 4.3.6, the quantities defined in Eqn(4.2.13), Eqn(4.3.3) - (4.3.5) using the Lanczos vectors and scalars satisfy

$$|\gamma_i - \gamma_i^*| / |\gamma_i^*| \leq \frac{2}{\beta^*} \left[M_i + \frac{2\varepsilon_1 \lambda_{max}(A)^2}{\lambda_{min}(A)} \right]. \qquad (4.3.32)$$

where $\gamma_i^* \equiv -p_i^T A r_{i+1}/p_i^T A p_i$. (See Eqn(4.2.7).) Moreover, for each i

$$|\omega_i - \omega_i^*| / \omega_i^* \leq N_i / [1 - N_i - N_{i-1} - |\varepsilon_i'| - \varepsilon_1 \lambda_{max}(A)/\lambda_{min}(A)]. \qquad (4.3.33)$$

where $\omega_i^* \equiv p_i^T A p_i/p_i^T r_i$. (See Eqn(4.2.6).) M_i and N_i were defined in Eqn(4.3.22) and (4.3.30).

Recall that $\beta^* \equiv \min_j (|\beta_j|, 2 \leq j \leq m+1)$ and $\lambda_{max}(A)$ is the largest eigenvalue of A. We note that if $\gamma_i = \gamma_i^*$, then $p_{i+1}^T A p_i = 0$. Also if $\omega_i = \omega_i^*$ then $v_{i+1}^T u_i = 0$, and we would be minimizing f exactly along p_i.

Proof of Theorem 4.3.5. First consider the relative error in γ_i. From the definitions given in Eqns(4.2.7), (4.2.13), (4.3.3) and (4.3.5) we have

$$\gamma_i - \gamma_i^* = \gamma_i^* [\gamma_i/\gamma_i^* - 1] = -\gamma_i^* [\sigma_i u_i^T A u_i / u_i^T A v_{i+1} + 1].$$

But from Lemma 4.3.3, Eqn(4.3.23) we have that $\sigma_i u_i = u_{i+1} - v_{i+1} + A^{-1}\delta w_{i+1}$ so that

$$\gamma_i - \gamma_i^* = -\gamma_i^*[u_{i+1}^T A u_i + u_i^T \delta w_{i+1}] / u_i^T A v_{i+1}$$

We compute a lower bound on the denominator. By the definitions in Eqns(4.3.4) and (4.3.16) we get that

$$|v_{i+1}^T A u_i| = |\beta_{i+1}(1 + \varepsilon'_{i+1}) + \frac{\omega_i \bar{\varepsilon}_i}{\beta_{i+1}}| = |\beta_{i+1} + \omega_i(\frac{\bar{\varepsilon}_i}{\beta_{i+1}} - \sigma_i \varepsilon'_{i+1})|.$$

But by Assumption 4.3.6, Eqn(4.3.15), and the fact that for all i $\omega_i < \alpha_i$ this expression is

$$> |\beta_{i+1}| - \beta^*/2 > |\beta_{i+1}|/2.$$

Eqn(4.3.32) follows immediately.

Now consider the relative error in ω_i. From the definition in Eqn(4.2.6) for ω_i^*, and the definition in Eqn(4.3.4) we have that

$$\omega_i - \omega_i^* = \omega_i - u_i^T A u_i / u_i^T v_i = \omega_i (\sigma_i u_i^T v_{i+1} / u_i^T v_i).$$

Dividing by ω^*, using Eqn(4.2.6) and then Theorem 4.3.4, we obtain

$$|\omega_i - \omega_i^*| / \omega_i^* = \omega_i \sigma_i |u_i^T v_{i+1}| / u_i^T A u_i \leq N_i \omega_i / u_i^T A u_i.$$

From Eqn(4.3.4) and (4.2.14),

$$u_i^T A u_i / \omega_i = (u_i^T \beta_{i+1} v_{i+1} + u_i^T \omega_i v_i) / \omega_i.$$

Then applying Lemma 4.3.3, Eqn(4.3.28), to the second term in the numerator we get that

$$u_i^T A u_i / \omega_i = (-\sigma_i u_i^T v_{i+1} + v_i^T v_i + \sigma_{i-1} u_{i-1}^T v_i - v_i^T A^{-1}\delta w_i).$$

Now if we apply Theorem 4.3.4 twice, Lemma 4.3.6, and Eqns(4.3.16) we get that this expression

$$\geq 1 - |\varepsilon'_i| - N_i - N_{i-1} - \varepsilon_1 \lambda_{max} (A)/\lambda_{min} (A).$$

$\square$

We want to prove that the associated 'conjugate gradient' values $f(x_i)$ decrease towards 0 as the number of Lanczos steps is increased. This tells us that as we increase i, the $\|r_i\|$ will 'converge' towards 0. To do this we need two more simple lemmas, one of which shows that each direction p_i has a significant projection on the corresponding gradient or residual r_i.

THEOREM 4.3.6 Under Assumptions 4.3.1 - 4.3.6 the associated values $f(x_i)$ obtained from the Lanczos quantities decrease monotonically , approaching zero as i is increased. Moreover, for each i

$$\| r_i \| \leq \sqrt{2f(x_i)\lambda_{max}(A)} \ . \tag{4.3.34}$$

LEMMA 4.3.7 Under Assumptions 4.3.1 - 4.3.5, the conjugate gradient residuals and directions obtained from the Lanczos quantities by using the definitions in Eqns (4.2.14) and Eqns(4.3.3)-(4.3.5) satisfy

$$| r_i^T p_i | > \rho_i^2 [1 - | \epsilon_i' | - N_{i-1} - \epsilon_1 \lambda_{max}(A)/\lambda_{min}(A)], \tag{4.3.35}$$

where N_{i-1} was defined in Theorem 4.3.4.

Proof. By Lemma 4.3.2 and the definitions of the p_i, r_i, and ρ_i, we obtain

$$r_i^T p_i = \rho_i^2 (1 + \epsilon_i') + \gamma_{i-1} r_i^T p_{i-1} - \rho_i^2 v_i^T A^{-1} \delta w_i.$$

But, by Theorem 4.3.4 and Eqn(4.3.3),

$$| \gamma_{i-1} r_i^T p_{i-1} | = | \rho_i^2 \sigma_{i-1} v_i^T u_{i-1} | \leq \rho_i^2 N_{i-1}.$$

If we invoke Lemma 4.3.6, then we get Eqn(4.3.35). $\qquad\square$

LEMMA 4.3.8 If Assumptions 4.3.1 - 4.3.6 are satisfied, then the associated conjugate gradient quantities satisfy

$$p_i^T A p_i < 2 r_i^T A r_i. \tag{4.3.36}$$

Proof. Using the definitions in Eqns(4.3.3)-(4.3.5) we get that

$$p_i^T A p_i = \rho_i^2 \omega_i [-\sigma_i u_i^T v_{i+1} + u_i^T v_i].$$

But, from Lemma 4.3.3 we have that

$$u_i^T v_i = v_i^T v_i + \sigma_{i-1} v_i^T u_{i-1} - v_i^T A^{-1} \delta w_i.$$

Therefore using Eqn(4.3.16), Lemma 4.3.6 and then Theorem 4.3.4 twice we obtain

$$\frac{p_i^T A p_i}{r_i^T A r_i} < \frac{\omega_i [1 + | \epsilon_i' | + N_i + N_{i-1} + \epsilon_1 \lambda_{max}(A)/\lambda_{min}(A)]}{v_i^T A v_i}$$

From Paige [1976] we have that

$$v_i^T A v_i = \alpha_i + \bar{\epsilon}_{i-1} + \delta \alpha_i \quad \text{where} \quad | \delta \alpha_i | \leq \epsilon n \lambda_{max}(A).$$

Furthermore, we know from the definition in Eqns(4.2.13) - (4.2.14) that

$$0 < \omega_i < \alpha_i.$$

Therefore the desired result follows from Assumption 4.3.6 and the fact that

$$[|\bar\varepsilon_{i-1}| + \varepsilon n \lambda_{max}(A)]/\alpha_i \le \varepsilon(3n+8)\lambda_{max}(A)/\lambda_{min}(A).$$

□

Proof of Theorem 4.3.6. We consider the values $f(x_i)$ of the function $f(x)$ in Eqn(4.2.3) at the iterates x_i. Observe that these values are completely defined from the residuals r_i. Since $f(x)$ is a quadratic function we have that

$$\Delta f_i / f_i \equiv [f(x_i) - f(x_{i+1})]/f(x_i) = [r_i^T \Delta x_i - \Delta x_i^T A \Delta x_i / 2] / [y_i^T A y_i / 2]$$

where as we demonstrated earlier $\Delta x_i = x_{i+1} - x_i = p_i/\omega_i$ and $y_i = x_i - x^*$. Since x^* is by definition the minimizing point of $f(x)$, $r^* = 0$, and we have therefore that $A y_i = -r_i$. Thus,

$$\frac{\Delta f_i}{f_i} = \frac{[2\omega_i r_i^T p_i - p_i^T A p_i]}{\omega_i^2 (r_i^T A^{-1} r_i)}.$$

From Theorem 4.3.5 and Assumption 4.3.6, we have that

$$\omega_i = (1 + a_i)\omega_i^* \quad \text{where}$$
$$|a_i| \le [N_i / (1 - N_i - N_{i-1} - |\varepsilon_i'| - \varepsilon_1 \lambda_{max}(A)/\lambda_{min}(A))] < 1/2.$$

This together with Eqn(4.2.6) for ω_i^* yields

$$\Delta f_i / f_i = \frac{[(2a_i + 1)(r_i^T p_i)^2]}{[(1 + a_i)^2 (p_i^T A p_i)(r_i^T A^{-1} r_i)]}.$$

Then from Lemmas 4.3.7 and 4.3.8 we obtain

$$\Delta f_i / f_i > \frac{[(1 - |\varepsilon_i'| - N_{i-1} - 1 - \varepsilon_1 \lambda_{max}(A)/\lambda_{min}(A))^2 \rho_i^4 (2a_i + 1)]}{[2(1 + a_i)^2 (r_i^T A r_i)(r_i^T A^{-1} r_i)]}.$$

But, $\rho_i^2 = r_i^T r_i$ so from Kantorovich's inequality, Luenberger [1973], we get

$$\frac{\rho_i^4}{[(r_i^T A r_i)(r_i^T A^{-1} r_i)]} > \frac{[4\lambda_{min}(A)\lambda_{max}(A)]}{[\lambda_{min}(A) + \lambda_{max}(A)]^2}.$$

Combining these relationships with Assumption 4.3.6, we get that for some $0 < \theta < 1$ and all i, $\Delta f_i \ge \theta f_i$. Therefore,

$$f_{i+1} \le (1-\theta) f_i \quad \text{so that } f_i \text{ decreases monotonically as i is increased.}$$

From this it is easy to show by contradiction that if we could let i increase indefinitely then the limiting value of f_i would have to be zero. Furthermore, since $f(x_i) = r_i^T A^{-1} r_i / 2$, we have that

$$| 2f(x_i) | \geq \| r_i \|^2 / \lambda_{max} (A),$$

which is Eqn(4.3.34). From the definition of $f(x)$ given in Eqn(4.2.3) we can get a similar inequality on the iterates x_i. ☐

The preceding results make certain assumptions about the sizes of the bounds obtained at intermediate points in the analysis, see in particular Assumptions 4.3.5 and 4.3.6. However, there are many matrices for which these assumptions are satisfied.

We have shown that (at least for some classes of matrices) the quantities generated by the Lanczos recursion yield quantities which satisfy the conjugate gradient recursions approximately. We then used the robustness of the conjugate gradient procedure to demonstrate that the associated values of $f(x_i)$ decrease monotonically as the number of Lanczos steps is increased. The bounds we have obtained depend upon i, and this limits the sizes of the Lanczos matrices which can be considered in the analysis. However, we note that for small matrices which are not too stiff and for which β^* is not too small, the limit on i is large. For such matrices the above analysis demonstrates that for large m we can expect the function values $f(x_i)$ to be essentially zero, and therefore the corresponding residual will be essentially 'zero'. In particular this means that the Lanczos vectors have essentially spanned the whole space. We would expect that whatever behavior the single-vector Lanczos recursion exhibits for small matrices should carry over to large matrices with some appropriate modifications in the estimates to acknowledge the effects of increasing the size of the given matrix. In practice this is precisely what we see.

The equivalence which we have demonstrated allows us to give meaning to taking more Lanczos steps than the order of the given matrix, since it is well-known that more than n steps of a minimization procedure may be required to accomplish the minimization of a function of n variables. It also gives us an interpretation of the significance of the local orthogonality of the Lanczos vectors. It is well-known that the directions of search which are used in optimization procedures need not be precisely the ones specified theoretically; they only have to have significant projections on the theoretical directions. In the next section we construct an argument which demonstrates a relationship between the 'convergence' of the associated conjugate gradient quantities and the appearance of the desired eigenvalues as eigenvalues of the Lanczos matrices.

SECTION 4.4   THE LANCZOS PHENOMENON

In this section we want to demonstrate a relationship between the 'convergence' of the associated conjugate gradient residuals (see Theorem 4.3.6) and the Lanczos Phenomenon (see Definition 4.3.1). In what we have presented thus far we have faithfully incorporated the effects of the finite precision arithmetic and we also do that here. We have the following relationship between the residuals $r_m = \rho_m v_m$ and the characteristic polynomial of the Lanczos matrix T_{m-1}.

LEMMA 4.4.1   For each m ≥ 2, the Lanczos vector v_m generated by the Lanczos recursion in Eqns(4.3.1) - (4.3.2) satisfies

$$\rho_m v_m = (-1)^{m+1} \frac{a_{m-1}(A)v_1}{a_{m-1}} + \sum_{j=1}^{m-1}(-1)^{m-j}\rho_j a_{j+1,m-1}(A)\,\delta w_j \bigg/ \prod_{k=j}^{m-1}\omega_k \quad (4.4.1)$$

where as defined in Chapter 3, $a_{m-1}(\mu)$ denotes the characteristic polynomial of the Lanczos matrix T_{m-1}, a_{m-1} denotes the determinant of T_{m-1}, and $a_{j+1,m-1}(\mu)$ denotes the characteristic polynomial of the submatrix of T_{m-1} obtained by deleting the first j rows and columns. We define $a_{m,m-1}(A) \equiv I$. δw_j is defined in Eqn(4.3.10).

We note that

$$\prod_{k=j}^{m-1}\omega_k = a_{m-1}/a_{j-1}.$$

Therefore, the error term in Eqn(4.4.1) is a sum of the effects of polynomial operators in A operating on the error terms δw_j obtained from the Lanczos recursion, where the polynomials are scaled characteristic polynomials of submatrices of T_{m-1}.

Proof of Lemma 4.4.1   A straight-forward argument using the Lanczos recursion in Eqns(4.3.1) - (4.3.2), the definitions in Eqn(4.2.14) and (4.3.3), the facts that

$$a_{m-1} = \prod_{k=1}^{m-1}\omega_k \text{ and that } \rho_j = \prod_{k=1}^{j-1}\sigma_k,$$

and induction on m, yields Eqn(4.4.1).   □

Theorem 4.3.6 tells us that the norms of the residuals converge towards zero as the number of Lanczos steps is increased. We would like to be able to use this theorem in Eqn(4.4.1) to obtain an explanation for the Lanczos Phenomenon. However, to do this rigorously we would have to obtain an appropriate bound on the summation term in Eqn(4.4.1), and we were not able to do this. However, the results of our numerical experiments make us believe that such a bound exists. See Section 4.7.

CONJECTURE 4.4.1 There are classes of matrices for which the following relationship is true. Let A be a positive definite real symmetric matrix. Let z_k be orthonormal eigenvectors of A corresponding respectively to the distinct eigenvalues λ_k, $1 \leq k \leq q$, of A. Under Assumptions 4.3.1 - 4.3.6, we have that given 'small' $\hat{\epsilon} > 0$, and any eigenvalue λ_k, $1 \leq k \leq q$, that for large enough m

$$
|v_1^T z_k| \, |a_m(\lambda_k)/a_m| \; \leq \; \hat{\epsilon} \; + \; \sum_{j=1}^{m} \rho_j |a_{j+1,m}(\lambda_k)| \, \|\delta w_j\| / \prod_{k=j}^{m} \omega_k \; \leq
$$
$$
\hat{\epsilon} \; + \; \epsilon_1 K_m \lambda_{\max}(A)/\lambda_{\min}(A)
$$
(4.4.2)

where $v_1^T z_k$ is the projection of the starting vector on z_k, and K_m is a slowly enough growing function of m such that for the m being considered the error term is dominated by the $\hat{\epsilon}$.

Partial Argument for Conjecture 4.4.1. By Lemma 4.4.1, we have that

$$
z_k^T r_{m+1} \; = \; \rho_{m+1} z_k^T v_{m+1} \; = \; a_m(\lambda_k) z_k^T v_1 / a_m \; + \; \sum_{j=1}^{m} \rho_j a_{j+1,m}(\lambda_k) z_k^T \delta w_j \; / \prod_{k=j}^{m} \omega_k
$$

But by Theorem 4.3.6 for large enough m, we have $\rho_{m+1} \equiv \|r_{m+1}\| < \hat{\epsilon}$, so the projection of r_{m+1} on z_k is bounded by $\hat{\epsilon}$ and we get Eqn(4.4.2). $\qquad \square$

Conjecture 4.4.1 states that for large enough m each distinct eigenvalue of A should be a near-zero of an appropriately-scaled characteristic polynomial of T_m. The scaling is the natural one. Scaling $a_m(\mu)$ by a_m is equivalent to scaling the characteristic polynomial of T_m so that its constant term is 1 or equivalently to scaling the characteristic polynomial of T_m^{-1} so that the coefficient of its highest order term is 1. The scaling of $a_{j+1,m}(\mu)$ by a_m/a_{j-1} is not so easy to justify directly. However, the principal submatrices of each Lanczos matrix mimic the large matrix w.r.t. being uniformly positive definite (see Lemma 2.3.7), their shapes and the sizes of their entries.

Once we see the equivalence in Section 4.3, the natural question to ask is why not just use conjugate gradients directly to compute eigenvalues? The problems with doing this are numerical. If we have a matrix whose spectrum consists of only a few well-separated clusters of eigenvalues, then when we apply a conjugate gradient procedure to Eqn(4.2.1) we get a numerically-small residual after only a few steps, and long before many of the individual eigenvalues of A have appeared as eigenvalues in the Lanczos matrices. Because of the numerical problems with dealing with very small residuals, it is not wise numerically to continue the conjugate gradient iterations beyond this point. The vectors generated by the Lanczos recursions are scaled on each step and therefore do not suffer from these numerical problems. Lanczos quantities can be generated indefinitely, as long as the β_{i+1} are not too small.

Lemma 4.4.1 and Conjecture 4.4.1 quantify the relationship between the Lanczos Phenomenon, the rate of convergence of the norms of the associated conjugate gradient residuals, and
the errors introduced by the finite precision arithmetic. We can view the use of a Lanczos
procedure without any reorthogonalization as a race, as the number of Lanczos steps is
increased, between the decrease in the norms of the residuals and the growth in the errors due
to the finite precision arithmetic.

The estimates which we obtained in Section 4.3 imply that the size of the A-matrix and the
sizes of the Lanczos matrices being considered are key issues in this race. However, our
experience has been that these sizes are secondary issues. As we will see in Section 4.7 the rate
of convergence depends primarily upon the distribution of the gaps between the eigenvalues of
A and the locations of the desired eigenvalues in the spectrum of A. The dependence of the
convergence rate upon the size of the original matrix, the matrix stiffness, and the size of the
Lanczos matrix being considered seems to be very weak.

Assuming that the solution of Eqn(4.2.1) has a nontrivial projection on each eigenspace,
then it is clear that the residual can be made arbitrarily small only if the set of Lanczos vectors
spans a subspace that contains at least one eigenvector from each eigenspace. Making the
residual arbitrarily small is equivalent to having minimized $f(x)$ over the entire space.

In order to use the Lanczos Phenomenon in a practical Lanczos algorithm for computing
eigenvalues of real symmetric matrices, it is necessary to devise an identification test for sorting
the 'spurious' eigenvalues of the Lanczos matrices from the 'good' ones. Conjecture 4.4.1 only
asserts that everything we want should be there if the Lanczos matrix is large enough. The
difficulty is that additional eigenvalues may also appear as eigenvalues of the Lanczos matrices.
These are caused by the losses in orthogonality of the Lanczos vectors, and they must be
identified and discarded.

SECTION 4.5   AN IDENTIFICATION TEST, GOOD VERSUS SPURIOUS EIGENVALUES

The Lanczos Phenomenon (see Definition 4.3.1) indicates that we should be able to
compute many of the eigenvalues of very large matrices, perhaps even all of them, by computing
eigenvalues of the associated real symmetric tridiagonal Lanczos matrices generated using the
single-vector Lanczos recursion with no reorthogonalization. However, because of the losses in
orthogonality of the Lanczos vectors, extra eigenvalues as well as extra copies of the desired
eigenvalues will appear as eigenvalues of the Lanczos matrices as the size m of these matrices is
increased. The numbers of these extra eigenvalues and their locations in the spectrum will vary
as m is varied. In order to use the Lanczos Phenomenon in a practical Lanczos algorithm for
computing eigenvalues of real symmetric matrices, it is necessary to devise an identification test
for sorting these extra or 'spurious' eigenvalues from the 'good' ones. We use the correspondence between the Lanczos tridiagonalization procedure and the conjugate gradient optimization

procedure for solving systems of equations Ax = b which was developed in Section 4.3 to justify the identification test which we use in our Lanczos procedures. Numerical experiments with this test have substantiated this choice.

We will be working with the eigenvalues of various submatrices of the Lanczos matrices generated by the Lanczos recursions. Recall the following definitions from Chapter 3. Given any tridiagonal mxm matrix T_m, we let $\hat{T}_2$ denote the $(m-1)x(m-1)$ tridiagonal matrix obtained from T_m by deleting the first row and the first column of T_m. We denote the characteristic polynomials of T_m and of $\hat{T}_2$, and the derivative of the characteristic polynomial of T_m by respectively, $a_m(\mu)$, $\hat{a}_2(\mu)$, and $a'_m(\mu)$. See Definitions 3.1.2 and 3.1.3 in Chapter 3. The eigenvalues of T_m and of the associated matrix $\hat{T}_2$ will be denoted respectively by

$$\mu_1^m \geq \mu_2^m \geq \ldots \geq \mu_m^m \quad \text{and} \quad \hat{\mu}_1 \geq \hat{\mu}_2 \geq \ldots \geq \hat{\mu}_{m-1}.$$

We have not included the superscript m in labelling the eigenvalues of $\hat{T}_2$. In practice we will also delete it from the μ_j^m. The value of m will typically be clear from the context. We have the following definition of spurious.

DEFINITION 4.5.1 Webster [1971]. Spurious $\equiv$ Outwardly similar or corresponding to something without having its genuine qualities.

The objective of this section is to describe the identification test which we use in our Lanczos procedures and to provide some justification for our choice. First however consider the following short table of eigenvalues computed from a Lanczos matrix which was generated by applying the Lanczos recursion with no reorthogonalization to one of our real symmetric test matrices. In Table 4.5.1 we see two types of 'spurious' eigenvalues. Type one is illustrated by eigenvalue μ_{44}. This eigenvalue is a less accurate copy or 'ghost' of the 'good' eigenvalue μ_{45}. Type two is illustrated by eigenvalue μ_{46}. This eigenvalue is genuinely spurious. If the size of the Lanczos matrix is increased, this eigenvalue 'disappears' only perhaps to reappear as a different eigenvalue elsewhere in the spectrum. We give a specific example of this type of behavior in Section 4.6.

TABLE 4.5.1 Spurious Eigenvalues

Classification	No.	Eigenvalue T_m
Accept	42	1.946284978373996
Repeat	43	1.946284978373996
Reject	44	1.947690935703804
Accept	45	1.947690935814651
Reject	46	1.948757755271330
Accept	47	1.949375014096254

We note that theoretically any irreducible, real symmetric tridiagonal mxm matrix has m distinct eigenvalues. (See Theorem 2.2.4 in Chapter 2.) However in practice, numerically-multiple eigenvalues occur. For example, look at the eigenvalues μ_{42} and μ_{43} in Table 4.5.1. Such multiplicities will occur with any matrix A if the corresponding Lanczos matrices are made large enough. How large the Lanczos matrix has to be before this type of phenomena occurs depends upon the eigenvalue distribution in the original matrix. Thus, for all practical purposes the Lanczos matrices can and do have multiple eigenvalues.

Lemma 4.5.1 which is just a restatement of Lemma 2.3.5 given in Chapter 2 states that the accuracy of any computed Ritz value as an approximation to some eigenvalue of A can be adequately measured by the size of the last component of the eigenvector u (of the Lanczos matrix) which corresponds to the computed μ, if the norm of the corresponding Ritz vector $z = V_m u$ is not small. Lemma 2.3.4 in Chapter 2 tells us that this norm is not small as long as the eigenvalue being considered is an isolated eigenvalue of the associated Lanczos matrix.

LEMMA 4.5.1 Paige [1980] Let T_m be generated by the Lanczos recursion in Eqn(4.3.1) - (4.3.2). Let $Tu = \mu u$ for some eigenvalue μ of T_m. Then there exists an eigenvalue λ of A such that

$$|\lambda - \mu| \leq \|AV_m u - \mu V_m u\| / \|V_m u\| \leq$$
$$[|\beta_{m+1} u(m)|(1 + 2\varepsilon_0) + 2\varepsilon_1 \sqrt{m}\lambda_{max}] / \|V_m u\| \qquad (4.5.1)$$

where ε_1 and ε_0 are defined in Eqn(2.3.15).

This lemma is central to the rest of the development. It is an easy consequence of the nontrivial machinery which Paige [1976,1980] develops in his error analysis of the Lanczos recursion. Using this lemma Paige then proves the following theorem for what he calls stabilized eigenvalues of the Lanczos matrices. These are eigenvalues for which the last component of a corresponding eigenvector of the Lanczos matrix is very small. Such eigenvalues are good approximations to eigenvalues of A. The bounds in Theorem 4.5.1 are weak, Paige [1980].

THEOREM 4.5.1 Paige [1980] Let T_m and V_m be the result of m steps of the Lanczos algorithm in Eqns(4.3.1) - (4.3.2). Paige defines a stabilized eigenvalue μ of T_m as one which satisfies

$$|\beta_{m+1} u(m)| \leq \sqrt{3}\varepsilon_2 m^2 \lambda_{max} (A). \qquad (4.5.2)$$

where ε_2 is defined in Eqn(2.3.24) as $\varepsilon_2 \equiv 2\sqrt{2}\max(6\varepsilon_0, \varepsilon_1)$. Then for any stabilized, isolated eigenvalue μ we have that there is an eigenvalue λ of A such that

$$|\lambda - \mu| \leq \varepsilon_2(m + 1)^3\lambda_{max} (A). \tag{4.5.3}$$

By the interlacing theorem we know that any numerically-multiple eigenvalue of a Lanczos matrix has stabilized in the sense that it must be an eigenvalue of all larger Lanczos matrices. This statement is not however the same statement as that made in the definition of stabilized eigenvalues given in Eqn(4.5.2). However, in practice one observes that shortly after an eigenvalue converges as an eigenvalue of the Lanczos matrices, the last components of its corresponding eigenvector become pathologically small, at which point in time that eigenvalue will have stabilized according to Paige's definition. Typically, this happens long before the eigenvalue becomes numerically-multiple. From this we infer that Paige's estimate in Eqn(4.5.3) is also applicable to any numerically-multiple eigenvalue. Therefore in each of our single-vector Lanczos procedures we accept numerically-multiple eigenvalues as 'accurate' approximations to eigenvalues of the original matrix A, and our identification test is applied only to the simple eigenvalues of the Lanczos matrices. By construction we have the following trivial lemma.

LEMMA 4.5.2 Let A be a positive definite, real symmetric matrix. Use the Lanczos recursion in Eqns(4.3.1) - (4.3.2) and the conjugate gradient - Lanczos tridiagonalization correspondence defined in Eqns(4.2.13) and (4.3.3) - (4.3.6) to define associated conjugate gradient quantities. Then we have for any m that

$$\prod_{j=2}^{m+1} \beta_j = (-1)^m \rho_{m+1}a_m. \tag{4.5.4}$$

Proof. The proof is a direct consequence of the facts (see Eqns(4.2.13) - (4.2.14)) that

$$\beta_{j+1} \equiv -\sigma_j\omega_j, \; \sigma_j \equiv \rho_{j+1}/\rho_j \text{ and that } \det(T_m) = a_m = \prod_{j=1}^{m} \omega_j.$$

$\square$

LEMMA 4.5.3 Under the hypotheses of Lemma 4.5.2, for any eigenvalue μ of T_m we have the following identity

$$\left[\frac{\hat{a}_2(\mu)}{a_{m-1}} \; \frac{a_{m-1}(\mu)}{a_{m-1}} \right] = \rho_m^2. \tag{4.5.5}$$

Proof. From Chapter 3, Corollary 3.2.4 we have that for any eigenvalue μ of T_m that

$$\hat{a}_2(\mu)a_{m-1}(\mu) = \prod_{k=2}^{m} \beta_k^2.$$

Applying Lemma 4.5.2 and rearranging the result yields Eqn(4.5.5). □

    Lemma 4.5.3 tells us that the only way that the associated conjugate gradient residuals can become arbitrarily small is for every eigenvalue of T_m to be either an eigenvalue of the submatrix T_{m-1} or of the submatrix $\hat{T}_2$, or to be simultaneously an eigenvalue of both of these submatrices. Thus, we conclude from Lemma 4.5.3 that there are at least two submatrices which are important in our analyses of the Lanczos recursion as a method for computing eigenvalues of large matrices. The first submatrix T_{m-1} is the natural one to consider when we are asking the question, "Which of the eigenvalues of the Lanczos matrices have converged?" Thus, we could track the progress of the convergence of the eigenvalues of the Lanczos matrices as their size is increased. This is in fact what van Kats and van der Vorst [1976,1977], Edwards et al [1979], and Parlett and Reid [1981] do in their Lanczos procedures with no reorthogonalization. Their identification tests rest purely on relationships between Lanczos matrices T_k as k is increased. See Chapter 2, Section 2.5 for more details on their alternative identification tests.

    Our approach differs from theirs in that we focus on a different submatrix, $\hat{T}_2$. It also differs in that we do not attempt to identify those eigenvalues which have converged. Rather at each stage we directly identify those eigenvalues which are spurious. All of the remaining eigenvalues are kept as 'good' eigenvalues. These 'good' eigenvalues may or may not have converged. Some of them may be accurate to only a few digits. Our approach has several distinct advantages over the more obvious approach. First the resolving power of our Lanczos procedures is increased. We do not require that the user provide a 'fuzzy' tolerance which is then used in determining convergence. Rather we use a tight machine-generated tolerance for deciding which eigenvalues are 'bad'. The remaining eigenvalues are then called 'good'. Our approach provides the user with early estimates of the entire spectrum, at least of that portion of the spectrum which is being computed, along with error estimates indicating the accuracy of each of the 'good' eigenvalues. This type of information is useful in deciding how much information about the given matrix can be computed in the amount of time and with the amount of storage available.

    The relationship in Eqn(4.5.5) indicates that the submatrix $\hat{T}_2$ can be expected to play an important although it be unintuitive role in the behavior of any Lanczos procedure without reorthogonalization. Numerical experiments computing all of the eigenvalues of the three matrices T_m, $\hat{T}_2$, and T_{m-1} for various values of m and comparisons of the quantities obtained led to the discover of the following identification test. An argument justifying this choice of test is given in Lemma 4.5.4.

IDENTIFICATION TEST 4.5.1 Let T_m, m = 1,2,... be Lanczos matrices generated by the Lanczos recursions in Eqns(4.3.1) - (4.3.2). Then for each m compare each computed simple eigenvalue μ_j^m of T_m with the eigenvalues of the corresponding submatrix $\hat{T}_2$. Any such eigenvalue which is also an eigenvalue (numerically) of the corresponding $\hat{T}_2$ matrix is labelled 'spurious' and is discarded from the list of computed eigenvalues. All remaining eigenvalues, including all numerically-multiple ones are accepted and labelled 'good'.

LEMMA 4.5.4 Let A be a positive definite, real symmetric matrix and let T_m be a corresponding Lanczos matrix obtained using the Lanczos recursions in Eqns(4.3.1) - (4.3.2). Given any eigenvalue μ of T_m and a corresponding unit eigenvector u with $Tu = \mu u$, there is an eigenvalue λ of A such that

$$|\lambda - \mu| \leq \frac{\|AV_m u - \mu V_m u\|}{\|V_m u\|} \equiv \frac{\|\varepsilon_m + \Delta W_m\|}{\|V_m u\|}. \tag{4.5.6}$$

where $\|\Delta W_m\| = O(\varepsilon\sqrt{m}\|A\|)$ and

$$\varepsilon_m \equiv \frac{\rho_m}{[(a'_m(\mu)/a_{m-1})(\hat{a}_2(\mu)/a_{m-1})]^{1/2}}. \tag{4.5.7}$$

Proof. This is a direct consequence of Lemma 4.5.2, the formulas for components of eigenvectors of T_m given in Eqn(3.2.11), Corollary 3.2.3, and the definitions of the associated conjugate gradient quantities. In particular we have that

$$[u(m)]^2 = \frac{a_{m-1}(\mu)}{a'_m(\mu)} = \frac{a_{m-1}(\mu)\hat{a}_2(\mu)}{a'_m(\mu)\hat{a}_2(\mu)} = \frac{\rho_m^2 a_{m-1}^2}{a'_m(\mu)\hat{a}_2(\mu)}.$$

$\square$

Observe that the expression obtained for ε_m is an identity. Thus we have obtained an equivalent expression for the residual bound, $\|AV_m u - \mu V_m u\|$. Practical experience has demonstrated repeatedly that this residual norm is a realistic and useful upper bound for the error in the computed eigenvalue, and it is used frequently in practical software.

Lemma 4.5.4 implies that if the associated conjugate gradient quantities are well-behaved in the sense that the norms of the conjugate gradient residuals are converging towards zero, then the only way that this bound cannot be 'small' is if the denominator becomes pathologically small. If we consider only 'isolated' eigenvalues of the Lanczos matrix so that the derivative of the characteristic polynomial, $a'_m(\mu)$, is not 'small', then pathological smallest can come only from μ being an eigenvalue (numerically) of $\hat{T}_2$. Thus, Lemma 4.5.4 indicates that those eigenvalues of T_m which are also eigenvalues of $\hat{T}_2$ may be problem eigenvalues.

Numerical tests verified that in fact these are the 'spurious' eigenvalues which we want to eliminate. In practice one sees this behavior exhibited by such eigenvalues matching eigenvalues of $\hat{T}_2$ to the accuracy with which the eigenvalues of the T_m have been computed. In Table

4.5.2 we give an example of this matching. Table 4.5.2 contains the eigenvalues given in Table 4.5.1 along with the corresponding eigenvalues of $\hat{T}_2$.

TABLE 4.5.2 Identifying Spurious Eigenvalues

Classification	No.	Eigenvalue T_m	No.	Eigenvalue $\hat{T}_2$
Accept	42	1.946284978373996	41	1.944310777494893
Repeat	43	1.946284978373996	42	1.946284978373996
Reject	44	1.947690935703804	43	1.947077133213947
Accept	45	1.947690935814651	44	1.947690935703804
Reject	46	1.948757755271330	45	1.948041036933613
Accept	47	1.949375014096254	46	1.948757755271330

Table 4.5.2 clearly illustrates the sharpness of our identification test. The T_m eigenvalue μ_{44} is spurious. It matches the $\hat{T}_2$ eigenvalue $\hat{\mu}_{44}$ to all digits. Eigenvalue μ_{46} is spurious and it matches eigenvalue $\hat{\mu}_{46}$ to all digits. Both kinds of spurious eigenvalues are identified by this test. Table 4.5.3 is a simple restatement of this test.

TABLE 4.5.3 Summary, Identification Test

Case	Evalue T_2	Multiple Evalue, T_m	Conclusion
1	Yes	Yes	Accept Evalue
2	Yes	No	Reject Evalue
3	No	No	Accept Evalue

For Case 1 we have Theorem 4.5.1 from which we infer that any numerically multiple eigenvalue of T_m must be a good approximation to an eigenvalue of **A**. Cases 2 and 3 rest upon Lemma 4.5.4 which implies that certain eigenvalues of $\hat{T}_2$ cause problems. There is an implicit assumption in our test that if an eigenvalue of T_m is simultaneously (numerically) an eigenvalue of T_{m-1} and of $\hat{T}_2$, then it will appear as a numerically-multiple eigenvalue of T_m. We do not have a proof that this is always true. This test can misclassify an eigenvalue whose eigenvector is very poorly represented in the starting vector. Eigenvalues with such small projections will appear (to within numerical accuracy) as eigenvalues of T_m and of $\hat{T}_2$ for all m and will therefore be misclassified until they begin to replicate. Our codes do however recoup from any such loss by using a check for such eigenvalues which is contained in the subroutine PRTEST. See Volume 2, Chapter 2, Section 2.5.

Subroutine PRTEST should be called only after convergence on some desired portion of the spectrum has been confirmed. PRTEST uses two empirically-observed 'facts'. First, if convergence is indicated on the approximations to the eigenvalues near any given eigenvalue λ of A, then the corresponding approximation to λ has almost surely converged. That is, if λ is reasonably well-separated from these other eigenvalues then it will have converged. If it were in a cluster and had not yet converged, then that should be reflected in the nonconvergence of its nearest neighbor. Second, not long after an approximation to a given eigenvalue of A has converged as an eigenvalue of the Lanczos matrices, T_m, the mth component of the corresponding T_m eigenvector will become pathologically-small. Therefore, for any σ the matrices

$$T_m - \sigma e_m e_m^T \qquad (4.5.8)$$

where e_m is the mth coordinate vector will also have μ as an eigenvalue. For each isolated eigenvalue μ of T_m which has been labelled spurious, PRTEST performs two Sturm Sequences plus a few multiplies to determine the number of eigenvalues of the matrix in Eqn(4.5.8) which are in the interval $[\mu - \varepsilon, \mu + \varepsilon]$ for 4 different values of σ scaled by T_m. Since σ enters only in the (m,m)-th position in this matrix, the Sturm sequencing does not have to be repeated for each value of σ used.

Our experience indicates that 'hidden' eigenvalues occur rarely. PRTEST was tested using the Poisson test matrices with zero boundary conditions where both the eigenvalues and eigenvectors were known a priori. See Section 4.6 for a description of these test matrices. Tests were performed by explicitly modifying the starting vector for the Lanczos recursions so that it had a zero or negligible projection on one or more of the eigenvectors of the original test matrix. Cullum, Willoughby and Lake [1983] contains a comparison of our identification test with the basic identification test proposed by Paige [1971]. That discussion illustrates the possible problems which can be encountered if one relies purely on the computed error estimates to determine which eigenvalues are 'good'.

The basic eigenvalue subroutine included with our programs, BISEC, is a Sturm sequencing procedure, Jennings [1977, Chapter 9], and can therefore be used to compute the eigenvalues of T_m in any subinterval. Three vectors of length m are needed for the α, β, and β^2 arrays. Vectors are also needed to store the computed eigenvalues of T_m and the running set of upper and lower bounds for the eigenvalues being computed on the subinterval being processed. The subintervals provided by the user are considered sequentially in algebraically increasing order. Starting at the lower end of a subinterval, the procedure computes successively the distinct eigenvalues of T_m in ascending algebraic order, locating eigenvalues to within a given tolerance set by the procedure using the machine epsilon.

It is easy to see that we do not need to compute the eigenvalues of $\hat{T}_2$ in order to carry out our identification test. For each computed simple, isolated eigenvalue μ of T_m we need only to determine if there is an eigenvalue of $\hat{T}_2$ which is pathologically close to μ. In order to make such a determination it is sufficient, given a measure of closeness $\bar{\varepsilon}_0$, to compute two Sturm sequences, one on the matrix $(\mu + \bar{\varepsilon}_0)I - \hat{T}_2$ and the other on the matrix $(\mu - \bar{\varepsilon}_0)I - \hat{T}_2$.

These tests are included directly in the BISEC subroutine. In fact the cost of these tests is negligible because the Sturm sequencing required for these tests is being done as part of the eigenvalue computations for T_m.

The Sturm sequencing is simply a count on the number of negative determinant ratios encountered as we progress through the recursion given in Eqn(3.5.2) in Chapter 3. However, we know that for any permutation matrix P, the matrices $P(\mu I - A)P^T$ and $(\mu I - A)$ have the same eigenvalues. Therefore, if we apply the permutation matrix whose jth column is the coordinate vector e_{m-j+1} to a Lanczos matrix, forming the matrix $P(\mu I - T_m)P^T$, then this matrix will have the same eigenvalues as A but its Sturm sequences will be identical to those obtained by using $(\mu I - T_m)$ but working up the diagonal from the bottom of the matrix rather than down from the top. Thus in BISEC we do our Sturm sequencing by counting ratios of determinants as we pass up the diagonal from the bottom of the matrix. But, this allows us to simultaneously determine whether or not a given number is both an eigenvalue of T_m and of the related $\hat{T}_2$ without any extra work.

In practice we use an $\bar{\epsilon}_0$ which depends upon the machine epsilon, the size of the Lanczos matrix being considered, and the matrix scale. For specific details the reader is referred to the BISEC subroutine in Volume 2, Chapter 2, Section 2.5. The same tolerance is used both for determining the numerical multiplicities of computed eigenvalues as eigenvalues of the Lanczos matrices and for determining spurious eigenvalues. The procedure computes eigenvalues by the successive bisection of intervals which are known to contain eigenvalues of the matrix in question. The tolerance $\bar{\epsilon}_0$ must be chosen to insure that a given eigenvalue of $\hat{T}_2$ is not allowed to label more than 1 eigenvalue of T_m as spurious. If an eigenvalue $\hat{\mu}$ of $\hat{T}_2$ were in the intersection of the $\bar{\epsilon}_0$-intervals for μ_j and μ_{j+1}, then both would be eliminated by that $\hat{\mu}$. Such multiple eliminations do not occur in BISEC because by construction (i) Each isolated μ_k is really the center of an interval

$$[l_k, u_k] \tag{4.5.9}$$

which defines that eigenvalue, (ii) The distance, $(l_k - u_{k-1})$, between successive eigenvalues of T_m is greater than $\bar{\epsilon}_0$, and (iii) spurious eigenvalues have the property that the corresponding eigenvalue of $\hat{T}_2$ is typically in the very small interval given in Eqn(4.5.9). Note that (ii) holds because we use the same tolerance in both the multiplicity and the spurious tests. (If μ_j is good then no eigenvalue of $\hat{T}_2$ is within $\bar{\epsilon}_0$ of μ_j). Combining the multiplicity and spurious tests simplifies the procedure considerably. Another way of stating (iii) is simply that if an eigenvalue of T_m is spurious, then the closest eigenvalue in $\hat{T}_2$ is within the tolerance of the eigenvalue computation.

If in fact, all of the numerically-multiple copies of a given eigenvalue μ of T_m were computed accurately, and the check for spuriousness were done before the check for numerical multiplicity, then not only would the genuinely spurious eigenvalues be discarded, but also all except one copy of each numerically-multiple eigenvalue would be rejected. Empirical results indicate that the one copy which would be left would be the most accurate approximation of the several copies. However, we do not do this. Not only would it be more expensive since each copy would have to computed fully, but we also would not be able to get meaningful error estimates. Meaningful error estimates can only be computed for those 'good' eigenvalues which are 'isolated' from other T_m eigenvalues. As programmed, BISEC computes 1 copy of each eigenvalue and then determines the multiplicity. Since this copy is the first copy encountered, it may not necessarily be the most accurate value for each multiple eigenvalue. However, it is accurate to within the tolerance $\bar{\varepsilon}_0$. If an eigenvalue is simple, then the $\hat{T}_2$ information is used to determine whether or not that eigenvalue is spurious.

In determining multiplicities, small ambiguities may occasionally occur because we can only estimate the accuracy of the tridiagonal eigenvalue computations. It is possible for 2 'good' eigenvalues that agree to 10 digits or more to be labelled as distinct. Therefore in the code in Volume 2, after the eigenvalue computations and the identification/multiplicity tests, and before computing the error estimates, we go through the list of 'good' eigenvalues and combine eigenvalues which differ from each other by less than a user-specified relative tolerance, RELTOL. We average the 'good' eigenvalues which have been lumped together, weighting the eigenvalues by their multiplicities. The numerical multiplicity is increased accordingly and this count includes any spurious eigenvalues which are within the tolerance. However, the values of the spurious eigenvalues are not used in determining the resulting averaged eigenvalue. For more details see subroutine LUMP in Volume 2, Chapter 2, Section 2.5. Error estimates are then computed only on the resulting simple, isolated, good eigenvalues.

Since there has been a great deal of controversy about the notion of a 'spurious' eigenvalue, in the next section we take the time to go through a detailed example showing how such an eigenvalue can appear in one portion of the spectrum, only to disappear from that portion as the size of the Lanczos matrix is increased, to reappear elsewhere.

SECTION 4.6   EXAMPLE, TRACKING SPURIOUS EIGENVALUES

The extra eigenvalues introduced because of the losses in orthogonality of the Lanczos vectors have been called 'redundant' or 'ghosts' by other authors. See for example Paige [1980]. These authors maintain that all of the extra eigenvalues are in some sense genuine because they believe that if if we were to continue to enlarge the Lanczos matrix each of these eigenvalues would become a duplicate of some previously converged eigenvalue of T_m. This however is not always the case, as we will illustrate below by taking an actual matrix and looking in detail at what happens to these extra eigenvalues as we enlarge the Lanczos matrices. There are actually two kinds of spurious eigenvalues, those which are truly replicating and those which are not. Those which are not do not behave in any predictable fashion as we will illustrate in Table 4.6.1. They may start out in one part of the spectrum, only to escape to some other part of the spectrum when we enlarge the Lanczos matrix. This escape is accomplished by a domino effect. The example which we consider is obtained by a discretization of the Laplace operator on the unit square. A brief description of the matrix we used is given below.

POISSON-TYPE MATRICES. We consider the Laplace equation on a rectangle,

$$u_{xx} + u_{yy} = \lambda'u, \quad R = \{(x,y) \mid 0 \leq x \leq X, \ 0 \leq y \leq Y\}. \tag{4.6.1}$$

for two types of boundary conditions: (1) Dirichlet conditions, the solution u=0 on the boundary of R, and (2) Neumann conditions, the normal derivative of the solution $\partial u/\partial n = 0$ on the boundary except that u = 0 when y = 0. We replace the differential operator in Eqn(4.6.1) by an algebraic equation, $Ax = \lambda x$, obtained by discretizing the Laplace operator on the rectangle. This discretization is uniform in the rectangle in each of the x and the y directions. The order of the resulting matrix is N = KX×KY where KX is the number of subdivisions in the x-direction and KY is the number of subdivisions in the y-direction. A is block tridiagonal. Each block B is KX×KX and there are KY blocks down the diagonal of A. Specifically we have

The parameter c is user-specified and must be chosen such that $0 < c \leq 0.5$ and the small matrix C = -(0.5 - c)I. For Dirichlet boundary conditions s=1, and for Neumann conditions $s = \sqrt{2}$. If we use the Dirichlet boundary conditions then we have a formula for the eigenvalues (and eigenvectors) of the test matrix. In this case for $1 \leq I \leq KX$, $1 \leq J \leq KY$, the eigenvalues are generated by the formula

$$\lambda(I,J) \equiv [1 - 2c \cos(\pi I/(KX + 1)) - 2(0.5 - c) \cos(\pi J/(KY + 1))]$$

If we use the Neumann boundary conditions then the eigenvalues and the eigenvectors are not known a priori. By varying the value of c we can obtain many different eigenvalue distributions.

We denote the particular test matrix which we used in this section by POIS1008. This was a Poisson matrix of order 1008 corresponding to a discretization of the Laplace operator on the unit square together with Dirichlet boundary conditions. KX was 56 and KY was 18. The codiagonal entries in the B blocks were -.44958715. The diagonal entries in the C blocks were

-.050412845. There were 18 blocks down the diagonal and each block was 56x56. We considered the subinterval [.088,.098], using Lanczos matrices of order m = 1068 to 1097. The eigenvalues were computed using the BISEC subroutine. The minimum off-diagonal β_{k+1} encountered was .126. The estimated norm of this matrix was 1.53. The tolerance

$$\bar{\epsilon}_0 = 1.4 \times 10^{-12}$$

was used in both the multiplicity and the spurious test. We observe that this is an easy matrix for the Lanczos procedure in that all of the eigenvalues of this matrix can be computed accurately using a Lanczos matrix of order m<2n. We know the true eigenvalues for this test problem and these values were used to verify that the identification test was rejecting the correct subset of eigenvalues.

The results of these computations are difficult to display. Table 4.6.1 is split into two pages. In Table 4.6.1 for each size Lanczos matrix considered, we list all of the distinct eigenvalues which were computed by BISEC in the interval [.08834,.09784]. The table entries are actually 10 times the computed eigenvalues. However, to simplify the discussion we will continue to refer to the table entries as eigenvalues. The columns labelled GOODJ for J = 1,2,...,6 contain the computed 'good' eigenvalues. Similarly, the columns labelled SPURJ for J = 1,2,...,5 contain those eigenvalues which were labelled spurious. If for a given Lanczos matrix of size M there is no entry in one of the SPURJ columns, then that simply means that there was no spurious eigenvalue between good eigenvalues J and J+1 for that particular value of M. We have used the interlacing properties to track the movement of the eigenvalues as we increased the order of the Lanczos matrix. Several examples of changes are emphasized by being labelled alphabetically. For example the reader can trace the path of the eigenvalue labelled J1 as we increase the size of M by simply scanning the table for succeeding labels J2, J3, etc.

If we look at the eigenvalues in column SPUR1 we see that there is a spurious eigenvalue in this location for almost every M considered. However, the 'good' eigenvalue which the SPUR1 eigenvalue approximates most closely changes as we increase M. In fact the error estimates oscillate. At M=1070 this spurious eigenvalue is closest to the good eigenvalue .09146. However, at a slightly larger M it is closest to the good eigenvalue .08834. 'Ownership' fluctuates between these two eigenvalues. Note also that although this spurious eigenvalue gets very close at various times to each of these two good eigenvalues, throughout the entire range of sizes considered in this Table, neither one of these two good ones becomes double. In fact none of the good eigenvalues in this interval becomes numerically-double during this range of sizes. This illustrates rather clearly that we cannot necessarily think of each of these extra eigenvalues as belonging to some converged eigenvalue and eventually converging to it.

We said earlier that it is possible for a spurious eigenvalue to appear in one part of the spectrum at some size M, but then to disappear from that location appearing elsewhere in the spectrum. We have been told that this is impossible. However, we invite the interested reader to consider the paths of eigenvalues E1, F1, and G1 in Table 4.6.1. E1 appears at M = 1082 as a spurious eigenvalue. At M = 1083 it has disappeared. Actually it has become the good eigenvalue E2= .09440. But in becoming E2 it forced the good eigenvalue F1 = .09940 at M = 1082 to move left at M = 1083 where it became the spurious eigenvalue F2 = .09432. At M = 1084 the spurious eigenvalue F2 becomes the good eigenvalue F3 = .09425, bumping G1 to become a spurious eigenvalue G2 = .09401. Thus, after two steps the spurious eigenvalue E1 = .09507 at M = 1082 has been transformed into the spurious eigenvalue G2 = .09401. Typically, in actual computations we would not perform such detailed computations. We would be comparing spurious eigenvalues from two different Lanczos matrices which varied more in size than those we considered here, and we could see a greater change in the locations of the spurious eigenvalues from one 'step' to the next in the Lanczos procedure. Something similar happens with the set of eigenvalues A1, B1, C1 and D1. This is a domino effect.

The spurious eigenvalue in column SPUR2 also persists but it also oscillates in terms of which of the two neighboring good eigenvalues it is trying to approximate. The spurious eigenvalues in columns SPURJ, J = 3,4,5 come and go. Table 4.6.2 provides a summary of the changes in the locations of the spurious eigenvalues in this small interval as we increase the size of the Lanczos matrix.

TABLE 4.6.1   POIS1008, Eigenvalues of T_M on $(.08834, .09784)$
for M = 1069 to 1090

M	GOOD1	SPUR1	GOOD2	SPUR2	GOOD3	SPUR3
1069	.8834	.8839	.9146	.9156	.9386	
					J1	I1
1070	.8834	.9144	.9146		.9386	.9396
				J2	I2	
1071	.8834	.9136	.9146	.9326	.9386	
				J3		
1072	.8834	.9124	.9146	.9271	.9386	
				J4		
1073	.8834	.9109	.9146	.9235	.9386	
				J5		
1074	.8834	.9077	.9146	.9204	.9386	
				J6		
1075	.8834	.8967	.9146	.9172	.9386	
				J7		
1076	.8834	.8846	.9146	.9157	.9386	
			K1	J8	D1	
1077	.8834		.9146	.9148	.9386	
		K2	J9	D2	C2	
1078	.8834	.9143	.9146	.9374	.9386	
		K3	J10	D3	C3	
1079	.8834	.9138	.9146	.9330	.9386	
		K4	J11	D4		
1080	.8834	.9129	.9146	.9283	.9386	
		K5	J12	D5		
1081	.8834	.9042	.9146	.9185	.9386	
		K6	J13	D6		
1082	.8834	.8838	.9146	.9153	.9386	
		K7	J14	D7		
1083	.8834	.8834	.9146	.9148	.9386	
	K8	J15	D8			G2
1084	.8834	.9146	.9146		.9386	.9401
		J16			H1	G3
1085	.8834	.9145	.9146		.9386	.9389
				H2	G4	
1086	.8834	.9143	.9146	.9377	.9386	
				H3		
1087	.8834	.9141	.9146	.9356	.9386	
1088	.8834	.9135	.9146	.9308	.9386	
1089	.8834	.8910	.9146	.9159	.9386	
1090	.8840	.9146	.9152		.9386	

TABLE 4.6.1   POIS1008, Eigenvalues of T_M on (.08834,.09784)
for M = 1069 to 1090

M	SPUR3	GOOD4	SPUR4	GOOD5	SPUR5	GOOD6
1069		.9425		.9440	.9632	.9784
1070	I1 .9396	.9425		.9440		.9784
1071		.9425		.9440		.9784
1072		.9425		.9440		.9784
1073		.9425		.9440		.9784
1074		.9425		.9440		.9784
1075		.9425		.9440		.9784
1076		.9425		B1 .9440	A1 .9605	.9784
1077		C1 .9425	B2 .9435	A2 .9440		.9784
1078		B3 .9425		.9440		.9784
1079		.9425		.9440		.9784
1080		.9425		.9440		.9784
1081		.9425		.9440		.9784
1082		.9425		F1 .9440	E1 .9507	.9784
1083		G1 .9425	F2 .9432	E2 .9440		.9784
1084	G2 .9401	F3 .9425		.9440		.9784
1085	G3 .9389	.9425		.9440		.9784
1086		.9425		.9440		.9784
1087		.9425		.9440		.9784
1088		.9425		.9440		.9784
1089		.9425		.9440	.9542	.9784
1090		.9425		.9440	.9473	.9784

TABLE 4.6.2 Locations of Spurious Eigenvalues

M	SPUR1	SPUR2	SPUR3	SPUR4	SPUR5
1069	x	x			x
1070	x		x		
1071	x	x			
1072	x		x		
1073	x	x			
1074	x		x		
1075	x	x			
1076	x	x			x
1077		x		x	
1078	x	x			
1079	x	x			
1080	x	x			
1081	x	x			
1082	x	x			x
1083	x	x		x	
1084	x		x		
1085	x		x		
1086	x	x			
1087	x	x			
1088	x	x			
1089	x	x			x
1090	x	x			x
1091	x	x			x
1092		x	x		
1093	x	x			
1094	x	x			

SECTION 4.7   LANCZOS PROCEDURE, EIGENVALUES

Our Lanczos eigenelement computations are carried out in two phases. First the desired eigenvalues are computed and then corresponding Ritz vectors are computed for some user-specified subset of the eigenvalues which were computed. In this section we describe our Lanczos procedure for computing eigenvalues of real symmetric matrices. In the next section we describe our companion procedure for computing eigenvectors of the computed eigenvalues. In both sections we present numerical results. We address the question of the rate of convergence of the overall procedure empirically.

It is essential to emphasize the importance of the matrix-vector multiply subroutine to the success of any Lanczos procedure. Lanczos procedures see the original A-matrix only through the results of its effect upon various Lanczos vectors v_i generated by the Lanczos recursions. What these procedures see must be consistent with our basic assumption that the original matrix is symmetric. The Lanczos procedures do not have any structural mechanism for forcing symmetry, as the EISPACK subroutines do. In addition, the speed of the matrix-vector multiply subroutine obviously affects the speed of the overall computations.

In order to determine the convergence properties of these procedures, families of test matrices were generated. Tests were also run on randomly-generated matrices. The effects of different starting vectors, of shifting the matrix, and of the definiteness or indefiniteness of the given matrix were all tested. Any Lanczos procedure requires that the starting vector have a nonzero projection on the invariant subspace associated with each eigenvalue of A that is to be computed. Single-vector Lanczos procedures cannot directly determine the multiplicities of the computed eigenvalues as eigenvalues of A. The multiplicity of a computed eigenvalue as an eigenvalue of a Lanczos matrix has no relationship to its multiplicity in the original A-matrix.

LANCZOS EIGENVALUE PROCEDURE FOR REAL SYMMETRIC MATRICES

Step 0.   Let A be an nxn real symmetric matrix. Set $v_0 \equiv 0$ and $\beta_1 \equiv 0$, and generate a random unit vector v_1 of length n.

Step 1.   For some suitable M generate the scalars α_j and β_{j+1} for $j \leq M$, using Eqns(4.3.1) - (4.3.2). Store the α_j, $1 \leq j \leq M$, the β_j, $2 \leq j \leq M + 1$, and the last two Lanczos vectors v_M and v_{M+1} off-line.

Step 2.   For some $m \leq M$, compute the eigenvalues of the Lanczos matrix T_m in the intervals of interest where convergence has not yet been established. Determine the numerical multiplicities of each of these eigenvalues. Accept numerically-multiple eigenvalues as converged approximations to eigenvalues of A. Test each simple computed eigenvalue of T_m to determine whether or not it is pathologically close to an eigenvalue of the corresponding $\hat{T}_2$ matrix. If so, reject that eigenvalue as spurious. Otherwise, accept it as an approximate eigenvalue of A.

Step 3.   Compute error estimates for the simple 'good' eigenvalues obtained in Step 2. If convergence is indicated, terminate. Otherwise increment m and if necessary enlarge T_M, and then go to Step 2.

In the BISEC program provided in Volume 2, Section 2.5, the multiplicity test and the test for rejection in Step 2 of the Eigenvalue Procedure are done simultaneously, and the extra cost incurred in doing this rejection test is negligible. See Section 4.5. In order to discuss the observed behavior of this procedure we need to introduce three definitions.

DEFINITION 4.7.1 Given a real symmetric matrix A with distinct eigenvalues $\lambda_{k_1} > \lambda_{k_2} > ... > \lambda_{k_q}$, define the minimal gap g_j for each eigenvalue λ_{k_j}, $1 \leq j \leq q$ as follows. Define $g_1 \equiv (\lambda_{k_1} - \lambda_{k_2})$ and $g_q \equiv (\lambda_{k_q} - \lambda_{k_{q-1}})$. Then for $j=2,...,q-1$ define

$$g_j \equiv \min [\lambda_{k_j} - \lambda_{k_{j+1}}, \lambda_{k_{j-1}} - \lambda_{k_j}]. \tag{4.7.1}$$

DEFINITION 4.7.2 Given the minimal gaps defined in Eqn(4.7.1), define the overall gap stiffness of the matrix A as

$$S_g(A) \equiv \max_j \ g_j \ / \ \min_j \ g_j. \qquad (4.7.2)$$

DEFINITION 4.7.3 The matrix stiffness of the matrix A is defined as

$$S(A) \equiv \max_j \ |\lambda_j| \ / \ \min_j \ |\lambda_j|. \qquad (4.7.3)$$

The numerical results indicate that the amount of computation required to compute the desired eigenvalues of a given matrix A depends directly upon the overall gap stiffness and how prevalent that stiffness is, upon the denseness of the eigenvalues of A in a neighborhood of the desired eigenvalue(s), and upon the relative locations of these eigenvalues in the spectrum of A. Typically, reasonably well-separated eigenvalues on the extremes of the spectrum of A will appear as eigenvalues of the Lanczos matrices for relatively small values of $m < n$. Well-separated interior eigenvalues can converge as fast or faster than clustered extreme eigenvalues. Tightly clustered eigenvalues in the middle of the spectrum would converge most slowly. The larger the overall gap stiffness $S_g(A)$ and the more persistent it is, the larger the value of m required to obtain all of the eigenvalues of A. See Examples 4.7.6 and 4.7.7 below. This procedure has no difficulty computing the individual members of any pair of eigenvalues which are very close together as long as the pair has a reasonable separation from the other eigenvalues.

Since there is no reorthogonalization, eigenvalues which have converged by a given size Lanczos matrix may begin to replicate as the size of the Lanczos matrix is increased further. The degree of replication which occurs depends upon the gap ratios, the larger the ratio, the more replication there will be. Replication does not affect the accuracy of any of the eigenvalue approximations. In practice it is usually not a problem. The storage gains achieved by not doing any reorthogonalization far outweigh the increases in the sizes of the Lanczos matrices needed to compensate for the replication.

COSTS OF LANCZOS RECURSION Each step of the basic single-vector Lanczos recursion requires two inner products of length n plus an additional 3n scalar multiplications/divisions and 2n additions/subtractions plus the cost of computing Av_i. The major storage requirements for the eigenvalue procedure are 2 vectors of length m for permanent storage of the Lanczos matrices, α_i and β_{i+1}, as they are generated and 2 vectors of length n for temporary storage of the current Lanczos vectors v_i and v_{i-1}. In the programs provided in Volume 2, at the end of the T_M generation, M and the vectors α, β, v_M, and v_{M+1} are stored off-line so that if the computations are rerun at a $m > M$, only the incremental scalars α_i, β_{i+1} have to be generated.

The overall costs of each step of the Lanczos matrix generation are summarized in Table 4.7.1. The primary cost is the computation of the Av_i. The original matrix A appears only in this computation. The user-supplied subroutine which computes Ax must be optimized to take advantage of any sparsity and/or structure in A. We emphasize that there is no requirement that A be sparse. The procedure only needs an Ax computation which is both accurate and economical.

TABLE 4.7.1   Cost of T_M Generation [a]

Computation	Multiplies/Divides	Adds/Subtracts
Av_i	pn	(p-1)n
$w_i = Av_i - \beta_i v_{i-1}$	n	n
$\alpha_i = w_i^T v_i$	n	n
$z_i = w_i - \alpha_i v_i$	n	n
$\beta_{i+1} = \| z_i \|$	n	n
$v_{i+1} = z_i / \beta_{i+1}$	n	
Cost/iteration	(p+5)n	(p+3)n

[a] p = average number of nonzero entries in each row of A. For some matrices the cost of computing Av_i may be less than (2p-1)n.

LOCAL ORTHOGONALITY   The plausibility arguments given in Section 4.4 for convergence of the Lanczos eigenvalue procedure require the nearest-neighbor, near-orthogonality of the Lanczos vectors. As Paige proved, see Eqn(2.3.14) in Theorem 2.3.1 in Chapter 2, this local orthogonality is preserved as long as the off-diagonal entries β_{i+1} are not too small. Numerical experiments indicate that it is quite reasonable to assume that until all of the distinct eigenvalues of the original matrix have been approximated well by eigenvalues of the Lanczos matrices, that all of the off-diagonal entries are uniformly bounded reasonably far away from zero. Some sample values taken from test runs are given in Table 4.7.2. These test problems can be found in Cullum and Willoughby [1979,1979b,1980,1981]. The minimal ratio $| \beta_{i+1} | / \| A \|$ is listed in the column marked Ratio.

The user must specify a value for M. Selecting a value larger than that which is required to obtain the desired eigenvalues accurately results in more computation than is necessary. However, as Lemma 2.3.8 in Chapter 2 states, it does not affect the accuracy achievable. The Lanczos Phenomenon asserts that for large enough m the eigenvalues which we want to compute will appear as eigenvalues of T_m. Paige's result, Lemma 2.3.5, suggests that any such accurate approximations to eigenvalues of A will be stabilized eigenvalues of such a Lanczos matrix. See Theorem 4.5.1, Eqn(4.5.2). In particular, they will also be eigenvalues of any larger Lanczos matrix. In the programs provided in Volume 2 the user specifies a priori the matrix sizes to be considered. It is possible to design a more intelligent program where the program uses the computed error estimates to decide whether or not a larger Lanczos matrix is required and to identify which subintervals should be included in the larger eigenvalue computations. However, without knowing a priori what the user wants to compute, it is not easy to design a general purpose program to do this.

TABLE 4.7.2   Minimal

$|\beta_{i+1}|/\|A\|$

Test Matrix	Order A	Order T_m	Minimum $|\beta_{i+1}|$	$\|A\|$	Ratio
POIS992	992	5952	.55	4.0	.14
POIS1008	1008	1764	.33	4.0	.08
KIRK250	250	2000	.85	102.4	.0083
KIRK567	567	1134	1.11	4.7	.24
KIRK992	992	2976	1.57	5.5	.29
KIRK1089	1089	4904	1.71	102.4	.017
KIRK1600	1600	4800	1.89	102.4	.018
KIRK4900	4900	29404	2.29	102.4	.022

ERROR ESTIMATES   The choice of error estimates for the computed 'good' eigenvalues was based upon Lemma 2.3.5. Numerically-multiple eigenvalues are accepted as converged approximations to eigenvalues of A. Moreover, in the implementation we label any 'good' eigenvalue which has a spurious eigenvalue sitting very close to it as converged. Error estimates are computed for the other simple, 'good' eigenvalues. For reasonable size m and $\|A\|$, we have by Lemma 2.3.5 that

$$\varepsilon_\mu \equiv |\beta_{m+1}u(m)| \qquad (4.7.4)$$

is the key to estimating the convergence of the computed 'good' eigenvalues. In practice this bound is a good but conservative reflection of the error in such eigenvalues. We use inverse iteration to compute the last components u(m) of the corresponding Lanczos matrix eigenvector required for these estimates. See Section 3.4. Each eigenvector computation for a mxm Lanczos matrix requires approximately 12m multiply-add operations. Subroutine INVERR in Volume 2, Chapter 2 was used for the inverse iteration computations. It is a slightly modified version of the EISPACK subroutine TINVIT. See Chapter 2 in Volume 2 for details.

In Table 4.7.3, for the Poisson test matrix of order n=1008 which was used in Section 4.6, we give a comparison of the computed error estimates and the true errors for computed 'good' eigenvalues. The tests we have run indicate that within any given subinterval of the spectrum, convergence occurs according to the sizes of the minimal gaps. See Definiton 4.7.1. Therefore, we do not need to compute error estimates on every computed eigenvalue. It is sufficient to compute estimates for some representative selection of the computed good eigenvalues with small gaps. The programs in Volume 2 however do compute error estimates for each isolated 'good' eigenvalue.

TABLE 4.7.3   POIS1008, Error Estimates vs. True Errors

Size T_m	'Good' Eigenvalue	Error Estimate	True Error
m=1008=n	1.6606618191	9.4×10^{-3}	1.1×10^{-4}
	1.6614072619	1.6×10^{-2}	6.4×10^{-4}
	1.6688217735	1.5×10^{-3}	2.6×10^{-6}
	1.6719607187	1.0×10^{-3}	2.7×10^{-5}
	1.6741447482	8.6×10^{-4}	1.7×10^{-6}
	1.6923622232	2.2×10^{-4}	5.2×10^{-8}
	1.6951062486	1.7×10^{-4}	2.9×10^{-8}
m=1126=1.12n	0.2690507887	8.6×10^{-13}	3.3×10^{-13}
	0.2824386303	1.0×10^{-8}	6.2×10^{-13}
	0.3171884350	1.7×10^{-5}	2.5×10^{-11}
	0.3190091584	6.7×10^{-5}	5.8×10^{-10}
	0.3228590426	5.0×10^{-4}	5.2×10^{-8}
	0.3576650427	2.2×10^{-3}	6.0×10^{-6}
	0.3628420891	3.6×10^{-3}	8.9×10^{-5}
	0.3633926493	7.8×10^{-3}	1.2×10^{-4}
	0.3640244896	2.5×10^{-2}	5.1×10^{-4}
	0.3688971202	9.7×10^{-5}	1.4×10^{-8}
	0.4839359213	5.6×10^{-2}	3.1×10^{-4}
	0.4842754360	9.2×10^{-4}	1.8×10^{-7}
	0.8475498551	7.9×10^{-4}	6.6×10^{-8}
	1.6719874543	5.9×10^{-4}	5.3×10^{-7}
	1.6741464513	1.8×10^{-5}	4.9×10^{-10}
	1.6923622747	4.3×10^{-7}	6.6×10^{-14}
	1.6951062773	3.0×10^{-7}	2.1×10^{-13}
m=1512=1.5n	1.6607681953	4.0×10^{-8}	9.8×10^{-13}
	1.6620674206	4.2×10^{-12}	2.4×10^{-13}
	1.6688243645	1.7×10^{-13}	2.1×10^{-13}
	1.6718896888	1.0×10^{-12}	2.3×10^{-13}
	1.6719879819	6.3×10^{-13}	1.8×10^{-13}

Convergence is said to have occurred when the error estimates on all of the isolated good eigenvalues in the desired subintervals of the spectrum are small enough. In practice, it is typically true that the actual error is a factor of 100 or more smaller than the error indicated by the error estimate, whenever the error estimate indicates convergence of a given eigenvalue to 8 or less digits. In Table 4.7.4 we illustrate how the error estimates can be used to tell the program what subintervals require a larger Lanczos matrix. The test matrix is a diagonally-disordered matrix of order 323. The estimates in Table 4.7.4 indicate that for the test matrix being considered, at m=486 the eigenvalues in the interval [99.79,100.23] have not yet converged. If the user wishes to compute these eigenvalues, then a larger Lanczos matrix must be used.

TABLE 4.7.4   KIRK323, Using Error Estimates to Identify Intervals Which Have Not Yet Converged, m = 486 = 3n/2

'Good' Eigenvalue	Estimate	'Good' Eigenvalue	Estimate
99.6934590	4.6×10^{-11}	99.7412735	3.4×10^{-12}
99.7864765	4.8×10^{-12}	100.004290	5.4×10^{-3}
100.006456	.014	100.008948	.023
100.015780	.029	100.020244	.037
100.026611	.058	100.030995	.083
100.038406	.050	100.041630	.037
100.059892	4.7×10^{-3}	100.064478	2.2×10^{-3}
100.066359	4.0×10^{-3}	100.070149	1.5×10^{-4}
100.230183	1.8×10^{-11}	100.279737	9.5×10^{-11}

NUMERICAL RESULTS   Tables 4.7.5 - 4.7.10 contain some of the results of our tests of the Lanczos eigenvalue procedure. The Poisson test matrices were described in Section 4.6. The other classes of matrices tested included matrices related to disordered systems, Kirkpatrick and Eggarter [1972], diagonal matrices and some smaller order matrices which have appeared in the literature, van Kats and van der Vorst [1976,1977]. Summarizing the results of such tests is difficult. First we present 3 specific examples to illustrate the overall behavior of the algorithm. Example 4.7.1 is a Poisson matrix of order 992 corresponding to the Laplace operator on a square with Neumann boundary conditions. We denote this matrix by POIS992. Example 4.7.2 is a disordered matrix of order 992 which we denote by KIRK992. Example 4.7.3 is a Poisson matrix of order 1008 corresponding to the Laplace operator on a square with Dirichlet boundary conditions. We denote this matrix by POIS1008. The eigenelements of POIS1008 were known a priori. However, neither the eigenvalues nor the eigenvectors of POIS992 and KIRK992 were known a priori. First we describe the disordered matrices considered.

DIAGONALLY-DISORDERED MATRICES.   Diagonally-disordered matrices arise in the study of two-dimensional $NX \times NY$ arrays of atoms in disordered systems. See for example, Kirkpatrick and Eggarter [1972]. The associated matrix A has order $N = NX \times NY$. If NX and NY are relatively prime, then all of the eigenvalues of A are distinct. A is almost block tridiagonal. Each B block is $NX \times NX$, and there are NY of these blocks down the diagonal of A.

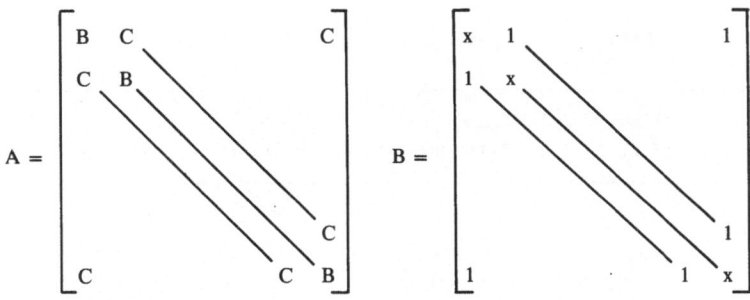

C is the unit matrix of order NX. The diagonal entries of A are randomly generated numbers. A scaling parameter bounds the magnitudes of these disorder terms. In the simplest case these entries are chosen from a uniform random distribution over an interval [− SCALE, SCALE] determined by the user-specified scaling parameter SCALE. KIRK992 in Table 4.7.5 is of this type. A second class of diagonally-disordered matrices is obtained by choosing the diagonal entries as either 0 or SCALE, as determined from the values f(j) obtained from a uniform random distribution on the interval [0,1]. If for some j, f(j)>CONC, then the jth diagonal entry is set to SCALE. Otherwise, it equals 0. The concentration CONC is specified by the user. In both classes of disordered matrices all of the nonzero, off-diagonal entries have value 1 and their pattern is that of the discretization of the Laplace operator on a square using only the nearest (x,y) neighbors and with doubly periodic boundary conditions.

POIS992 is defined by KX=31, KY=32, and c = .484131. POIS1008 was defined in Section 4.6. KIRK992 is defined by NX= 31, NY=32 with the scaling parameter 6 so that the diagonal entries are uniformly distributed over the interval [− 6, 6]. The spectrums of POIS992 and of POIS1008 fall in the interval [0,2]. The spectrum of KIRK992 is in the interval [− 5.4, 5.4]. The $|\beta_i|$ for all i considered were larger than .55, .33 and 1.57 respectively, for POIS992, POIS1008 and KIRK992. See Table 4.7.2.

RATE OF CONVERGENCE Table 4.7.5 summarizes the observed convergence of the eigenvalues of these three examples as the size of the Lanczos matrices is increased. Results are given for a large number of values of m to illustrate the overall convergence in detail. For POIS992, the eigenvalue computations on the larger Lanczos matrices were only done on the subintervals in the center of the spectrum where convergence had not yet occurred. For each value of m considered and each test matrix we have listed the number of the eigenvalues of A being approximated, the number of these eigenvalues which are accurate to at least 10 significant digits and the number which are accurate to at least 6 significant digits. For POIS992 and KIRK992 these numbers are estimates obtained using the error estimates. Thus for these two matrices, the numbers in the table are conservative lower bounds on the true values. Typically, the computed eigenvalues have several digits more accuracy than indicated by the error estimates whenever these estimates are 'large', $\geq 10^{-6}$. For POIS1008 the numbers of eigenvalues were computed by comparisons of the computed 'good' eigenvalues with the true eigenvalues.

TABLE 4.7.5   Convergence Patterns, 3 Test Matrices

Matrix	Order of T_m	Eigenvalues Approximated	Accuracy ≥ 10 Digits	Accuracy ≥ 6 Digits
POIS992				
	1984(2n)	945	667	774
	2480(2.5n)	974	830	919
	2976(3n)	981	896	956
	3472(3.5n)	986	917	958
	4000(4n)	988	943	958
	4960(5n)	992	949	992
	5952(6n)	992	992	
KIRK992				
	496(.5n)	432	67	88
	922(n)	728	215	256
	1488(1.5n)	921	467	500
	1984(2n)	992	957	992*
	2480(2.5n)	992	992	
POIS1008				
	504(.5n)	478	62	70*
	756(.75n)	681	113	123*
	1008(n)	867	171	307
	1260(1.25n)	989	852	878*
	1512(1.5n)	1006	991	997*
	1764(1.75n)	1008	1008	

The most rapid convergence was exhibited by POIS1008. By only m = 1.75n = 1764, all of the eigenvalues of POIS1008 had been computed to at least 10 significant digits. The diagonally-disordered matrix KIRK992 also exhibited rapid convergence. By m=2n, the error estimates indicated that all of the eigenvalues were good to at least 8 digits. (They were in fact good to at least 10 significant digits.) Similar convergence has been observed on other simple diagonally-disordered test matrices of order 567, 927 and 1505, respectively.

POIS992 exhibited the slowest terminal convergence. Observe however that by m=2.5n, 974 eigenvalues of POIS992 were being approximated and the error estimates indicated that 919 of these were good to at least 6 digits. Only a few eigenvalues of POIS992 were not being approximated by the time m = 2.5n. It is not until somewhere between m = 4n and m = 5n that every eigenvalue of POIS992 is represented in the spectrum of the corresponding Lanczos matrix. The few missing eigenvalues were in clusters directly in the center of the spectrum of POIS992. The asterisk * on some of the entries in Table 4.7.5 indicates that these numbers of eigenvalues were accurate to at least 8 digits or more.

The difference in the terminal convergence between POIS992 and KIRK992 can be explained by the differences in the gap structure in the middle of the spectrums of these two matrices. Only 6 interior eigenvalues of KIRK992 have gaps less than $1/10$ of the average gap. POIS992 has 104 interior eigenvalues with gaps less than this. In the middle of the spectrum of KIRK992 we see there are no eigenvalues with gaps less than average gap/10. For POIS992 however, there are long chains of eigenvalues with small gaps on either side of the middle of the spectrum. The difficulties with the convergence occur in these two chains. Every other eigenvalue of POIS992 is present at $m = 2.5n$. The last ones to converge are in clusters in these chains. In practice, when the Lanczos matrix is enlarged to obtain eigenvalues which have not yet converged at smaller size Lanczos matrices; we could compute the eigenvalues of the larger Lanczos matrix only in those small subintervals of the spectrum where convergence has not yet been obtained.

Tests were also run on 2 large diagonally-disordered matrices where the diagonal entries were either 0 or SCALE. All of the diagonally-disordered test matrices were obtained from

TABLE 4.7.6
KIRK1600, Convergence of Representative Eigenvalues
from the Desired Intervals $(.1,.9)$ and $(1.1,1.9)$.

No. T_m	'Good' Eigenvalue	Amingap	Error Estimates $m=2n$	Error Estimates $m=3n$
1	.108538583112	7.88×10^{-3}	6.2×10^{-5}	1.7×10^{-8}
2	.116416158091	2.16×10^{-3}	8.4×10^{-4}	1.5×10^{-11}
4	.125695041582	2.89×10^{-4}	5.7×10^{-3}	4.7×10^{-9}
5	.125984202306	2.89×10^{-4}	8.4×10^{-2}	2.9×10^{-9}
6	.131418112867	1.35×10^{-3}	2.7×10^{-2}	2.1×10^{-10}
8	.140417847900	4.55×10^{-3}	1.2×10^{-1}	1.3×10^{-11}
10	.150454046930	1.83×10^{-3}	4.4×10^{-2}	1.9×10^{-11}
12	.159800239400	5.49×10^{-3}	6.4×10^{-2}	6.8×10^{-11}
16	.177438347611	4.22×10^{-3}	1.3×10^{-2}	1.7×10^{-11}
19	.190218830700	4.28×10^{-4}	1.2×10^{-2}	3.6×10^{-11}
20	.190646629966	4.28×10^{-4}	1.4×10^{-1}	1.6×10^{-10}
195	1.23316155483	3.46×10^{-3}	1.4×10^{-4}	7.3×10^{-12}
196	1.24376265893	9.50×10^{-3}	2.0×10^{-6}	1.2×10^{-10}
197	1.25326700799	8.10×10^{-3}	1.3×10^{-6}	4.9×10^{-11}
208	1.31796912319	2.01×10^{-3}	2.3×10^{-9}	1.3×10^{-11}
209	1.31997989839	2.01×10^{-3}	7.2×10^{-10}	1.5×10^{-10}
210	1.33085004976	1.09×10^{-2}	3.6×10^{-11}	7.5×10^{-11}
211	1.35121009484	6.86×10^{-3}	4.2×10^{-11}	1.2×10^{-8}
212	1.35807260268	5.11×10^{-3}	8.4×10^{-11}	5.2×10^{-10}
224	1.42478180911	1.84×10^{-3}	4.8×10^{-8}	3.1×10^{-11}
228	1.44227149920	2.85×10^{-3}	1.1×10^{-6}	2.6×10^{-10}
229	1.44737404957	3.81×10^{-3}	4.3×10^{-8}	4.0×10^{-11}

Kirkpatrick [1979]. One of these matrices, Example 4.7.4, was of order $n=1600$, and we denote it by KIRK1600. The other one, Example 4.7.5 was of order $n = 4900$, and we denote it by KIRK4900. Some of the results obtained for these matrices are included in Tables 4.7.6

and Table 4.7.7. For KIRK1600, NX = NY = 40, CONC = .7, and SCALE = 100. For KIRK4900, NX = NY = 70, CONC = .7, and SCALE = 100.

TABLE 4.7.7
KIRK4900, m=3n, Convergence of Representative Eigenvalues on (.1,.9) [a]

No. T_m	'Good' Eigenvalue at m=3n	Computed Amingap m=4n	Error Estimates at m=3n	$\dfrac{\text{Estimate}}{\text{Amingap}}$
1	.100184031023	9.286×10^{-4}	1.3×10^{-8}	1.4×10^{-5}
2	.101112591037	9.286×10^{-4}	9.2×10^{-12}	9.9×10^{-9}
3	.102424974865	1.312×10^{-3}	1.2×10^{-10}	9.2×10^{-8}
4	.105801742503	5.928×10^{-4}	8.8×10^{-11}	1.5×10^{-7}
5	.106394577433	5.928×10^{-4}	3.6×10^{-11}	6.1×10^{-8}
6	.111238683033	4.844×10^{-3}	2.7×10^{-8}	5.6×10^{-6}
7	.117388719646	5.052×10^{-3}	1.6×10^{-8}	3.2×10^{-6}
18	.138395951209	3.499×10^{-4}	6.1×10^{-10}	1.7×10^{-6}
19	.138745818377	3.499×10^{-4}	2.2×10^{-9}	6.3×10^{-6}
20	.139454780546	6.850×10^{-4}	2.6×10^{-9}	3.8×10^{-6}
21	.140139782448	6.850×10^{-4}	5.7×10^{-10}	8.3×10^{-7}
180	.401048914269	1.774×10^{-3}	2.8×10^{-9}	1.6×10^{-6}
219	.462776965800	1.499×10^{-4}	1.4×10^{-6}	9.3×10^{-3}
220	.464964879878	6.365×10^{-4}	7.7×10^{-8}	1.2×10^{-4}
221	.465601349313	6.365×10^{-4}	3.0×10^{-7}	4.7×10^{-4}
222	.466617916442	1.017×10^{-3}	7.0×10^{-9}	6.9×10^{-6}
223	.469426316630	7.622×10^{-5}	7.7×10^{-7}	1.0×10^{-2}
224	.469502542959	7.623×10^{-5}	3.1×10^{-7}	4.1×10^{-3}
225	.470921967811	1.161×10^{-3}	2.5×10^{-8}	1.6×10^{-5}
226	.472082454128	1.161×10^{-3}	1.8×10^{-7}	1.6×10^{-4}
227	.474707818128	1.354×10^{-3}	1.4×10^{-10}	1.0×10^{-7}
228	.476061429528	1.168×10^{-3}	2.2×10^{-10}	1.9×10^{-7}

[a] All eigenvalues computed at m=3n differ from the corresponding eigenvalues computed at m=4n in at most the 10th digit.

KIRK1600 was analyzed in detail. We computed all of the eigenvalues of T_{4800}. The eigenvalues of KIRK1600 range from -3.511 to 102.37. There are no eigenvalues in the interval (3.499, 97.67). At m = 3n = 4800, 1468 eigenvalues were being approximated. Convergence to at least 9 digits was observed on all of the eigenvalues except those in 6 small subintervals. One of these subintervals contained 0 and convergence was not indicated on any of the eigenvalues in this particular subinterval. However, in the other 5 subintervals, all of which were contained in the interval [99,101.03], many of the eigenvalues were computed to 6 digits or more. These subintervals were identified by examining the error estimates as was done for example in Table 4.7.4.

For KIRK1600 Table 4.7.6 contains representative computed 'good' eigenvalues from two subintervals of interest [.1, .9] and [1.1, 1.9], together with the corresponding computed minimal A-gaps which were computed at m=3n using the computed 'good' eigenvalues, and the error estimates computed at m=2n=3200 and m=3n=4800. All of these eigenvalues were computed accurately by m=3n. The eigenvalues in the interval [1.1,1.9] converged slightly more quickly than those in the interval [.1,.9]. This is to be expected since the gap structure in these two subintervals is very similiar, but the subinterval [1.1,1.9] is closer to a very large gap in the spectrum. The storage requirements for the KIRK1600 eigenvalue computations at m=4800 were approximately 260K bytes. This can be compared with the EISPACK requirements of approximately 4M bytes of storage. It is interesting to note in Table 4.7.6 that for the eigenvalues μ_{211} and μ_{212} that the error estimates at m = 3n are worse than those computed at m = 2n. This is caused by the presence of nearby spurious eigenvalues.

For KIRK4900, we computed the eigenvalues on these same two subintervals. This computation was repeated at m=3n and at m=4n. The latter computation was just to verify (for unbelievers) that all of the eigenvalues on these two intervals had converged. All of the good eigenvalues computed at m=3n agreed to at least 10 digits with the good eigenvalues computed at m=4n. Some of the results of this computation for the harder subinterval [.1,.9] are given in Table 4.7.7. We again list representative good eigenvalues, their computed minimal A-gaps, and the corresponding error estimates obtained using inverse iteration on T_m at m = 3n = 14,700. The storage requirements at m = 3n = 1 4,700 were approximately 750K bytes compared to the approximately 10M bytes which would be needed for the corresponding EISPACK program. The results for KIRK4900 demonstrate conclusively that the proposed procedure can be used on very large matrices.

We can illustrate very vividly the effect of the gap distribution upon the rate of convergence by considering two simple examples. Example 4.7.6 has a gap distribution such that as we move through the spectrum of A from the smallest to the largest eigenvalue, the gaps increase similarly. That is the smallest (largest) eigenvalues have the smallest (largest) gaps, and the gaps increase monotonically from the lower end of the spectrum to the upper end. Table 4.7.8 illustrates the convergence obtained for Example 4.7.6. The upper end of the spectrum is obtained readily. The lower end, even though these are extreme eigenvalues is obtained only after a great amount of work. The gap stiffness for this example is 100 million.

Example 4.7.7 given in Table 4.7.9 has a gap distribution which is ideal for a Lanczos procedure if the user wishes to compute many or even all of the eigenvalues of the given matrix. For this example all of the minimal eigenvalue gaps g_j are in the range

$$g_{av}/10 \leq g_j \leq 50g_{av} \quad \text{where } g_{av} \equiv \sum_{j=1}^{q} g_j/q. \qquad (4.7.5)$$

By only m = n all of the eigenvalues of this matrix have been computed to at least 10-digit accuracy.

In practice, the gap structure of the particular matrix being considered lies somewhere between the extremes of these last two examples. In general most of the eigenvalues can be computed readily, typically with $m < 3n$, for any matrix which does not have extreme differences in its gap structure.

With our Lanczos eigenvalue procedure it is suggested that if the user has no a priori knowledge of the eigenvalue distribution in the given matrix A, then the eigenvalues of the Lanczos matrix T_{2n} in the user-specified subintervals, be computed along with error estimates of the accuracy of the computed 'good' eigenvalues. An examination of the 'good' eigenvalues together with their error estimates will yield a reasonable picture of the degree of difficulty that exists in computing the remaining desired eigenvalues. Typically, at such a value of m reasonably well-separated eigenvalues will have converged and 'clusters' of eigenvalues can be identified.

EXAMPLE 4.7.6 Let $A_1(i,j) \equiv$ min (i,j), $1 \leq i \leq j \leq n$. We note that $A_1 = L_1^{-1} L_1^{-T}$ where L_1 is bidiagonal with $L_1(i,i) = 1$ and $L_1(i,i-1) = -1$, $1 \leq i \leq n$. We consider n = 150.

TABLE 4.7.8   Convergence, Example 4.7.6

Order T_m	Number of Eigenvalues Approximated	Eigenvalues Accurate to $\geq$ 10 Digits	Eigenvalues Accurate to $\geq$ 5 Digits
50	30	14	15
300	79	43	48
750	119	70	75
1200	141	98	104
1500	150	119	150

At n=150, A_1 has eigenvalues that increase in magnitude from $\lambda_1 = .25$ to $\lambda_{150} = 9180$. More importantly the gaps between successive eigenvalues increase monotonically from 8×10^{-5} at λ_1 to 8000 at λ_{150}. Thus, this matrix has a gap stiffness of 100 million. As m is increased convergence of the eigenvalues of the Lanczos matrices occurs monotonically from the largest eigenvalue $\lambda_1 = 9180$ down to the smallest eigenvalue $\lambda_{150} = .2500$. In this example the smallest eigenvalues, even though they are at the lower extreme of the spectrum, converge last. Convergence to at least 5 digit accuracy on all of the 150 eigenvalues occurred by $m = 1500 = 10n$, (m was blindly incremented by > 300).

EXAMPLE 4.7.7 $A_2 \equiv A_1^{-1}$, n = 150 and n = 300.

For n = 150, A_2 has eigenvalues that vary in magnitude from $\lambda_{150} = 1.089 \times 10^{-4}$ to $\lambda_1 = 3.99956$. The corresponding minimal gaps increase monotonically from 8.7×10^{-4} at λ_1

to a maximum of 4.17×10^{-2} at λ_{75} and then decrease monotonically to 1.307×10^{-3} at λ_{150}. This matrix has a gap stiffness less than 50. We have the following very interesting results.

TABLE 4.7.9   Convergence, Example 4.7.7

Order A_2 n	Order T_m	Number of Eigenvalues Approximated	Eigenvalues Accurate [a] to ≥ 10 Digits
150	50	50	0
	90	90	0
	150	150	150
300	100	100	0
	200	200	0
	300	300	300

[a] These counts are based upon error estimates computed using Eqn(4.7.1).

Table 4.7.9 indicates that all of the eigenvalues of Example 7.4.5 converge more or less simultaneously. The extreme eigenvalues of A_2 converge at approximately the same rate as the interior eigenvalues. By $m = n$ all of the eigenvalues of A_2 have been approximated to at least 10 digits. In this example the effect of location of a given eigenvalue in the spectrum is balanced by the gap structure. The minimal gaps increase monotonically as we move from the lower extreme of the spectrum to the center. These gaps decrease monotonically as we continue to move on out of the spectrum at the upper extreme. With all of the eigenvalues being approximated at essentially the same rate, there is not any significant loss in the global orthogonality of the Lanczos vectors. In this situation, if we want to compute many or even all of the eigenvalues of A, then the Lanczos procedure can do this easily. If however, only a few extreme eigenvalues are required then we see that the cost of this computation for this particular example is as much as computing all of the eigenvalues of this example.

These tests and others indicate that the matrix gap stiffness, as defined in Eqn(4.7.2), is the primary key to understanding the rate of convergence of the Lanczos eigenvalue procedure when there is no reorthogonalization of the Lanczos vectors. Tests verify that to within the limits of the accuracy of the eigenvalue subroutine used, the matrix stiffness (as defined in Eqn(4.7.4)) is not a primary factor in the rate of convergence of the Lanczos procedure. This is not totally unexpected since (at least theoretically) the matrices A and $A + \tau I$ generate the same Lanczos vectors and thus have the same convergence rates, but for an appropriate choice of τ the matrix stiffness of $A + \tau I$ will be small even if A is very stiff.

These last two examples illustrate the possible tradeoff between using a given matrix A and in using its inverse. There is an increased cost incurred for replacing the Ax computations by the repeated solution of the equation $Ax = v_i$. However, if the equation $Ax = b$ can be solved cheaply; and if in fact the gap distribution in A^{-1} is radically superior to that in A, then it is possible that the overall cost of the Lanczos computation can be reduced significantly by using A^{-1} instead of A.

TABLE 4.7.10

Observed Convergence on Randomly-Generated Matrices, n=100.
Tests 1-5, Uniform Distribution, Tests 6-9, Gaussian Distribution

Test No.	Order T_m	Number Good Eigenvalues	Error Estimates $<10^{-8}$	Number Multiple Eigenvalues
1	50	50	5	0
RANDOM1	100	87	31	5
Nnzero=5	150	100	87	23
Rseed=35613297	200	100	100	36
3	50	50	3	0
RANDOM3	100	89	24	5
Nnzero=11	150	100	100	19
Rseed=215372912				
4	50	50	1	0
RANDOM4	100	90	25	4
Nnzero=35	150	100	100	18
Rseed=93262178				
6	50	50	7	0
RANDOMG1	100	84	29	7
Nnzero=5	150	100	77	23
Rseed=35613297	200	100	100	32
8	50	50	1	0
RANDOMG3	100	90	26	4
Nnzero=11	150	100	100	18
Rseed=512743219				

To eliminate the possibility that the observed convergence of our procedure was due to some hidden structural properties of our test matrices, we ran some tests on matrices whose sparsity patterns and entries were determined randomly. Tests 1 through 5 used a uniform random distribution. Tests 6 through 9 used a Gaussian random distribution. The results of several of these tests are given in Table 4.7.10. In this table Nnzero is the average number of nonzero entries per row and column in the test matrix, and Rseed is the seed used by the random number generator.

For a given matrix A and a corresponding Lanczos matrix T_m, the question of whether or not a given subset of the eigenvalues of A can be approximated by eigenvalues of T_m, depends primarily upon the gap structure in the matrix A and upon the relative locations of these eigenvalues in the spectrum of A. Tests verify that the effects of the choice of the Lanczos starting vector v_1, of scaling the matrix (replacing A by αA) or of shifting the matrix, (replacing A by A + σI) are all secondary. Different implementations of the matrix-vector multiply subroutine lead to very different α , β sequences (Lanczos matrices), but on the tests which we ran there were no significant differences in the actual convergence of the eigenvalues of the Lanczos matrices as the sizes of these matrices was increased. In exact arithmetic we would expect no difference in convergence, but this insensitivity also seems to be valid in practical computations. Without this insensitivity, the Lanczos procedure would not be very attractive.

SECTION 4.8   LANCZOS PROCEDURE, EIGENVECTORS

The Lanczos eigenelement computations are carried out in two phases. In the previous section we discussed the eigenvalue computations. In this section we describe the complementary eigenvector computations. For each converged 'good' eigenvalue of the Lanczos matrices we can compute a corresponding approximate eigenvector of the original matrix.

LANCZOS EIGENVECTOR PROCEDURE FOR CONVERGED EIGENVALUES

Step 0.   The matrices T_m, m $\leq$ M, generated in the Lanczos eigenvalue computations are read from storage. The program decides whether or not to enlarge this history and enlarges it if necessary.

Step 1.   For each eigenvalue μ being considered, the first value $m1(\mu)$ of m for which μ appears as an eigenvalue of T_m to within a specified accuracy $\varepsilon(\mu)$ is determined using Sturm sequencing. Also if possible the first value $m2(\mu)$ of m for which μ appears as a double eigenvalue of T_m to within $\varepsilon(\mu)$ is also determined.

Step 2.   Using the information obtained in Step 1 for each μ, an initial guess is made at an appropriate size $ma(\mu)$ for the Lanczos matrix to be used in the corresponding Ritz vector computation. The corresponding unit eigenvector of that size Lanczos matrix is computed and the size of its last component is used to estimate the accuracy of that Ritz vector. If this error estimate is small enough, then the procedure returns to the beginning of Step 2 with the next eigenvalue in the list of eigenvalues being considered. (If this list has been exhausted, then the procedure goes to Step 3.) Otherwise, the size of the Lanczos matrix is increased and the eigenvector computation is repeated on the larger Lanczos matrix. A maximum of 10 such Lanczos matrix eigenvector computations are allowed per eigenvalue considered. Typically, only 1 or 2 such computations are required. The associated expensive Ritz vector computations are not performed until Step 3.

Step 3.   For each eigenvalue μ being considered and the corresponding Lanczos matrix eigenvector u computed in Step 2, the corresponding normalized Ritz vector is calculated.

$$y \equiv Vu/\|Vu\|, \qquad (4.8.1)$$

where $V \equiv \{v_1, v_2, \ldots v_m\}$ and $m = ma(\mu)$ is the size determined for μ in Step 2. A single pass through the Lanczos vectors is all that is required to generate a whole set of Ritz vectors. The number of Ritz vectors which can be computed on such a pass depends upon the amount of storage available. The number of Lanczos vectors actually used for a particular Ritz vector computation varies with the particular eigenvalue μ being considered.

Step 4. Compute the corresponding error estimates for each Ritz vector $\|Ay-\mu y\|$ and terminate.

The FORTRAN code for this procedure is contained in Chapter 2 of Volume 2. In order for this procedure to work properly each eigenvalue for which a Ritz vector is to be computed must have converged as an eigenvalue of the Lanczos matrices, in the sense that the last component of the corresponding eigenvector of the Lanczos matrix is pathologically small. As in the eigenvalue computations, the amount of computation required to compute the desired Ritz vectors depends directly upon the gap stiffness, the denseness of the desired eigenvalues, and the relative locations of these eigenvalues in the spectrum of A. This cost can be estimated a priori without any eigenvector computations for the Lanczos matrices or Ritz vector computations for the original matrix. This estimate can be obtained using Sturm sequencing, and it is provided in the FORTRAN code in Volume 2, Chapter 2.

We know that if μ is an isolated eigenvalue of a Lanczos matrix T_m such that $\mu = \lambda_j(A) + \varepsilon$ for some j and small ε, then

$$\|Ay - \mu y\| \; / \; \min_{k \neq j} |\lambda_k(A) - \lambda_j(A)| \qquad (4.8.2)$$

is a good estimate of the error in the associated Ritz vector $y \equiv Vu/\|Vu\|$ where u is a unit eigenvector of T_m corresponding to μ. In practice the λ_i in the denominator of Eqn(4.8.2) are replaced by the corresponding computed 'good' eigenvalues $\mu_{k(i)}$. See Theorem 1.3.2 in Chapter 1. We have the following lemma which states that Ritz vectors corresponding to different eigenvalues of A are nearly orthonormal. These Ritz vectors are actually computed using different size Lanczos matrices.

LEMMA 4.8.1 Let μ_j, $j = 1,\ldots,q$ be eigenvalues of corresponding Lanczos matrices $T_{m(j)}$. Assume that the corresponding unit Ritz vectors $y_j \equiv V_{m(j)}u_j/\|V_{m(j)}u_j\|$ satisfy $\|Ay_j - \mu_j y_j\| = \|\varepsilon_j\| < \varepsilon$. Then for $\mu_i \neq \mu_j$,

$$|y_i^T y_j| < 2\varepsilon/|\mu_i - \mu_j|. \qquad (4.8.3)$$

Proof. Since

$$\mu_i y_i^T y_j = y_i^T A y_j - \varepsilon_i^T y_j = \mu_j y_i^T y_j - \varepsilon_i^T y_j + y_i^T \varepsilon_j,$$

we obtain Eqn(4.8.3) immediately. □

Given a 'good' eigenvalue μ (of some Lanczos matrix) which is a good approximation to some eigenvalue of A, the corresponding Ritz vector should be a good approximation to an eigenvector of A, if we compute the Ritz vector at a size Lanczos matrix ma(μ) at which μ has stabilized as an eigenvalue of the Lanczos matrices. An appropriate size can be obtained by selecting an appropriate tolerance $\varepsilon(\mu)$ and then using Sturm Sequencing to determine the number of eigenvalues of the Lanczos matrices in the interval $[\mu - \varepsilon(\mu), \mu + \varepsilon(\mu)]$ as we let m increase. We can continue to increment m after μ first appears to attempt to determine when μ replicates as an eigenvalue of the Lanczos matrices. The cost of the Sturm sequencing for each eigenvalue is less than $6m(\mu)$ where $m(\mu)$ is a function of μ. See Step 1.

Inverse iteration on the Lanczos matrices is inexpensive and is used to determine the 'goodness' of the size ma(μ) selected. See Chapter 3, Section 3.4. Typically, not more than one iteration of inverse iteration is required. If this is the case, then using T_m requires approximately 12m multiply-add operations per eigenvalue. An additional 2mn arithmetic operations are required to compute the corresponding Ritz vector plus of course the cost of computing the Lanczos vectors. However, if we consider a set of eigenvalues simultaneously, then the cost of a single pass through the Lanczos vectors can be averaged over the set of eigenvalues used.

In order for the inverse iteration procedure to work well, the eigenvalue being considered should be reasonably well-separated from the other eigenvalues of the Lanczos matrix being used. Therefore, we must also be concerned with the gaps between eigenvalues of the Lanczos matrices. The gaps in the Lanczos matrices are not however controlled by the gaps between the eigenvalues in the original matrix. As we saw in Section 4.6, spurious eigenvalues may appear anywhere in the spectrum of a Lanczos matrix. However, if a spurious eigenvalue is very close to a 'good' eigenvalue which we are considering, then the presence of such an eigenvalue would typically be reflected through a poorer error estimate for that 'good' eigenvalue. A poor estimate on Step 2 of the eigenvector procedure would cause an increase in the size of the Lanczos matrix being considered for the Ritz vector computation. Looping would continue until a Lanczos matrix eigenvector with a small enough last component was obtained. If for some eigenvalue μ the program is unable to find such an eigenvector, then it will not compute the corresponding Ritz vector unless the user has set a flag which indicates that the Ritz vectors are to be computed using the best Lanczos matrix eigenvectors that the program was able to find.

Table 4.8.1 demonstrates the sensitivity of the accuracy of the computed Ritz vector to the choice of Lanczos matrix, ma(μ). Table 4.8.1 summarizes the results from computing Ritz vectors for three of the eigenvalues of a 100×100 diagonally-disordered matrix (KIRK100) as we varied the size of the Lanczos matrix from $m1(\mu)$ to $m2(\mu)$. For each eigenvalue μ these sizes were computed as the first value of m at which μ appears as an eigenvalue of T_m to within a specified tolerance and the first value of m at which μ appears as a double eigenvalue to within that tolerance. In each case there is a range of Lanczos matrix sizes for which the Aerror is essentially constant. This range always includes a broad band of values in the middle region between $m1(\mu)$ and $m2(\mu)$. Large errors occur near $m1(\mu)$ because μ has not yet stabilized in the sense that the last component of the corresponding eigenvector of the Lanczos matrix has not yet become pathologically small. Large errors occur near $m2(\mu)$ because there are spurious eigenvalues near μ.

TABLE 4.8.1   KIRK100, Computing Ritz Vectors for Varying M
CONC= .7, SCALE= 100., SEED= 123456789

Evalue μ	M	Amingap	Aerror	$\dfrac{\text{Aerror}}{\text{Amingap}}$	$\beta_{M+1}u(M)$
-1.7496969	125	0.03265	5.2×10^{-5}	1.6×10^{-3}	5.2×10^{-5}
	145		6.5×10^{-9}	2.0×10^{-7}	6.5×10^{-9}
	155		9.1×10^{-11}	2.8×10^{-9}	4.1×10^{-11}
	165		7.1×10^{-11}	2.2×10^{-9}	3.0×10^{-11}
	175		7.1×10^{-11}	2.2×10^{-9}	7.6×10^{-12}
	185		7.1×10^{-11}	2.2×10^{-9}	1.1×10^{-12}
	195		7.1×10^{-11}	2.2×10^{-9}	5.9×10^{-12}
	205		7.1×10^{-11}	2.2×10^{-9}	1.0×10^{-13}
	215		7.1×10^{-11}	2.2×10^{-9}	3.0×10^{-13}
	225		7.1×10^{-11}	2.2×10^{-9}	2.6×10^{-12}
	235		9.2×10^{-11}	2.8×10^{-9}	5.9×10^{-11}
	258		1.4×10^{-7}	4.2×10^{-6}	1.4×10^{-7}
100.05124	92	0.00873	4.9×10^{-3}	5.6×10^{-1}	4.9×10^{-3}
	108		2.2×10^{-6}	2.5×10^{-4}	2.2×10^{-6}
	116		1.4×10^{-8}	1.6×10^{-6}	1.4×10^{-8}
	124		4.8×10^{-9}	5.5×10^{-7}	4.8×10^{-9}
	132		6.8×10^{-10}	7.8×10^{-8}	6.8×10^{-10}
	140		9.9×10^{-11}	1.1×10^{-8}	3.6×10^{-11}
	148		9.4×10^{-11}	1.1×10^{-8}	5.4×10^{-12}
	156		9.3×10^{-11}	1.1×10^{-8}	7.8×10^{-12}
	164		1.1×10^{-10}	1.3×10^{-8}	4.7×10^{-11}
	172		2.8×10^{-10}	3.2×10^{-8}	2.7×10^{-10}
	180		1.3×10^{-9}	1.5×10^{-7}	1.3×10^{-9}
	199		4.6×10^{-7}	5.3×10^{-5}	4.6×10^{-7}
100.05997	106	2.2×10^{-5}	1.2×10^{-5}	5.7×10^{-1}	1.2×10^{-5}
	124		1.2×10^{-8}	5.6×10^{-4}	1.2×10^{-8}
	133		2.1×10^{-10}	9.6×10^{-6}	1.7×10^{-10}
	142		1.3×10^{-10}	6.3×10^{-6}	8.5×10^{-11}
	151		9.7×10^{-11}	4.6×10^{-6}	1.2×10^{-11}
	160		9.7×10^{-11}	4.6×10^{-6}	1.4×10^{-12}
	169		9.9×10^{-11}	4.7×10^{-6}	1.8×10^{-12}
	178		1.9×10^{-10}	9.2×10^{-6}	1.6×10^{-10}
	187		5.5×10^{-10}	2.6×10^{-5}	5.5×10^{-10}
	196		1.1×10^{-10}	5.4×10^{-6}	4.3×10^{-11}
	205		1.1×10^{-10}	5.0×10^{-6}	9.0×10^{-12}
	235		4.2×10^{-7}	2.0×10^{-2}	4.2×10^{-7}

In Tables 4.8.1 and 4.8.2 we use the following notation. M denotes the size of the Lanczos matrix used in the Ritz vector computation.

$$\text{Aerror} \equiv \| AVu - \mu Vu \| / \| Vu \|,$$

$$\text{Terror} \equiv \| T_M u - \mu u \| \text{ with } \| u \| = 1, \text{ and}$$

$$\text{Amingap} \equiv \min_{k(i) \neq k(j)} | \mu_{k(i)} - \mu_{k(j)} | \text{ for } \mu_{k(i)} \text{ a 'good' eigenvalue}$$

Table 4.8.1 demonstrates that the quantities $\beta_{M+1} u(M)$ are good predictors of the actual residual norms (Aerror) of the computed Ritz vectors.

We also observe that if we enlarge the Lanczos matrix beyond $m2(\mu)$, determining the size $m3(\mu)$ where μ has a multiplicity of 3 as an eigenvalue of the Lanczos matrix, then there is a similar span of sizes between $m2(\mu)$ and $m3(\mu)$ which yield good Ritz vectors. We observe further that if the 'good' eigenvalue μ being considered is a simple eigenvalue of the original A-matrix, then such a Ritz vector will be essentially the same as the Ritz vector obtained for smaller M. If however the eigenvalue μ is a multiple eigenvalue of the original matrix, then a Ritz vector obtained using this larger Lanczos matrix is typically independent of the one obtained for the smaller value of M. Thus, by continuing to enlarge the Lanczos matrix we can obtain linearly independent Ritz vectors which span the entire invariant subspace corresponding to such multiple eigenvalues. The amount of additional work required depends upon the rate of convergence of the given eigenvalue. If we have an eigenvalue which replicates easily, then it is not costly to compute such additional eigenvectors. Additional eigenvectors can be computed by simply continuing to enlarge the Lanczos matrices and obtaining the sizes $mj(\mu)$, $j = 1, 2, ..., s$ where s is the multiplicity of μ as an eigenvalue of the original matrix.

NUMERICAL EXAMPLE To demonstrate the effectiveness of this procedure on a large matrix we considered a diagonally-disordered matrix KIRK1089 of order 1089 whose diagonal entries were either 0 or 100. We first computed the eigenvalues of this matrix using the eigenvalue procedure described in Section 4.7 and a Lanczos matrix of size 3300. At this value of m, 1036 eigenvalues of KIRK1089 were being approximated. There were very tight clusters of eigenvalues near 0, 99, 100 and 101 containing eigenvalues which were not yet being approximated by m=3300. The eigenvector computations were all made using eigenvalues which had converged to at least 10 digits as eigenvalues of the Lanczos matrices.

It was not practical to store the Lanczos vectors for reuse in the eigenvector computations. They were regenerated for the eigenvector computations. KIRK1089 had eigenvalues ranging from $\lambda_{1089} = -3.604$ to $\lambda_1 = 102.435$. The eigenvalue gaps ranged from 10^{-7} or less to .186. There were no eigenvalues in the interval [3.6, 97.6]. Approximate eigenvectors associated with eigenvalues of KIRK1089 which were scattered throughout the spectrum and which had various size Amingaps were computed as Ritz vectors using inverse iteration on

$(T_{ma(\mu)} - \mu I)$, for the sizes $ma(\mu)$ determined in Step 2 of the Eigenvector Procedure. Some of these results are summarized in Table 4.8.2. They clearly indicate the power of this eigenvector procedure.

TABLE 4.8.2   KIRK1089, Computing Ritz Vectors
CONC= .7, SCALE= 100., SEED= 123456789

Evalueμ	M	Amingap	Aerror	$\dfrac{\text{Aerror}}{\text{Amingap}}$	Terror
-3.6046373	169	0.03216	1.2×10^{-9}	3.8×10^{-8}	7.8×10^{-13}
-3.0483198	767	0.00817	1.1×10^{-10}	1.3×10^{-8}	1.6×10^{-12}
-2.9042581	803	0.00392	2.8×10^{-10}	7.2×10^{-8}	1.8×10^{-11}
-2.7834770	990	0.00308	2.5×10^{-10}	8.0×10^{-8}	4.2×10^{-11}
-2.5554347	1211	0.00327	9.7×10^{-10}	3.0×10^{-7}	1.3×10^{-11}
-2.3008228	1672	0.00430	1.4×10^{-10}	3.2×10^{-8}	4.9×10^{-11}
-2.0915135	1681	0.00568	5.0×10^{-10}	8.8×10^{-8}	1.1×10^{-10}
-1.9024592	1992	0.00172	1.1×10^{-10}	6.2×10^{-8}	1.8×10^{-11}
-1.7328250	2285	0.00207	1.5×10^{-9}	7.2×10^{-7}	2.9×10^{-11}
-1.5512152	2384	0.00261	1.1×10^{-10}	4.1×10^{-8}	1.9×10^{-11}
-1.5476645	2379	0.00355	2.0×10^{-10}	5.5×10^{-8}	1.6×10^{-11}
-1.3571523	2250	0.00665	1.1×10^{-11}	1.7×10^{-9}	1.2×10^{-11}
-1.1860977	2652	0.00197	3.5×10^{-10}	1.8×10^{-7}	1.3×10^{-10}
-1.0237748	2592	0.00481	9.7×10^{-11}	2.0×10^{-8}	2.2×10^{-11}
$-.88306080$	3294	0.00278	2.8×10^{-10}	1.0×10^{-7}	2.9×10^{-10}
$-.73430204$	3214	0.00560	3.8×10^{-10}	6.7×10^{-8}	3.4×10^{-10}
$-.59786176$	3523	0.00267	1.2×10^{-10}	4.6×10^{-8}	6.7×10^{-11}
$-.59519365$	3524	0.00267	2.4×10^{-10}	9.1×10^{-8}	2.0×10^{-10}
$-.46753395$	3577	0.00188	1.3×10^{-9}	6.8×10^{-7}	1.0×10^{-9}
$-.21995797$	3346	0.00640	8.3×10^{-11}	1.3×10^{-8}	8.0×10^{-11}
0.1828204	3542	0.00048	2.5×10^{-10}	5.2×10^{-7}	1.8×10^{-10}
0.8528662	3292	0.00271	2.9×10^{-10}	1.1×10^{-7}	6.5×10^{-11}
0.9913656	2662	0.00754	5.8×10^{-11}	7.7×10^{-9}	6.5×10^{-11}
2.5308241	1224	0.00287	9.5×10^{-10}	3.3×10^{-7}	5.5×10^{-12}
2.6188205	1097	0.00514	1.4×10^{-10}	2.7×10^{-8}	5.7×10^{-11}
2.7569963	1018	0.00474	1.3×10^{-10}	2.7×10^{-8}	1.3×10^{-11}
3.0457472	820	0.00199	2.6×10^{-10}	1.3×10^{-7}	2.3×10^{-11}
98.178376	363	0.00117	2.6×10^{-10}	2.3×10^{-7}	7.6×10^{-12}
98.610657	1173	5.9×10^{-5}	8.7×10^{-10}	1.5×10^{-5}	6.7×10^{-12}
99.017506	845	0.00227	1.3×10^{-9}	5.5×10^{-7}	1.2×10^{-11}
99.501541	567	0.00013	6.6×10^{-10}	4.9×10^{-6}	2.3×10^{-11}
101.02235	1104	0.00211	1.7×10^{-9}	8.2×10^{-7}	1.6×10^{-12}
101.93034	325	0.02918	3.0×10^{-10}	1.0×10^{-8}	1.0×10^{-11}
102.19677	266	2.0×10^{-5}	2.2×10^{-10}	1.1×10^{-5}	2.1×10^{-12}
102.25791	175	0.00848	7.6×10^{-10}	8.9×10^{-8}	1.4×10^{-12}
102.43574	111	0.13879	3.8×10^{-10}	2.7×10^{-9}	1.1×10^{-12}

With respect to the implementation of this procedure, it is essential to point out that the Lanczos vectors which are generated for the Ritz vector computations must be identical to those which were used in the eigenvalue computations. The success of this procedure requires this match between the quantities computed in the eigenvalue computations and those computed in the eigenvector computations. The storage requirements for this program depend upon the number of Ritz vectors to be computed and upon the degree of difficulty in computing them.

The degree of difficulty affects the size of the Lanczos matrix needed for the Ritz vector computation. The program is designed so that it can be used to determine a priori good estimates of these sizes and can therefore determine a good estimate of the overall amount of storage required for any particular set of eigenvalues specified by the user. Typically, this determination requires no more storage than that required for the original eigenvalues computations. The procedure as programmed keeps both the computed eigenvectors of the Lanczos matrices in main storage as well as the Ritz vectors which are being computed. Please see Chapter 2, Volume 2 for more details. In the next section we consider two other classes of real symmetric problems, the Hermitian matrices and certain real symmetric generalized eigenvalue problems.

SECTION 4.9   LANCZOS PROCEDURES, HERMITIAN, GENERALIZED SYMMETRIC

The basic Lanczos procedures discussed in the preceding sections can be modified to handle three other classes of 'real' symmetric problems. Chapters 3, 4, and 5 of Volume 2 contain respectively Lanczos procedures for (1) Hermitian matrices; (2) Factored inverses of positive definite, real symmetric matrices; and (3) Certain real symmetric generalized problems.

The modifications which are required to handle class (2) are trivial and occur primarily in the matrix-vector multiply subroutine. That subroutine becomes an equation solver which uses the factorization of the original matrix or of a shifted, scaled, and permuted version of A. If the original matrix A is sparse, it is sometimes possible to use permutation matrices to obtain a matrix whose factors are also sparse, SPARSPAK [1979].

The modifications of the real symmetric procedure required to handle Hermitian matrices or to handle real symmetric generalized problems include changes in the Lanczos recursion. We first consider Hermitian matrices and then consider real symmetric generalized problems where one of the two matrices is positive definite.

HERMITIAN MATRICES   The reader is referred to Chapter 1, Section 1.5 for a summary of some of the basic properties of Hermitian matrices. In this section we define a Lanczos procedure for such matrices. Actually we need only to modify the basic Lanczos recursion used to generate the Lanczos matrices. With this change the statements of the Lanczos procedures for real symmetric matrices given in Sections 4.7 and 4.8 become descriptions of the corresponding Hermitian Lanczos procedures. We can obtain a suitable Lanczos recursion for such matrices by considering the general Lanczos recursions given in Wilkinson [1965, Chapter 6] and specializing them to Hermitian matrices. We write any Hermitian matrix A as $A = B + iC$ where B and C are real matrices and i is the square root of -1.

LEMMA 4.9.1   Let A be a $n \times n$ Hermitian matrix. An appropriate Lanczos recursion is given by the following equations.

$$v_{i+1}\beta_{i+1} = Av_i - \alpha_i v_i - \bar{\beta}_i v_{i-1}. \tag{4.9.1}$$

$$\alpha_i \equiv v_i^H A v_i \text{ and}$$
$$|\beta_{i+1}| \equiv \|Av_i - \alpha_i v_i - \bar{\beta}_i v_{i-1}\|. \tag{4.9.2}$$

Proof. Wilkinson [1965, Chapter 6] presents the following two-sided recursions for a general nxn matrix A. Let v_1 and w_1 be randomly-generated starting vectors scaled so that $v_1^T w_1 = 1$. Then for i = 1,2,...,m use the following recursion to define vectors v_{i+1}, w_{i+1} such that

$$\beta_{i+1}v_{i+1} = Av_i - \alpha_i v_i - \gamma_i v_{i-1}$$
$$\gamma_{i+1}w_{i+1} = A^T w_i - \alpha_i w_i - \beta_i w_{i-1} \tag{4.9.3}$$

where the scalars α_i, β_i, and γ_i are chosen so that for each m, $V_m^T W_m = I$ where $W_m \equiv \{w_1,...,w_m\}$ and $V_m \equiv \{v_1,...,v_m\}$.

Since A is Hermitian if we take the complex conjugates of the quantities in the A^T recursion in Eqn(4.9.3), we obtain a recursion for the conjugates of the w_i vectors.

$$\bar{\gamma}_{i+1}\bar{w}_{i+1} = A\bar{w}_i - \bar{\alpha}_i\bar{w}_i - \bar{\beta}_i\bar{w}_{i-1},$$

where bar denotes the complex conjugate. But this says that if we set $v_1 \equiv \bar{w}_1$, then $\bar{w}_i$ and v_i satisfy the same recursion and thus for each i we have that

$$\alpha_i = \bar{\alpha}_i, \ \gamma_i = \bar{\beta}_i.$$

But this yields Eqn(4.9.1). Eqn(4.9.2) follow directly from Eqn(4.9.1) and the fact that the coefficients are chosen so that

$$W_m^T V_m = \bar{V}_m^T V_m = I_m.$$

□

Thus, given a Hermitian matrix, we can use the Lanczos recursions to generate a family of Hermitian, tridiagonal matrices. Note however, that the off-diagonal entries of the Lanczos matrices are not uniquely specified by Eqn(4.9.2). Eqn(4.9.2) controls only the magnitudes of these quantities. The phase angle of each of these entries is not specified. We would like to set each phase angle to 0 degrees so that the Lanczos matrices would be real symmetric. The following lemma demonstrates that we can do this because the set of eigenvalues of any resulting Lanczos matrix is invariant with respect to the particular phase angles chosen.

Moreover, the corresponding Lanczos vectors differ at most by a unitary diagonal matrix. Therefore, the convergence rate of the Hermitian Lanczos procedure as defined by Eqns(4.9.1) - (4.9.2) is independent of which particular phase angles are selected for the off-diagonal entries, and we are free to set

$$\beta_{i+1} \equiv \| Av_i - \alpha_i v_i - \beta_i v_{i-1} \| \tag{4.9.4}$$

LEMMA 4.9.2 (Exact Arithmetic) Let A be a Hermitian matrix. Use Recursions (4.9.1) - (4.9.2) to generate two families of Lanczos matrices. The sequence T_m, m = 1,2,... corresponds to setting the phase angle of each off-diagonal entry generated to 0. The sequence T_m^θ, m = 1,2,... corresponds to selecting these phase angles in some other fashion. The corresponding sets of Lanczos vectors are denoted by V_m and V_m^θ, respectively. Then for any m we have that

$$T_m^\theta = D_m T_m D_m^{-1} \quad \text{and} \quad V_m^\theta = V_m D_m^{-1} \tag{4.9.5}$$

where D_m is the mxm diagonal matrix whose kth diagonal entry is $\exp(\sum_{j=2}^{k} \theta_j)$.

Proof. The proof is by induction. First set $\beta_1^\theta \equiv \beta_1 \equiv 0$ and $v_1^\theta \equiv v_1$. Then clearly, $\alpha_1^\theta \equiv \alpha_1$, and therefore $v_2 \beta_2 = v_2^\theta \beta_2^\theta$. Both v_2 and v_2^θ have unit norm, so that we easily get that for some angle θ_2,

$$\beta_2^\theta = e^{i\theta_2}\beta_2 \quad \text{and therefore} \quad v_2 = e^{i\theta_2} v_2^\theta.$$

But these relationships yield

$$\alpha_2^\theta \equiv (v_2^\theta)^H (Av_2^\theta - \bar{\beta}_2^\theta v_1^\theta) = \alpha_2.$$

The use of Eqns(4.9.1) - (4.9.2) then yields $\beta_3^\theta = \exp(i\theta_3)\beta_3$.

Now assume that for all $j \leq k$ that

$$\alpha_j = \alpha_j^\theta, \quad \beta_j^\theta = \exp(i\theta_j)\beta_j \quad \text{and} \quad v_j = [\exp(i\sum_{\ell=2}^{j} \theta_\ell)]v_j^\theta$$

But from Eqns(4.9.1) - (4.9.2) we get that

$$v_{k+1}^\theta \beta_{k+1}^\theta = [\exp(-i\sum_{j=2}^{k} \theta_j)](A - \alpha_k I)v_k - \exp(-i\theta_k)\beta_k[\exp(-i\sum_{j=2}^{k-1} \theta_j)]v_{k-1}$$

Therefore,

$$v_{k+1}^\theta \beta_{k+1}^\theta = \exp(-i\sum_{j=2}^{k} \theta_j)[\beta_{k+1}v_{k+1}].$$

Therefore, we have that

$$(\bar{\beta}_{k+1}^{\theta})\beta_{k+1}^{\theta} \; = \; \beta_{k+1}^2 \;\; \text{so that} \;\; \beta_{k+1}^{\theta} \; = \; \exp(i\theta_{k+1})\beta_{k+1}$$

for some angle θ_{k+1}. Furthermore, we get by the direct substitution of these relationships into the formula for α_{k+1}^{θ} that $\alpha_{k+1}^{\theta} \; = \; \alpha_{k+1}$.

These equalities together with the definition of the diagonal matrix D_m can now be used to prove (by direct substitution) that the matrix T_m^{θ} can be obtained from the real matrix T_m by the diagonal similarity transformation defined by D_m. Therefore, these two matrices have the same eigenvalues. □

Lemma 4.9.2 states that given a Hermitian matrix and a starting vector that all of the possible Lanczos matrices of size m which can be generated using the Eqns(4.9.1) - (4.9.2) have the same set of eigenvalues. Furthermore, the corresponding Lanczos vectors generated are identical to within scaling factors of magnitude 1. Therefore, we have the following Corollary which tells us that we can use the Lanczos recursion to generate real symmetric tridiagonal matrices directly from a given Hermitian matrix. This simplifies the computations considerably since all of the eigenvalue and eigenvector computations for the Lanczos matrices can be done in real arithmetic.

COROLLARY 4.9.2 (Exact Arithmetic) Under the hypotheses of Lemma 4.9.2, the rate of convergence of the eigenvalues and of the Ritz vectors of the Hermitian Lanczos matrices generated using Eqns(4.9.1) - (4.9.2) is independent of the particular choice of the phase angles for the off-diagonal entries in these matrices.

An alternative approach to computing eigenelements of Hermitian matrices would be to apply the Lanczos procedure for real symmetric matrices directly to the associated 2n x 2n real symmetric matrix

$$M \; \equiv \; \begin{bmatrix} B & -C \\ C & B \end{bmatrix}. \tag{4.9.6}$$

LEMMA 4.9.3 The real symmetric eigenvalue problem

$$Mw \; = \; \lambda w \tag{4.9.7}$$

is equivalent to the Hermitian problem in Eqn(4.9.1). In particular each eigenvalue of A appears as an eigenvalue of M with twice the multiplicity it has in A. Any eigenvector $z = x + iy$ of A yields corresponding eigenvectors

$$\begin{bmatrix} x \\ y \end{bmatrix} \; \text{and} \; \begin{bmatrix} y \\ -x \end{bmatrix}$$

of M. Conversely any eigenvector $w = (w_1, w_2)$ of M yields a corresponding eigenvector $z = w_1 + iw_2$ of A.

Proof. Let λ and $z = x + iy \neq 0$ be respectively an eigenvalue and an eigenvector of A. Then the relationship $Az = \lambda z$ immediately yields the equivalent relationships

$$Bx - Cy = \lambda x \text{ and } Cx + By = \lambda y.$$

These are, however, the eigenvalue - eigenvector relationships for M. Therefore the transposes of the two vectors $[x^T, y^T]$ and $[y^T, -x^T]$ are eigenvectors of M. The converse follows trivially. $\square$

The FORTRAN code for the Hermitian Lanczos procedures which use the recursions given in Eqns(4.9.1) - (4.9.2) with each $\beta_{i+1} \geq 0$ are contained in Chapter 3 of Volume 2 of this book. The reader is referred to Volume 2 for the documentation and listing for that program. The arithmetic in this procedure is a mixture of double precision real and double precision complex. The Lanczos matrices are generated using complex arithmetic for computing the Lanczos vectors. The Lanczos matrices are however real symmetric. Therefore the eigenvalue and eigenvector computations for the Lanczos matrices use double precision real arithmetic. The Ritz vector computations use the real eigenvectors of the Lanczos matrices together with the complex Lanczos vectors to generate complex eigenvectors of the original Hermitian matrix. The modified Gram-Schmidt formulation for the diagonal entries α_i was not used in the Hermitian case. Observe that the modified form would be

$$\alpha_i = v_i^H(Av_i - \beta_i v_{i-1}).$$

See Eqns(4.3.2) for the real symmetric case. The correction term is obviously complex-valued whereas the α_i must be real. Therefore, we could use such a correction term only if we were willing to accept the real part of the correction and then reject the imaginary part.

One cute but perhaps useless observation is the following. If the original matrix is actually real symmetric we can apply the Hermitian procedure to it to determine which eigenvalues of A are multiple. We have the following Lemma.

LEMMA 4.9.4 Let A be a real symmetric matrix. Apply the Hermitian Lanczos procedure given in Eqns(4.9.1) - (4.9.2) to A. Let μ be an eigenvalue of an associated Lanczos matrix T_m which has stabilized in the sense that the mth component of the corresponding eigenvector of T_m has the property that $u(m) \approx 0$. Define the associated Ritz vector $y \equiv V_m u \equiv y_1 + iy_2$. Then if the ratios of the components

$$y_1(j)/y_2(j) \neq \text{constant over } 1 \leq j \leq n,$$

and $\|y_1\|$ and $\|y_2\|$ are not 'small', then μ is an approximation to a multiple eigenvalue of A.

Proof.

$$\| Ay - \mu y \|^2 = \| Ay_1 - \mu y_1 \|^2 + \| Ay_2 - \mu y_2 \|^2 .$$

But the Hermitian analog of Lemma 4.5.1 tells us that this residual norm will be small. Therefore, each of the individual residual norms for y_1 and y_2 is small. But since the ratios of the corresponding components of these two vectors are not equal to a constant, we have that these two vectors must be independent. Therefore, the multiplicity of μ as an eigenvalue of A is at least two. □

REAL SYMMETRIC GENERALIZED EIGENVALUE PROBLEMS We consider the question of devising a Lanczos procedure for the real symmetric generalized eigenvalue problem defined in Chapter 1, Section 1.6. We are given two real symmetric matrices A and B and we are asked to determine scalars λ and vectors $x \neq 0$ such that

$$Ax = \lambda Bx. \tag{4.9.8}$$

In this section we assume that either A or B is a positive definite matrix. W.l.o.g we assume that B is positive definite. Furthermore, we assume that the Cholesky factorization $B = LL^T$ is available for use in the Lanczos procedure. The matrix L is lower triangular. It is easy to demonstrate that there is a basis of eigenvectors X which is B-orthogonal. That is, $X^T BX = I$, and $AX = X\Lambda$ where Λ is the diagonal matrix whose entries are the corresponding eigenvalues. We saw in Section 1.6 that this problem is equivalent to the real symmetric problem

$$L^{-1}AL^{-T}y = \lambda y \text{ where } y = L^T x. \tag{4.9.9}$$

Using this relationship we can obtain the following lemma.

LEMMA 4.9.5 Let A and B be real symmetric matrices. Then if B is positive definite and $B = LL^T$ is the Cholesky factorization of B, then an appropriate Lanczos recursion for the generalized eigenvalue problem given in Eqn(4.9.8) is given by the following equations. Define $\beta_1 \equiv 0$, $v_0 \equiv 0$. Choose v_1 randomly and normalized so that $v_1^T B v_1 = 1$. Then for i=2,... we have that

$$\beta_{i+1} B v_{i+1} = Av_i - \alpha_i B v_i - \beta_i B v_{i-1}. \tag{4.9.10}$$

$$\alpha_i \equiv v_i^T (Av_i - Bv_{i-1}) \text{ and}$$
$$|\beta_{i+1}| \equiv \| L^{-1}(Av_i - \alpha_i v_i - \beta_i v_{i-1}) \| . \tag{4.9.11}$$

Proof. Simply substitute the matrix $C \equiv L^{-1}AL^{-T}$ into the Lanczos recursions for real symmetric matrices given in Eqns(4.3.1) and then multiply both sides of the resulting equation by L to obtain

$$LL^Tw_{i+1}\beta_{i+1} = Aw_i - \alpha_i LL^Tw_i - \beta_i LL^Tw_{i-1} \equiv r_i$$

where $w_i \equiv L^{-T}v_i$. Making this same substitution in Eqn(4.3.2) and rearranging the terms we obtain Eqn(4.9.11). $\qquad\qquad\qquad\qquad\qquad\qquad\qquad\qquad\qquad\qquad\qquad\qquad\quad$ $\square$

The choice of β_{i+1} in Eqn(4.9.11) corresponds to normalizing each Lanczos vector w.r.t. the B-norm. That is for each i, $v_i^TBv_i = 1$. Observe that this procedure generates real symmetric tridiagonal matrices. Also observe the effect of the numerical condition of B. This approach cannot handle the case when B is only semi-definite. In that case if we factored B we would get $B = WW^T$ where W is rectangular. If A is nonsingular and can be factored as $A = LDL^T$, where L is lower triangular and D is diagonal, then we could apply the procedure for real symmetric matrices to this corresponding inverse problem.

$$W^TL^{-T}D^{-1}L^{-1}Wz = \theta z \text{ where } z \equiv W^Tx \text{ and } \theta = 1/\lambda. \qquad (4.9.12)$$

This is the approach used by Ericsson and Ruhe [1980]. They also incorporate shifts σ, factoring $(A-\sigma B)$.

The Lanczos procedures for these types of real symmetric generalized problems are contained in Chapter 5 of Volume 2. These procedures allow the user to work with permutations of the A and B matrices. Such permutations are often useful in obtaining sparse factorizations when the original matrix B is sparse, SPARSPAK [1979].

CHAPTER 5

REAL RECTANGULAR MATRICES

SECTION 5.1   INTRODUCTION

If A is a ℓxn matrix then the singular value decomposition of A has the following form.

$$A = Y\Sigma X^H \tag{5.1.1}$$

where Y is a ℓxℓ unitary matrix, X is a nxn unitary matrix and for $\ell \geq n$, Σ is a real rectangular diagonal matrix of the form

$$\Sigma = \begin{bmatrix} \Sigma_1 \\ 0 \end{bmatrix}. \tag{5.1.2}$$

Σ_1 is an nxn diagonal matrix with nonnegative entries and 0 denotes the $(\ell-n)$xn matrix of zeros. For $\ell \leq n$, Σ would be of the form $[\Sigma_1 \ 0]$.

To simplify the discussion in this Chapter we will assume that $\ell \geq n$. This is typical in the applications. For example in fitting data, we would typically have more measurements than fitting parameters. The corresponding statements for the case $\ell \leq n$ can be obtained directly from the arguments given, if these arguments are applied to A^H rather than to A. If A is real then Y and X are orthogonal matrices. We have that $A^H A = X\Sigma^T\Sigma X^H$ and $AA^H = Y\Sigma\Sigma^T Y^H$. Two equivalent forms of Eqn(5.1.1) which are easily obtained from that equation are

$$AX = Y\Sigma \ \text{ and } A^H Y = X\Sigma^T. \tag{5.1.3}$$

Thus, the Y span the range or column space of A and the X span the range of A^H.

In this chapter we present a single-vector Lanczos procedure with no reorthogonalization for computing singular values and corresponding singular vectors of real rectangular matrices. First in Section 5.2 we summarize some of the properties of singular values and briefly discuss their relationship to the eigenvalues of the given matrix. Second in Section 5.3 we briefly discuss several applications. Finally in Section 5.4 we describe our Lanczos procedure. FORTRAN code for this procedure is provided in Chapter 6 of Volume 2 of this book. Theorem 5.1.1 states that the matrix decomposition given in Eqn(5.1.1) exists for any matrix. Proofs of this Theorem can be found in many places, see for example Stewart [1973, p. 319] or Lawson and Hanson [1974, p. 19].

THEOREM 5.1.1 Let A be a ℓxn matrix, then there are unitary matrices Y and X and a rectangular ℓxn diagonal matrix Σ satisfying Eqn(5.1.1).

Proof. Assume $\ell \geq n$. Let σ_j, $1 \leq j \leq n$, be the nonnegative square roots of the eigenvalues of $A^H A$ where $\sigma_1 \geq \sigma_2 \geq \ldots \geq \sigma_r > 0$ and $\sigma_{r+1} = \ldots = \sigma_n = 0$. Then $r = \text{rank}(A)$. Define $\Sigma_1 = \text{diag}\{\sigma_1, \ldots, \sigma_n\}$ and $\Sigma_{11} = \text{diag}\{\sigma_1, \ldots, \sigma_r\}$. Let $X = (X_1, X_2)$ be a unitary basis of eigenvectors of $A^H A$ where X_1 is nxr and corresponds to the nonzero singular values σ_j, $1 \leq j \leq r$, and X_2 is nx(n-r) and corresponds to the zero singular values σ_j, $r + 1 \leq j \leq n$. Define the ℓxr matrix

$$Y_1 \equiv AX_1 \Sigma_{11}^{-1} \tag{5.1.4}$$

and let Y_2 be any ℓx$(\ell-r)$ matrix whose columns form a unitary set of vectors that makes the set $Y \equiv (Y_1, Y_2)$ a complete unitary basis for C^ℓ. Clearly, by construction

$$AA^H Y_1 = A(A^H AX_1 \Sigma_{11}^{-1}) = Y_1 \Sigma_{11}^2$$

so that the columns of Y_1 are eigenvectors of the matrix AA^H corresponding to the nonzero eigenvalues of AA^H. Therefore, $AA^H Y_2 = 0$ since Y_2 must span the unitary complement of Y_1 which is the space spanned by the eigenvectors of AA^H corresponding to the zero eigenvalues. But this implies that $Y_2^H AA^H Y_2 = 0$ so that $A^H Y_2 = 0$ and therefore Y_2 spans the null space of A^H. Thus, we have from Eqn(5.1.4) that $A^H Y_1 = X_1 \Sigma_{11}$ and $A^H Y_2 = 0$. But combining these two equations, we obtain Eqn(5.1.3). Note that if $\ell \leq n$ then the above argument applies to A^H. $\square$

Y_2 spans the null space of A^H and since $A^H AX_2 = 0$, X_2 spans the null space of A. Therefore from Eqn(5.1.3), A can be written as

$$A = Y_1 \Sigma_{11} X_1^H. \tag{5.1.5}$$

In this representation we used only those singular vectors corresponding to the nonzero singular values of A. In fact in many computations which use a singular value decomposition, for example solving systems of equations or in least squares computations, the null spaces of A and of A^H are not used. It is sufficient to have the singular vectors Y_1 and X_1 corresponding to the nonzero singular values. Clearly the sets of singular vectors Y and X are not unique if $A^H A$ or AA^H has any multiple eigenvalues. We have the following Corollary to Theorem 5.1.1 which will prove useful in Section 5.4.

COROLLARY 5.1.1 Let $A = Y\Sigma X^H$ be a singular value decomposition of a ℓxn matrix A with $\ell \geq n$. Set $Y = (\tilde{Y}_1, \tilde{Y}_2)$ where $\tilde{Y}_1$ is ℓxn. Then

$$AX = \tilde{Y}_1 \Sigma_1, \quad A^H \tilde{Y}_1 = X\Sigma_1 \quad \text{and} \quad A^H \tilde{Y}_2 = 0. \tag{5.1.6}$$

If A has rank $r < n$, set $X = (X_1, X_2)$ where X_1 is nxr and X_2 is nx(n-r), then $AX_2 = 0$. In fact, the singular vectors $\tilde{Y}_1$ and X corresponding to the singular values, including any zero singular values, are characterized by the following conditions:

$$Ax = \sigma_j\, y, \quad A^H y = \sigma_j\, x, \quad y \neq 0, \quad \text{and} \quad x \neq 0. \tag{5.1.7}$$

Thus, A has a zero singular value if and only if A has a non-trivial null space. The remaining left singular vectors $\tilde{Y}_2$ can be any unitary set of $(\ell - n)$ vectors such that $(\tilde{Y}_1, \tilde{Y}_2)$ is a unitary set.

Proof. The first part of the proof follows directly from Eqns(5.1.6). The characterization in Eqn(5.1.7) follows directly from the fact that the vectors $\{X, \tilde{Y}_1\}$ are respectively, linearly independent sets of eigenvectors of $A^H A$ and of AA^H. $\qquad\qquad\qquad\square$

SECTION 5.2   RELATIONSHIPS WITH EIGENVALUES

Since the singular values are the positive square roots of the eigenvalues of a Hermitian matrix and the singular vectors are the corresponding eigenvectors of these Hermitian matrices, all of the perturbation theorems that apply to Hermitian matrices are directly applicable to these quantities. For details see for example, Chapter 1 and Stewart [1973]. Thus, for small perturbations in the given matrix A we can expect small perturbations in the singular values and in the subspaces spanned by the singular vectors. This behavior is in sharp contrast to the sensitivity of the eigenvalues and eigenvectors of a non-Hermitian matrix to such perturbations, as the following simple example clearly demonstrates. This example has appeared in many different places and is attributed to Wilkinson [1965].

EXAMPLE 5.2.1 For any positive number $\varepsilon > 0$, let A(n) be the nxn matrix with 1's on the superdiagonal and 0's elsewhere. Let B(n) be the matrix obtained from A(n) by perturbing the first element in the nth row by ε. For example

$$A(4) = \begin{bmatrix} 0 & 1 & 0 & 0 \\ 0 & 0 & 1 & 0 \\ 0 & 0 & 0 & 1 \\ 0 & 0 & 0 & 0 \end{bmatrix} \quad \text{and} \quad B(4) = \begin{bmatrix} 0 & 1 & 0 & 0 \\ 0 & 0 & 1 & 0 \\ 0 & 0 & 0 & 1 \\ \varepsilon & 0 & 0 & 0 \end{bmatrix} \tag{5.2.1}$$

The eigenvalues of A(n) are all 0, but the eigenvalues of B(n) are the nth roots of ε. If $n=10$ and $\varepsilon = 10^{-10}$, then .1 is an eigenvalue of B(10). Thus, a small perturbation 10^{-10} gives a magnitude change in the eigenvalues which is larger than the perturbation by a factor of one billion.

Because of the stability of singular values under perturbations it may be advantageous to use the singular value decomposition rather than the corresponding eigenvalue-eigenvector relationships in analyzing certain physical systems. A key consideration in making this replacement is whether or not we can obtain the desired information by looking only at the singular values and vectors instead of the eigenvalues and eigenvectors.

A natural question to ask is whether or not there is any direct relationship between the singular values and the eigenvalues of a square matrix. The answer unfortunately is that there is seldom a simple relationship. Example 5.2.1 can also be used to illustrate the large differences which can exist between the eigenvalues of a matrix and its singular values. The eigenvalues of A(n), for any n, are all 0. However, n-1 of the singular values of A(n) are equal to 1 and only one singular value is zero. The rank of A(n) is therefore n-1. This is easily verified by forming $[A(n)]^T[A(n)]$ and looking at the resulting matrix. For this particular example the smallest singular value and the smallest eigenvalue are identical. However, that is not typically true.

Theorem 5.2.1 is given in Marcus [1960]. A proof for part (ii) was given in Weyl [1949]. Part (i) follows from the fact that the singular values are the nonnegative square roots of the eigenvalues of A^HA and therefore if we take any unit right eigenvector p_i of A corresponding to λ_i, we have that $\sigma_n^2(A) \leq p_i^H A^H A p_i \leq \sigma_1^2(A)$. This follows since the algebraically-largest and the algebraically-smallest eigenvalues of any Hermitian matrix B are respectively, the maxima and the minima of corresponding Rayleigh quotients $p^H B p$. Part (iii) follows directly from the definitions of the spectral and the Frobenius matrix norms, $\|A\|_2$ and $\|A\|_F$, and the fact that the trace of any matrix equals the sum of its eigenvalues.

THEOREM 5.2.1 Let A be a square nxn matrix. Let $\sigma_1(A) \geq \dots \geq \sigma_n(A)$ be the singular values of A. Let $|\lambda_1(A)| \geq \dots \geq |\lambda_n(A)|$ be the eigenvalues of A. Then

(i) For all j, $\sigma_n(A) \leq |\lambda_j(A)| \leq \sigma_1(A) \leq \|A\|_F \equiv \left[\sum_{i,j=1}^{n} |a_{ij}|^2 \right]^{1/2}$

(ii) For all k, $\prod_{j=1}^{k} |\lambda_j(A)| \leq \prod_{j=1}^{k} \sigma_j(A)$, with equality at $k=n$ (5.2.2)

(iii) $\|A\|_2 = \sigma_1$, $\|A\|_F^2 = \sum_{j=1}^{n} \sigma_j^2$. If $\sigma_n > 0$, then $\|A^{-1}\|_2 = 1/\sigma_n$.

Theorem 5.2.1 tells us that the largest and the smallest singular values are respectively greater than or equal to the magnitude of the eigenvalue which is largest in magnitude and less than or equal to the magnitude of the eigenvalue which is smallest in magnitude. In Section 5.3 we will see that this means that we cannot predict the sensitivity of the solution of a system of equations Ax = b to changes in either A or b (when A is not normal) by looking only at the eigenvalues of A.

Normal matrices have the property that the singular values and the eigenvalues have the same magnitudes (when the eigenvalues are ordered by decreasing magnitude). A is a normal matrix if and only if $A^H A = AA^H$. We have the following well-known equivalent characterization of normality.

THEOREM 5.2.2 The nxn matrix A is normal if and only if there exists a unitary transformation X that diagonalizes A. That is, $A = X \Lambda X^H$ where Λ is diagonal and X is unitary.

COROLLARY 5.2.2 If A is a normal matrix and if $|\lambda_1| \geq \ldots \geq |\lambda_n|$, then $\sigma_j(A) = |\lambda_j(A)|$.

Proof of Corollary 5.2.2. We have that $A = X \Lambda X^H$, where X is a unitary matrix. Write $\lambda_j = e^{i\theta_j}\sigma_j$ where $\sigma_j \equiv |\lambda_j|$. Define $D_\theta = \text{diag}\{e^{i\theta_1}, \ldots, e^{i\theta_n}\}$, and $Y \equiv XD_\theta$. Then $A = Y\Sigma X^H$ is a singular value decomposition of A where $\Sigma = D_\theta^{-1}\Lambda = \text{diag}\{\sigma_1,\ldots,\sigma_n\}$. $\square$

Ruhe [1975] proves that a matrix is almost normal, see Henrici [1962], if and only if its singular values are close to the magnitudes of its eigenvalues, and that the closeness of all of the singular values of a matrix to all of its eigenvalues implies that the eigenvalue problem for that matrix is well-conditioned.

In the rest of this Chapter we will assume that the matrix that we are given is real. The ratio of the largest to the smallest singular value provides a measure of the sensitivity of the solution of Ax = b to perturbations in the problem parameters. An alternative procedure for measuring the condition of a system of linear equations, has been devised by Cline, Moler, Stewart and Wilkinson [1979]. An approximation to the condition of the given matrix is obtained by solving 2 cleverly-chosen systems of equations involving A and A^T. This procedure however, will not tell you how persistent the ill-conditioning is. Ill-conditioning can be caused by only one or a few or by many small or exceptionally-large singular values.

LOWER BOUNDS ON SINGULAR VALUES Since the ratio of the largest singular value to the smallest is of such interest, it is important to be able to obtain bounds on these singular values. It is easy to obtain upper bounds on the largest singular value of a matrix, any matrix norm of A will provide such a bound, but very difficult to obtain a lower bound on the smallest singular value. Varah [1975] obtained such a lower bound for the very special case of a matrix which is strictly diagonally dominant both by rows and by columns. As a slight digression we derive an extension of this result to a different class of matrices. A related paper by Anderson and Karasalo [1975] looks at another special case, triangular matrices. Matrices are seldom triangular. However, it is true that if we form a QR factorization of A, that is, A = QR where Q is an orthogonal matrix and R is upper triangular, then the singular values of R are the same as the singular values of A. This statement follows from Lemma 5.2.6. Thus, any lower bound on the singular values of R is a lower bound on the singular values of A.

DEFINITION 5.2.1 Given any nxn matrix A define $\hat{A} \equiv (\hat{a}_{ij})$ where $\hat{a}_{ii} \equiv |a_{ii}|$ and $\hat{a}_{ij} \equiv -|a_{ij}|$ for $i \neq j$. Any matrix $B \equiv (b_{ij})$ is a nonsingular M-matrix if and only if $b_{ij} \leq 0$ for $i \neq j$, and $\det(B_k) > 0$ for $1 \leq k \leq n$ where B_k is the kxk leading principal submatrix of B. Any matrix A is called an H-matrix if and only if $\hat{A}$ is a nonsingular M-matrix.

We note that much has been written about such matrices, see for example Ostrowski [1937], Fiedler and Ptak [1962], Varga [1962, 1976] and Varah [1975].

LEMMA 5.2.1   Let A be an H-matrix. Then $\sigma_n(A) \geq \sigma_n(\hat{A})$.

Proof.   A straight-forward induction argument demonstrates that $|A^{-1}| \leq \hat{A}^{-1}$. Choose z such that $\|z\|_2 = 1$. Then we have that

$$\sigma_1^2(A^{-1}) \leq |z|^T |A^{-1}|^T |A^{-1}| |z| \leq |z|^T \hat{A}^{-T} \hat{A}^{-1} |z| \leq \sigma_1^2(\hat{A}^{-1}).$$

But, for any matrix B, $\sigma_1(B^{-1}) = (\sigma_n(B))^{-1}$. $\square$

We need the following well-known Lemma relating the $\| \ \|_\infty$, $\| \ \|_1$, and the $\| \ \|_2$ norms. These relationships are used in the proof of Lemma 5.2.3.

LEMMA 5.2.2   For any matrix A we have that $\|A\|_1 = \|A^H\|_\infty$, and $\|A\|_2^2 \leq \|A\|_1 \|A\|_\infty$. In particular for any nonsingular matrix B,

$$\sigma_n(B) = \|B^{-1}\|_2^{-1} \geq \left(\|B^{-H}\|_\infty^{-1} \|B^{-1}\|_\infty^{-1} \right)^{1/2}.$$

We need the following Lemma of Varah [1975]. It provides a lower bound on the smallest singular value of any matrix that is both row and column dominant, and the lower bound is given in terms of the size of the dominance.

LEMMA 5.2.3   Varah [1975]. For any nxn matrix A that is strictly diagonally dominant

$$\|A^{-1}\|_\infty^{-1} \geq \delta_R \equiv \min_{1 \leq i \leq n} \{ |a_{ii}| - \sum_{j \neq i}^n |a_{ij}| \} > 0.$$

If in addition, A^H is also strictly diagonally dominant with corresponding row dominance δ_C, then

$$\sigma_n(A) = \|A^{-1}\|_2^{-1} \geq [\delta_R \delta_C]^{1/2}.$$

We also need the following Lemma which relates the singular values of a diagonally-scaled matrix of full rank to those of the original matrix.

LEMMA 5.2.4   Let A be any real ℓxn matrix of full rank with $\ell \geq n$. Let $u > 0$ be a positive n-vector with $\|u\|_\infty = 1$. If $B \equiv AD_u$, then $\sigma_n(A) \geq \sigma_n(B)$.

Proof.   There exists a vector z_0 with $\|z_0\|_2 = 1$ and $\sigma_n^2(A) = z_0^T A^T A z_0$. Define $y_0 \equiv D_u^{-1} z_0$. Then $\|y_0\|_2 \geq 1$ and

$$\sigma_n^2(A) = y_0^T D_u A^T A D_u y_0 \geq y_0^T B^T B y_0 / \|y_0\|_2^2 \geq \sigma_n^2(B).$$

COROLLARY 5.2.4 Let A be any real nonsingular square matrix. Let $u > 0$ and $v > 0$ be positive n-vectors with $\|u\|_\infty = \|v\|_\infty = 1$. If $C \equiv D_v A D_u$, then $\sigma_n(A) \geq \sigma_n(C)$.

Proof. Observe that $C = D_v B$ where $B = A D_u$. Apply Lemma 5.2.4 to C^T, obtaining $\sigma_n(B) \geq \sigma_n(C^T)$. But the singular values of C^T are the same as those of C, so the desired result is obtained by invoking Lemma 5.2.4 for B. $\qquad\square$

Thus, if for a given matrix we could identify positive vectors u and v for which the matrix $C \equiv D_v A D_u$ was strictly diagonally dominant w.r.t rows and w.r.t columns, then we could apply Lemma 5.2.3 and obtain a lower bound on the smallest singular value of A. If we make one more assumption about A, we can in fact obtain such a bound. We assume that A is irreducible.

DEFINITION 5.2.2 A matrix A is irreducible in the eigenvalue sense if and only if there is no permutation matrix P such that

$$PAP^T = \begin{bmatrix} A_1 & 0 \\ A_2 & A_3 \end{bmatrix}, \quad \text{where } A_1 \text{ is square.}$$

Whether or not a given matrix is an irreducible H-matrix can be verified by the triangular factorization of $\hat{A}$. Let

$$\hat{A} = \hat{L}\hat{U}, \quad \hat{L} = (\hat{\ell}_{ij}), \quad \hat{U} = (\hat{u}_{ij})$$

where $\hat{L}$ is lower triangular and $\hat{U}$ is unit upper triangular. If $\hat{\ell}_{ii} > 0$, for $1 \leq i \leq n$, then $\hat{A}$ is a non-singular M-matrix . Moreover, if for $1 \leq k \leq n - 1$ there exist $i,j > k$ such that $\hat{\ell}_{ik} \hat{u}_{kj} \neq 0$, then $\hat{A}$ is irreducible.

A major consequence of $\hat{A}$ being irreducible is that $\hat{A}^{-1} > 0$. For such a matrix the eigenvalue of maximum modulus, λ_*, is unique, positive, real and simple. Moreover, there exist positive vectors u , $v > 0$ such that

$$\|u\|_\infty = \|v\|_\infty = 1 \text{ and } \hat{A}^{-1}u = \lambda_* u, \quad \hat{A}^{-T}v = \lambda_* v.$$

These quantities are called the Frobenius root and vectors, see for example Householder [1964]. Combining the above Lemmas and comments we obtain the following Lemma.

LEMMA 5.2.5 If A is an irreducible H-matrix, and λ_*, u, v, are respectively, the Frobenius root and vectors of $\hat{A}^{-1}$, then we have the following lower bound on the smallest singular value of A.

$$\sigma_n(A) \geq \sigma_n(\hat{A}) \geq \min_i [u(i)v(i)] / \lambda_*.$$

Proof. First we prove that $B \equiv D_v A D_u$ is diagonally dominant both by rows and by columns. This is easily done by direct computation. Note that $\hat{b}_{ij} = \hat{a}_{ij} v(i) u(j)$ so that the row dominance of the ith row is given by the following expression.

$$\delta_{R_i} = |b_{ii}| - \sum_{k \neq i}^{n} |b_{ik}| = v(i)[\hat{A}u](i) = [v(i)u(i)] / \lambda_*.$$

A similar computation demonstrates that the column dominance $\delta_{C_i} = \delta_{R_i}$. Thus, B satisfies the conditions of Lemma 5.2.3 and we obtain the desired result. $\qquad\square$

Thus, we have obtained a lower bound on the smallest singular value of any matrix A that is an irreducible H-matrix. The computation of this bound requires the Frobenius root of $\hat{A}^{-1}$ and the corresponding left and right Frobenius vectors.

The preceding arguments can also be used to obtain a lower bound for the smallest singular value of any nonsingular, irreducible triangular matrix. If A is a nonsingular upper triangular matrix, define $\hat{A}$ as before and define the upper triangular matrix $\tilde{A} \equiv (\tilde{a}_{ij})$ such that for $j > i$, $\tilde{a}_{ij} = - \max_{k>i} |a_{ik}|$ and $\tilde{a}_{ii} = |a_{ii}|$. It is trivial to demonstrate that any nonsingular triangular matrix is an H-matrix. This is obviously the case since the principal minors of $\hat{A}$ are just products of its diagonal entries. It is also clear that $|A^{-1}| \leq \hat{A}^{-1} \leq \tilde{A}^{-1}$, and that therefore $\sigma_n(A) \geq \sigma_n(\hat{A}) \geq \sigma_n(\tilde{A})$. A lower bound on the smallest singular value of $\tilde{A}$ can be obtained from an estimate of the largest eigenvalue of the matrix $\tilde{A}^{-T} \tilde{A}^{-1}$. But, such an estimate is easily obtainable. Given $v > 0$ we solve for $u > 0$ such that $\tilde{A}^T \tilde{A} u = v$. Then, we have that $\min_i [u(i)/v(i)] \leq \sigma_1(\tilde{A}^{-1}) \leq \max_i [u(i)/v(i)]$, see for example Varga [1962, p32]. For an alternative approach to determining bounds, see Jennings [1982].

EFFECTS OF MATRIX SCALINGS ON SINGULAR VALUES We now return to the main discussion to consider the effects of certain types of scalings on the singular values of a given matrix. In solving practical matrix problems we may introduce some type of balancing transformations or scalings. Row or column scalings of a matrix can radically affect the singular values of the matrix. If we are trying to solve a system of equations Ax=b, then we may introduce such scalings to improve the numerical condition of the given problem. Column scaling, where all of the elements of A in a given column are multiplied by a scalar, corresponds to scaling the corresponding component of the unknown vector x. If we are solving a system of equations by least squares then row scaling, where all of the elements in each row of A are multiplied by a scalar, is the same as assigning new weights to the residuals in the sum of squares.

Certain types of transformations do not alter the singular values. One such important family of transformations is the set of unitary transformations. These transformations preserve the lengths of columns (rows) and the singular values of a given matrix. This is stated in the following Lemma.

LEMMA 5.2.6 Let P and Q be unitary transformations of size $\ell x \ell$ and nxn respectively. Then the matrices A and $B \equiv PAQ^H$ have the same singular values. If Y and X are the left and right singular vectors of A, then PY and QX are the left and right singular vectors of B.

The proof of Lemma 5.2.6 is trivial since if $A = Y\Sigma X^H$ then $B = (PY)\Sigma(X^HQ^H)$. The singular value decomposition procedure in LINPACK [1979] makes extensive use of this result. There the given matrix is transformed via orthogonal transformations into a bidiagonal matrix whose singular values are the same as those of the original matrix.

SECTION 5.3   APPLICATIONS

SOLVING SYSTEMS OF EQUATIONS First we consider the problem of solving a system of equations $Az = b$, and show how the sensitivity of the solution of such a system to perturbations in the matrix is related to the singular values of the matrix A. An excellent reference for this part of the discussion is Lawson and Hanson [1974]. If A is square and A^{-1} exists, then the solution z^* exists, is unique and $z^* = A^{-1}b$. However, if A is singular (or nearly singular) or $\ell \neq n$, then how do we obtain a 'solution'? To be specific we assume that A is real and that $\ell \geq n$. We have the following definition of the pseudoinverse of a matrix.

DEFINITION 5.3.1 The pseudoinverse A^+ of an ℓxn matrix A is that matrix such that $A^+AA^+ = A^+$, $AA^+A = A$, and AA^+ and A^+A are Hermitian.

Instead of trying to find a vector z such that $Az = b$, we determine the vector z^* of minimum norm that solves the following minimization problem.

$$\min_{z \epsilon E^n} \| Az - b \|^2. \tag{5.3.1}$$

If $\ell > n$, then it is very unlikely that there is a z such that $Az = b$, and the minimum in Eqn(5.3.1) will not be 0. Substituting the singular value decomposition of A into Eqn(5.3.1), using the definitions of rank and of Y_1, Y_2, and of X_1, X_2 given in Theorem 5.1.1, and setting $z = z_1 + z_2$ where $z_1 = X_1\gamma_1$ and $z_2 = X_2\gamma_2$, we obtain the equivalent minimization problem

$$\min_{z_1 \epsilon E^r} \| \Sigma_{11}X_1^Tz_1 - Y_1^Tb \|^2 + \| Y_2^Tb \|^2. \tag{5.3.2}$$

Observe that only the first term in statement (5.3.2) depends upon z and this term can be made to vanish by setting

$$\Sigma_{11}X_1^Tz_1 = Y_1^Tb. \tag{5.3.3}$$

Therefore any z which minimizes expression (5.3.1) is of the form $z = z_1 + z_2$ where z_1 is given by Eqn(5.3.3) as $X_1(\Sigma_{11}^{-1}Y_1^Tb)$ and z_2 is orthogonal to z_1 but otherwise arbitrary. But since $\| z \|^2 = \| z_1 \|^2 + \| z_2 \|^2$ the solution of minimum norm must have $z_2 = 0$.

The pseudoinverse A^+ of A is the transformation which maps a given vector b into the vector $z^*(b)$ of minimum norm which minimizes the expression in Eqn(5.3.1). We obtain the following representation for A^+ from the singular value decomposition of A.

$$A^+ = X_1 \Sigma_{11}^{-1} Y_1^T. \tag{5.3.4}$$

It is not difficult to demonstrate directly that A^+ as defined in Eqn(5.3.4) satisfies the requirements of Definition 5.3.1. We note also that we could write

$$A^+ = X \Sigma^+ Y^T \text{ where } \Sigma^+ = [\Sigma_{11}^{-1} \ 0]. \tag{5.3.5}$$

Eqns(5.3.4) and (5.3.5) clearly point out the fact that Y and X vectors corresponding to either the zero singular values of A or to $\ell \neq n$, play no role in the pseudoinverse so they do not have to be computed. Weighted least squares computations appear in many applications, see for example Lawson and Hanson [1974].

Pseudoinverses should, however, be used with care as the following well-known trivial example illustrates. A small perturbation in the matrix A may cause very large changes in the pseudoinverse of A. Let

$$A(\varepsilon) = \begin{bmatrix} 1 & 0 \\ 0 & \varepsilon \end{bmatrix} \text{ then } A^+(0) = \begin{bmatrix} 1 & 0 \\ 0 & 0 \end{bmatrix} \text{ but } A^+(10^{-10}) = \begin{bmatrix} 1 & 0 \\ 0 & 10^{10} \end{bmatrix}. \tag{5.3.6}$$

In fact $A^+(\varepsilon)$ is clearly discontinuous at $\varepsilon = 0$. For numerical computations with the pseudoinverse it is necessary to incorporate certain modifications based upon the effective rank of A which we discuss briefly below.

Now consider the question of how the minimum norm solution of the least squares problem in Eqn(5.3.1) varies with perturbations (errors) in the matrix A. If we knew a priori that A had rank $r<n$, then we would set the n-r smallest singular values of A to zero. A worst case estimate of the variations could then be obtained using the corresponding condition number

$$K_r(A) \equiv \sigma_1/\sigma_r. \tag{5.3.7}$$

We have the following Theorem from LINPACK [1979, Section 11.4], Stewart [1977] which provides such a worst case estimate.

THEOREM 5.3.1 (Stewart [1977]) Let E be a perturbation of A satisfying

$$\sigma_{r+1} < \|E\|_2 \leq 10 \|E\|_2 \leq \sigma_r.$$

Let z_0 be the solution of $Az = b$ and let $z_0 + \delta z_0$ be the solution of $(A+E)z = b$. Then the relative change in the solution satisfies

$$\frac{\| \delta z_0 \|}{\| z_0 \|} \leq 9 \left[K_r(A) + K_r^2(A) \frac{\| w \|}{\| A \| \ \| z_0 \|} \right] \frac{\| E \|}{\| A \|} \qquad (5.3.8)$$

where $w \equiv b - Az_0$ is the corresponding residual.

In practice we do not know the rank of A a priori. An effective rank is often determined by simply looking at the computed singular values, deciding that some of them are 'negligible', and then treating these negligible singular values as zeroes. Unfortunately if A has not been scaled properly, the corresponding effective rank obtained in this manner may give misleading results. Proper choices for scalings are not readily available; in fact this topic is not well-understood. Golub and Kahan [1965] and LINPACK [1979, Introduction] recommend scaling the rows and columns of A so that the estimates of the errors in each of the entries of the resulting matrix are all approximately equal to a common value ε . The matrix A becomes $A_s = D_R A D_C$ where D_R and D_C are diagonal scaling matrices and the equation $Az = b$ becomes $A_s y = b_s$ where $b_s = D_R b$ and $z = D_C y$.

When determining the rank of a matrix by looking at the sizes of the singular values of that matrix, we must take into account the relative sizes of the singular values which are treated as 'zero' and the remaining nonzero singular values. The singular values that are not suppressed must all be significantly larger than the errors in the entries of the matrix A (or in the vector b). To have some real understanding of the sensitivity of a given system of equations to perturbations, we should have not only an estimate of the 'condition number' σ_1/σ_n, but of the set of condition numbers σ_1/σ_k for $k = n, n-1,...,s$ for some appropriate choice of s.

If for some given ill-conditioned matrix computation we know the above set of condition numbers, then it may be possible to modify the required computation so that it becomes numerically stable. We find examples of this form of stabilization in many papers, see for example, Lawson and Hanson [1974], Ekstrom and Rhoads [1974], Fisher and Howard [1980] and Barakat and Buder [1979]. Below we give an example of an ill-conditioned problem where we need some type of numerical stabilization. In practice these types of computations are further complicated by the presence of measurement errors (noise) in the data. In its original continuous formulation this type of problem would typically be called an ill-posed problem. A problem is ill-posed (or ill-conditioned) if small changes in the data can yield large changes in the solution. For more discussion of these kinds of ill-posed problems see e.g. Tikhonov and Arsenin [1977].

EXAMPLE, ILL-POSED PROBLEM   Consider the following type of ill-posed problem. Given a kernel K and an experimentally-measured function g, compute f such that

$$g(t) = \int_0^1 K(t,s)f(s)ds, \qquad 0 \leq t \leq 1. \qquad (5.3.9)$$

In practice data is available only at a finite number of points and we have to use a discrete matrix approximation to Eqn(5.3.9) of the form Az=b. A will be an invertible but ill-conditioned matrix. Some specific applications where such equations arise are the numerical differentiation of tabulated data, the deconvolution of data obtained in spectroscopy experiments, inverse problems in geophysics, and signal processing problems, such as in radar and sonar. For examples of some of these applications see the following references, Anderssen and Bloomfield [1974], Huang and Parrish [1978], Backus and Gilbert [1970], and Andrews and Hunt [1977].

The objective is to compute a solution f over the entire interval. Since the data is determined experimentally, it will almost certainly contain errors. We assume that we do not have an a priori model for f and that we do not know the covariance of the measurement errors. We make two basic assumptions. (A-1) The desired solution f is smooth, possessing one or more derivatives. (A-2) The noise in the data is random, additive and highly oscillatory.

A numerical procedure for computing f in Eqn(5.3.9) should have the following objectives: (O-1) to replace the original problem by a better-conditioned problem; (O-2) to reduce the influence of the noise on the computed solution; (O-3) to provide a physically-meaningful solution that approximates the true solution in some reasonable sense. Clearly, the errors in the data impose a limitation on the achievable accuracy.

Modified singular value decompositions have been used to solve such problems. There is singular value decomposition with truncation and singular value decomposition with damping. (More details on some of the comments made here can be found in Cullum [1979,1980] and in the above references.) We will use the notation $Kf = g$ to denote Eqn(5.3.9). We assume that K is a 1-1 mapping whose kernel is in L_2 and we obtain an equation of the form Az=b by using some quadrature rule. Picard's Theorem, Smithies [1958], gives necessary and sufficient conditions for the existence of a solution to Eqn(5.3.9).

THEOREM 5.3.2 Let K be in $L_2([0,1] \times [0,1])$ and $g \in L_2[0,1]$. Let y_k, k=1,2,... be a complete, orthonormal set of eigenfunctions for the operator KK^T and let x_k, k=1,2,... be a complete, orthonormal set of eigenfunctions for the operator K^TK where K^T is the adjoint of K. Let σ_k^2 be the corresponding set of eigenvalues. Then Eqn(5.3.9) has an L_2 solution f if and only if g is in the closure of the range of K and

$$\sum_{k=1}^{\infty} (g^Ty_k)^2 / \sigma_k^2 < \infty . \qquad (5.3.10)$$

THEOREM 5.3.3 Under the hypotheses of Theorem 5.3.2, Eqn(5.3.9) has the following unique solution

$$f = \sum_{k=1}^{\infty} (g^T y_k) x_k / \sigma_k \qquad (5.3.11)$$

The system $\{x_k\}$, $\{y_k\}$, $\{\sigma_k\}$, k=1,2,... is the singular value decomposition of the operator K, Smithies [1958]. That is,

$$K = \sum_{k=1}^{\infty} \sigma_k y_k x_k^T. \qquad (5.3.12)$$

Similarly we have for any corresponding nxn matrix approximation A of K that $A = Y\Sigma X^T$, where the of columns of Y and X are orthonormal sets of eigenvectors $\{y_j\}$ and $\{x_j\}$, $1 \leq j \leq n$ of AA^T and $A^T A$ respectively, and Σ is the diagonal matrix whose entries are the singular values σ_j, j=1,2,...,n of A. Since A is real, Y and X are orthogonal matrices. We will assume that A is square and invertible.

The method of singular value decomposition with truncation (SVDT) considers the sizes of the singular values and decides which ones are 'not significant'. 'Insignificant' singular values, for example $\sigma_{k+1},...,\sigma_n$, are set equal to zero and the equation

$$\tilde{A}z = \tilde{b} = \tilde{Y}\tilde{Y}^T b \qquad (5.3.13)$$

is solved instead of Az = b. In Eqn(5.3.13) $\tilde{A} = Y\tilde{\Sigma}X^T$ with $\tilde{\Sigma} = \text{diag}\{\sigma_1,...,\sigma_k,0,...,0\}$ and $\tilde{Y} = (y_1, y_2,...,y_k, 0,...,0)$. This is equivalent to replacing the inverse of A by the pseudoinverse of $\tilde{A}$.

SVDT implicitly assumes that the projection of the data b on those singular vectors which correspond to the singular values which have been set to zero is small. In fact it assumes more, namely that $|b^T y_j| / \sigma_j \ll 1$ for all j > k. There is no a priori reason why this would be true for an arbitrary matrix A. However, Theorem 5.3.2 tells us that this is not an unreasonable assumption when the matrix A is obtained by the discretization of a continuous operator K of the type given in Eqn(5.3.9) and n is 'large'. Theorem 5.3.2 requires that $|(g^T y_j)| / \sigma_j \rightarrow 0$ as $j \uparrow \infty$. Varah [1973,1979] derived some error estimates for SVDT, but he considered only the effect of roundoff errors in the computations and not the effects of measurement errors in the data. One problem for which the SVDT approach is not suitable, is the numerical differentiation of data. The spectrum of the operator K for this problem varies continuously and there is no clearly defined 'significant' part. In fact for many problems, deciding which singular values should be set to 'zero' is not a straightforward exercise, and as in the differentiation problem, there may not even be an 'insignificant' subspace. See LINPACK [1979, Introduction pp 7-12] for a discussion of possible difficulties.

Alternatively we could use a singular value decomposition with damping (SVDD) instead of one with truncation, see for example Ekstrom and Rhoads [1974]. Actually they use the eigenvector decomposition of a particular real symmetric matrix associated with their given problem but their ideas extend readily to the singular value decomposition. Using damping instead of truncation allows the small singular values and corresponding singular vectors to remain in the solution, but mollifies their effect by introducing damping factors on each term in the discrete analog of Eqn(5.3.11). The user does not have to arbitrarily set some of the singular values to zero.

For some problems the smallest singular values correspond to those singular vectors which have the maximum number of oscillations (when the magnitudes of the components of the vector are plotted versus the component number). In such problems modifying the terms in the solution which correspond to the smallest singular values also mollifies the effects of the errors in the data whenever these errors are highly oscillatory.

DATA REDUCTION In some applications such as image processing, the amount of data which is generated cannot be transferred reasonably and it is necessary to do data reduction. Sometimes this can be accomplished by using a few of the singular vectors of a matrix A associated with the given information. Typically, these would be the singular vectors associated with the largest singular values. In each case the amount of information retained must be small enough to be processed readily but still sufficient to regenerate the essence of the original information. The following well-known theorem of Eckart and Young [1936] tells us that for any prespecified rank r, the singular value decomposition of a given matrix can be used to obtain the 'best' matrix of that rank which approximates A. See for example Stewart [1973, p.322].

THEOREM 5.3.4 (Eckart and Young [1936]) Let A be a ℓxn real matrix with singular value decomposition $A = Y\Sigma X^T$. Then the matrix B_k of rank k which best approximates A, in the sense that

$$\| A - B_k \| \leq \| A - C \| \qquad (5.3.14)$$

for all possible C of rank k, is given by

$$B_k = Y\Sigma_k X^T \quad \text{and} \quad \| A - B_k \|^2 = \sum_{j=k+1}^{n} \sigma_j^2 . \qquad (5.3.15)$$

where Σ_k is obtained from Σ by replacing the singular values σ_j, $j \geq k + 1$ by zeros. This result holds for any norm which is unitarily invariant, that is for any matrix norm for which $\| UA \| = \| A \|$ when U is a unitary matrix.

Theorem 5.3.4 tells us that if we wish to approximate A by another matrix of smaller rank k, then the best we can do (in a least squares sense) is to compute the singular vectors

x_j, y_j $j = 1,...,k$ corresponding to the k largest singular values and approximate A by

$$B_k \equiv \sum_{j=1}^{k} \sigma_j y_j x_j^T. \tag{5.3.16}$$

Much of the above discussion appeared in Golub [1968]. Examples of the use of the k singular vectors corresponding to the k largest singular value to achieve data reduction schemes are given in Andrews and Hunt [1977].

SECTION 5.4   LANCZOS PROCEDURE, SINGULAR VALUES AND VECTORS

A reliable computer program for computing the singular values and singular vectors of small to medium size matrices (order less than several hundred) is contained in the LINPACK Library [1979]. This program is based upon the algorithm given in Golub and Kahan [1965]. It is a modification of the program given in Golub and Reinsch [1970]. A sequence of orthogonal transformations is applied to the given $\ell \times n$ matrix A to explicitly transform it into a bidiagonal matrix with the same singular values. The singular values and singular vectors of the bidiagonal matrix are computed. The orthogonal transformations which were used to obtain the bidiagonal matrix are then used to transform these singular vectors into singular vectors for A.

Typically, the orthogonal transformations do not preserve the zero-nonzero structure of A. Fill-in occurs. (See Example 3.1 in Chapter 3.) Thus even if the original matrix is sparse, the transformed matrices probably will not be sparse and the locations in the matrix at which the fill-in occurs cannot be specified a priori. Therefore, it is necessary to provide enough computer storage for a full $\ell \times n$ matrix. The number of arithmetic operations required is a function of the product $[\min(\ell,n)] [\max(\ell,n)]^2$. This program computes all of the singular values of the given matrix A in a magnitude size order which cannot be determined a priori. See LINPACK [1979, Chapter 11] and Wilkinson [1978] for more details.

Many applications involve large matrices, and the large storage requirements and arithmetic operations counts required by the LINPACK singular value algorithm deter its use on large matrices. For example, if $n = \ell = 1000$, then more than 4 million bytes of computer storage are required. Moreover, the user usually wants only some subset of the singular value spectrum not the whole spectrum and only a few of the corresponding singular vectors. Which few are required of course depends upon the particular application.

One obvious approach to the problem of computing singular values is the direct computation of the eigenvalues of the associated Hermitian matrix $A^H A$ (or AA^H if $\ell \leq n$). However, as the following well-known example illustrates, this approach may lead to erroneous results because of the squaring effect. Let

$$A^H = \begin{bmatrix} 1 & 0 & \mu \\ 1 & \mu & 0 \end{bmatrix} \quad \text{then } A^H A = \begin{bmatrix} 1+\mu^2 & 1 \\ 1 & 1+\mu^2 \end{bmatrix} \quad .$$

If $|\mu| < \sqrt{\varepsilon_{mach}}$ where ε_{mach} is the machine epsilon, then numerically $\mu^2 = 0$ and $[A^H A](i,j) = 1$, for $1 \leq i,j \leq 2$. Therefore, numerically $A^H A$ would have rank 1 and its eigenvalues would be $\sigma_2^2 = 0$ and $\sigma_1^2 = 2$. However, it is clear that numerically A has rank 2, and its singular values are $|\mu|$ and $[2 + \mu^2]^{1/2}$.

Thus, very small singular values may appear numerically as zero eigenvalues in $A^H A$ or AA^H. The squares of any singular values less than 1 are also closer together than the singular values themselves are. At the other end of the spectrum large singular values become larger eigenvalues and as eigenvalues they have bigger gaps between them. The smallest and the largest singular values of A are on the extremes of the spectrum of either $A^H A$ or AA^H.

There is however an alternative transformation Lanczos [1961] which preserves the sizes of the singular values. This transformation has also been used in the same context elsewhere. See in particular, Golub and Kahan [1965]. Paige [1974] uses this transformation in a procedure for solving least squares problems. O'Leary and Simmons [1981] and Bjorck and Duff [1980] use it in procedures for solving ill-posed problems. We restrict the discussion to real matrices, and without loss of generality in the following discussions we assume that $\ell \geq n$ unless we specifically state otherwise.

Associated with any real $\ell \times n$ matrix A is the real symmetric $(\ell + n) \times (\ell + n)$ matrix

$$B = \begin{bmatrix} 0 & A \\ A^T & 0 \end{bmatrix}. \tag{5.4.1}$$

It is straightforward to demonstrate that the eigenvalues of B are the $n \pm$ pairs of the singular values of A plus $(\ell - n)$ additional zero eigenvalues if A is rectangular. The overall multiplicity of the zero eigenvalue of B is $\ell + n - 2r$ where r is the rank of A. We have the following Lemma which states that the singular value decomposition of A can be used to generate an eigenvalue-eigenvector decomposition of B. The proof follows directly from Definition 5.1.1 and the definition of B. In the remainder of this section we will use the following notation.

DEFINITION 5.4.1 Let y be a $\ell \times 1$ vector and let x be a $n \times 1$ vector then we use $(y,x)^t$ to denote the transpose of the $(\ell + n) \times 1$ column vector $\begin{bmatrix} y \\ x \end{bmatrix}$.

LEMMA 5.4.1 Let A be a real $\ell \times n$ matrix with $\ell \geq n$, and let B be defined by Eqn(5.4.1). Let $A = Y \Sigma X^T$ where $\Sigma \equiv [\Sigma_1, 0]^T$ with $\Sigma_1 \equiv \text{diag}\{\sigma_1,...,\sigma_n\}$ be the singular value decomposition of A. Let r be such that $\sigma_r > \sigma_{r+1} = ... = \sigma_n = 0$. Partition $Y = (Y_1, Y_2)$ such that Y_1 is $\ell \times n$ and Y_2 is $\ell \times (\ell - n)$. Let Z be the following $(\ell + n) \times (\ell + n)$ matrix

$$Z = \frac{1}{\sqrt{2}} \begin{bmatrix} Y_1 & Y_1 & \sqrt{2} Y_2 \\ X & -X & 0 \end{bmatrix}. \tag{5.4.2}$$

Then Z is orthogonal and if

$$\Lambda \equiv \begin{bmatrix} \Sigma_1 & 0 & 0 \\ 0 & -\Sigma_1 & 0 \\ 0 & 0 & 0 \end{bmatrix} \quad \text{then} \quad BZ = Z\Lambda. \tag{5.4.3}$$

That is, Z is an eigenvector basis for B. Furthermore, zero singular values of A appear in the spectrum of B as eigenvalues which have eigenvectors of the form $(y,x)^t$ with $x \neq 0$ and $y \neq 0$.

Proof. The orthogonality of Z follows directly from the orthogonality of Y and the orthogonality of X. Eqn(5.4.3) follows directly from Eqns(5.4.2) and (5.1.3). Finally, the characterization of the eigenvectors corresponding to zero singular values follows from Corollary 5.1.1. □

Therefore, each singular value σ_j of A, including any zero singular values, appears in the symmetrized matrix B as the pair of eigenvalues $\pm \sigma_j$ with a corresponding pair of eigenvectors $(y,x)^t$ and $(y,-x)^t$. For computational purposes we need the converse of Lemma 5.4.1. We want to compute the singular values of A by computing the nonnegative eigenvalues of the larger but real symmetric matrix B, and the singular vectors by computing the corresponding eigenvectors of B. An obvious difficulty arises when $\ell \neq n$. In this case any zero singular values of A will mix with the zero eigenvalues of B arising from the rectangular shape of A. Lemma 5.4.2 below, however, states that any eigenvalue of B arising from a true singular value of A will have an eigenvector of the form $(y,x)^t$ with $x \neq 0$ and $y \neq 0$, whereas those corresponding to eigenvalues arising from the rectangular shape of B have only eigenvectors of the form $(y,0)^t$. We will capitalize on this characterization in our procedure (see Lemma 5.4.4) by attempting to force our singular value procedure to ignore vectors of the form $(y,0)^t$. (If $\ell < n$, then the extraneous zero eigenvalues of B would have only eigenvectors of the form $(0,x)^t$.)

We have the following converse of Lemma 5.4.1 which allows us to obtain suitable sets of singular vectors of A from the computed eigenvectors of B. Recall we are assuming that $\ell \geq n$. Since the vectors Y_2 perform no real function in the singular value decomposition of A, we do not attempt to compute these singular vectors. We only compute right singular vectors X and a suitable set of left singular vectors Y_1.

LEMMA 5.4.2 Let A be a real $\ell \times n$ matrix with $\ell \geq n$ and B be defined by Eqn(5.4.1). (i) For any positive eigenvalue λ_i of B let $(y_i,x_i)^t$ denote a corresponding eigenvector of norm $\sqrt{2}$. Then λ_i is a singular value of A and x_i and y_i are respectively, right and left singular vectors of A corresponding to λ_i. (ii) For $\lambda_i = 0$, if B has corresponding orthogonal eigenvectors $(y_j,x_j)^t$ with $x_j \neq 0$ and $y_j \neq 0$ for $j = 1,...,t$ and some $t \geq 1$, then 0 is a singular value of A, and left singular vectors can be obtained by orthogonalizing these y_j, and right singular vectors can be obtained by orthogonalizing these x_j. Otherwise, A has full rank.

Proof. (i) Consider any positive eigenvalue λ_i of B and a corresponding eigenvector $(y_i,x_i)^t$ with norm $\sqrt{2}$. By Eqn(5.4.1) we have that $Ax_i = \lambda_i y_i$ and $A^T y_i = \lambda_i x_i$, but these are just the singular value/vector equalities given in Eqns(5.1.3). Furthermore, from these equalities we get that $x_i^T x_i = y_i^T y_i = 1$. (ii) Suppose that for $\lambda_i = 0$ and for $j = 1, \ldots, t$ where $t \geq 1$, we have a set $\{(x_j, y_j)^t\}$ of orthonormal eigenvectors of B corresponding to 0 such that $x_j \neq 0$ and $y_j \neq 0$. Then by Lemma 5.4.1 these vectors correspond to zero singular values, so that A does not have full rank. For these vectors we have that $A^T y_j = 0$ and $A x_j = 0$, $x_j \neq 0$ and $y_j \neq 0$ from which it is clear that we can take any orthogonalized combinations of the $x_j(y_j)$ vectors as singular vectors. By Lemma 5.4.1, if no such vectors exist then A has full rank. □

Therefore, we can generate a singular value decomposition of a given ℓxn matrix A by computing the eigenvectors of the related real symmetric matrix B given in Eqn(5.4.1). Golub and Kahan [1965] pointed out that the single-vector Lanczos recursions together with the matrix B in Eqn(5.4.1) could be used to describe an algorithm for computing a singular value decomposition of the matrix A. However, they did not develop this idea because they believed that the numerical difficulties associated with these recursions made such a straight-forward approach too unreliable unless total reorthogonalizations of all of the Lanczos vectors were incorporated. Instead they developed an algorithm based upon Householder transformations. This algorithm has become the universal standard for matrices of order less than a few hundred. In their discussion however, they did make several suggestions with regard to a Lanczos procedure for B. In particular they noted that if the starting vector for the recursion were chosen properly, the nominal amount of work required to perform two iterations of the Lanczos recursion on B could be reduced by half. Lemma 5.4.3 is an extension of these observations.

In the context of developing a Lanczos procedure for solving the linear least squares problem for large sparse matrices, Paige [1974] elaborated upon this Lanczos bidiagonalization algorithm proposed in Golub and Kahan [1965]. In passing he then stated, without presenting any numerical results, that he had tried using this algorithm for computing singular values and that it worked very well. In Cullum, Willoughby and Lake [1983] we presented some comparisons between our proposed procedure and the corresponding procedure implemented as proposed in that paper. There we demonstrated that our procedure provides a unique look-ahead capability that allows the user to obtain early estimates of the spectrum of the given matrix, superior resolution power, and eliminates the ambiguity associated with selecting a tolerance which is used to determine which eigenvalues of the Lanczos matrices are 'good'. See Chapter 4, Section 4.5.

In many practical applications the user wants only some small portion of the singular value spectrum. This portion may or may not be on the extremities of the spectrum, that depends upon the particular application. Golub, Luk and Overton [1981] applied an iterative block Lanczos version of the above Lanczos bidiagonalization procedure to B in Eqn(5.4.1) to compute a few of the largest singular values of a real matrix A and corresponding singular vectors. As demonstrated in Chapter 7, Section 7.3, iterative block Lanczos procedures are maximization procedures for sums of Rayleigh quotients of B, and are therefore suitable for computing a few of the extreme eigenvalues of B. Thus, they are suitable for computing the largest singular values.

The single-vector Lanczos procedure with no reorthogonalization which we describe in this section can be used to compute either a few or many of the distinct singular values at either end of the spectrum or on interior portions.

We obtain a Lanczos singular value procedure for any real ℓxn matrix A by applying the Lanczos procedure for real symmetric matrices described in Chapter 4 to the corresponding matrix B in Eqn(5.4.1) with the particular choice of starting vector suggested by Golub and Kahan [1965]. See Lemma 5.4.3. The Lanczos procedure which we obtain can be used to compute the distinct singular values of many matrices. We emphasize again that we get only the distinct singular values and not their A-multiplicities. As for the real symmetric eigenvalue procedure, getting the true multiplicities would require additional computation. We also emphasize that this procedure cannot be used to determine whether or not a general ℓxn matrix with $\ell \neq n$ has full rank. We cannot guarantee that a zero eigenvalue of B caused by $\ell \neq n$ will not appear in our computations. However, see Lemma 5.4.4, our procedure is designed to suppress such eigenvalues. First in Lemma 5.4.3 we state the recursions which result from applying the Lanczos recursions given in Eqns(4.3.1) - (4.3.2) to the matrix B in Eqn(5.4.1) with the special starting vector.

LEMMA 5.4.3 Let A be a real $\ell \times n$ matrix with $\ell \geq n$, and let B be the associated real symmetric matrix defined in Eqn(5.4.1). Apply the Lanczos tridiagonalization recursion specified in Eqns(4.3.1) - (4.3.2) to B with a unit starting vector $(v,0)^t$ where v is ℓx1 or with a unit starting vector $(0,u)^t$ where u is nx1 . Then in either case the diagonal entries of the real symmetric tridiagonal Lanczos matrices generated are all identically zero, the Lanczos vectors generated alternate between the 2 forms $(v,0)^t$ or $(0,u)^t$, and the Lanczos recursions in Eqns(4.3.1) and (4.3.2) reduce to the following:

(i) For a starting vector of the form $(v,0)^t$, define $v_1 \equiv v$, $u_0 \equiv 0$, and $\beta_1 \equiv 0$. For i=1,...,m we obtain the Lanczos recursions

$$\begin{aligned}
\beta_{2i}\, u_i &= A^T v_i - \beta_{2i-1}\, u_{i-1} \\
\beta_{2i+1}\, v_{i+1} &= A u_i - \beta_{2i}\, v_i
\end{aligned} \tag{5.4.5}$$

(ii) For a starting vector of the form $(0,u)^t$ define $u_1 \equiv u$, $v_0 \equiv 0$, and $\beta_1 \equiv 0$. For i=1,...,m we obtain the Lanczos recursions

$$\begin{aligned}
\beta_{2i}\, v_i &= A u_i - \beta_{2i-1}\, v_{i-1} \\
\beta_{2i+1}\, u_{i+1} &= A^T v_i - \beta_{2i}\, u_i
\end{aligned} \tag{5.4.6}$$

In either case the β_j are chosen to make the u_i and v_i unit vectors.

Proof. The proof consists of directly writing down the Lanczos recursions for B and then observing that they reduce to either Eqns(5.4.5) or (5.4.6). □

Lemma 5.4.3 tells us that with a starting vector of the form $(v,0)^t$ or $(0,u)^t$, the total storage required for 2 successive Lanczos vectors is only $\ell + n$ words. Since each diagonal entry of the Lanczos matrices is 0 only one vector array is needed to store the scalars β_i. At each iteration only either Au or $A^T v$ is computed, not both. Thus, although B is twice the size of A, the computational requirements per iteration of the Lanczos recursion do not increase proportionately when we replace A by B in Eqn(5.4.1).

However, B has twice as many distinct nonzero eigenvalues as A has distinct nonzero singular values since each singular value of A (including also any zero singular values) appears as a ± pair of eigenvalues of B. Moreover, the ± pairs of the small singular values of A appear at the center of the spectrum of B as the most interior eigenvalues. These small singular values are typically the most difficult for our Lanczos procedure to compute. Therefore the number of iterations required to get the desired eigenvalues of B may be considerably more than what would be required for a real symmetric matrix whose eigenvalue distribution matched that of the singular values of A. We present one such specific comparison.

An obvious question to consider is whether or not it matters which choice of starting vector $(v,0)^t$ or $(0,u)^t$ that we use, and whether or not we will lose any part of the desired eigenspace of B if one of these special forms is used. A related question for the case $\ell \neq n$ is what effect does one of these choices have on our ability to compute the zero singular values of A (if there are any) without being confused by zero eigenvalues of B which occur simply because $\ell \neq n$. The following Lemma provides an answer to these questions.

LEMMA 5.4.4 (Exact Arithmetic) Let A be a real ℓxn matrix. Let B be the associated real symmetric matrix given in Eqn(5.4.1).

(i)   If $\ell > n$ then any eigenvector of B corresponding to a zero eigenvalue of B which occurs because $\ell \neq n$, has no projection on vectors of the form $(0,u)^t$ where u is nx1.

(ii)   If $\ell < n$ then statement (i) holds for vectors of the form $(v,0)^t$ where v is ℓx1.

(iii)   If a nx1 vector u is generated randomly, then the expectation for each distinct singular value of A that u will have a projection on a corresponding right singular vector of A is the same as the expectation that for the corresponding distinct eigenvalue of $A^T A$ that u will have a projection on a corresponding eigenvector of $A^T A$. We have a similar statement for the case when $\ell < n$, using v and AA^T.

Proof. The proof of (i) is immediate from Lemma 5.4.1 which states that for $\ell > n$, any eigenvector of B which corresponds to an extraneous 0 eigenvalue caused by $\ell \neq n$ must be of the form $(y,0)^t$ where y is of length ℓ. Similarly, when $\ell < n$, such eigenvectors must be of the form $(0,x)^t$. Part (iii) follows from the fact that the singular vectors of A are eigenvectors of $A^T A$ and of $A A^T$, and eigenvectors of these matrices are respectively, right and left singular vectors of A. $\square$

W.r.t. part (iii) of Lemma 5.4.4, we can actually show that in exact arithmetic the odd-numbered Lanczos vectors generated by applying the Lanczos recursion in Eqn(5.4.6) to B are identical to the Lanczos vectors generated when the real symmetric Lanczos procedure is applied to $A^T A$, with u as the starting vector. In particular by combining the two equations in Eqns(5.4.6) we get that for each i

$$\beta_{2i}\beta_{2i+1}u_{i+1} \;=\; A^T A u_i \;-\; (\beta_{2i}^2 + \beta_{2i-1}^2)u_i \;-\; \beta_{2i-1}\beta_{2i-2}u_{i-1}$$

where $\beta_1 \equiv 0$. Using the fact that in exact arithmetic the vectors u_i are orthonormal, it is straightforward to show that the coefficients in the above recursion are precisely the corresponding α and β coefficients that would be obtained by applying the Lanczos recursions in Eqns(4.3.1) - (4.3.2) to $A^T A$.

COROLLARY 5.4.4 Under the hypotheses of Lemma 5.4.4, if we apply the Lanczos recursions in Eqns(5.4.5) or (5.4.6) to B in Eqn(5.4.1) then (i) if $\ell \geq n$, an appropriate starting vector has the form $(0,u)^t$ where u is nx1 and (ii) if $\ell \leq n$, an appropriate starting vector has the form $(v,0)^t$ where v is ℓx1. These choices minimize the chances of generating eigenvectors of B which are associated with any zero eigenvalues of B arising from $\ell \neq n$.

The real symmetric tridiagonal matrices T_m, m = 1,2,... generated by recursions (5.4.5) or (5.4.6) are of a very simple form. They have zero diagonal entries and the super and the subdiagonal entries are the coefficients β_j generated in the recursions. That is for each m,

$$T_m(i,i) \;=\; 0, \quad \text{and} \quad T_m(i,i+1) \;=\; T_m(i+1,i) \;=\; \beta_{i+1}. \tag{5.4.7}$$

These matrices have some of the properties of the B matrix. We have the following Lemma.

LEMMA 5.4.5 (i) For even orders, m = 2j, j=1,2,... , the eigenvalues of T_m occur in $\pm$pairs. (ii) For odd orders, m = 2j-1, j=1,2,..., the eigenvalues of T_m occur in $\pm$ pairs except for an extraneous '0' eigenvalue. (iii) Given any positive eigenvalue μ of T_m with eigenvector w then the vector $\widetilde{w}$ obtained from w by setting $\widetilde{w}(i) = w(i)$ for i odd and $\widetilde{w}(i) = -w(i)$ for i even is an eigenvector of T_m corresponding to the eigenvalue $-\mu$. (iv) Any eigenvalue μ of T_m with $\mu = O(\varepsilon)$ where $\varepsilon \approx 0$, has an eigenvector w (normalized by setting w(1) = 1) whose even numbered components are $O(\varepsilon)$. If in fact $\mu = 0$, then the even numbered components are 0.

Proof. For any γ , let $d_i(\gamma) \equiv \det(T_i - \gamma I)$. We prove by induction that for any γ and for any j

$$d_{2j}(\gamma) = d_{2j}(-\gamma) \text{ and } d_{2j-1}(\gamma) = -d_{2j-1}(-\gamma). \tag{5.4.8}$$

That is for any order m, if γ is an eigenvalue of T_m then $-\gamma$ is also an eigenvalue of T_m. In addition for odd orders m, $d_m(0) = -d_m(0)$, so therefore 0 must always be an eigenvalue of such a T_m. Set $d_0(\gamma) \equiv 1$. Clearly, $d_1(\gamma) = -\gamma = -d_1(-\gamma)$ and $d_2(\gamma) = \gamma^2 - \beta_2^2 = d_2(-\gamma)$. For any symmetric tridiagonal matrix with 0 diagonal entries we have the following recursion for the principal determinants $d_i(\gamma) = -\gamma d_{i-1}(\gamma) - \beta_i^2 d_{i-2}(\gamma)$. Now assume that (i) and (ii) are true for j=k, then using this determinant recursion we get that $d_{2k+1}(-\gamma) = \gamma d_{2k}(-\gamma) - \beta_{2k+1}^2 d_{2k-1}(-\gamma) = \gamma d_{2k}(\gamma) + \beta_{2k+1}^2 d_{2k-1}(\gamma) = -d_{2k+1}(\gamma)$. Similarly, we get that $d_{2k+2}(-\gamma) = d_{2k+2}(\gamma)$. (iii) is proved by direct substitution. (iv) follows easily from Eqn(3.2.9) in Corollary 3.2.3 in Chapter 3 for the components of eigenvectors of tridiagonal matrices, using mathematical induction, and the relationships used to prove (i) and (ii). □

Lemma 5.4.5 tells us that we need only compute the nonnegative eigenvalues of each T_m and that to avoid confusion with an extraneous zero eigenvalue which occurs only because of the special structure of T_m, we should use only even ordered Lanczos matrices.

Before proceeding we make one more observation related to part (iii) of Lemma 5.4.4 and to our comments about the matrix $A^T A$. Consider for example Eqns(5.4.6). For any m we can write the first equation in matrix form as

$$AU_m = V_m J_m \tag{5.4.10}$$

where $U_m = (u_1,...,u_m)$, $V_m = (v_1,...,v_m)$ and J_m is the mxm bidiagonal matrix with $J(j,j) = \beta_{2j}$ and $J(j,j+1) = \beta_{2j+1}$. The nonzero singular values of J are the same as the positive eigenvalues of

$$K_m \equiv \begin{bmatrix} 0 & J_m \\ J_m^T & 0 \end{bmatrix}.$$

A simple permutation maps K_m into T_{2m}. Since singular values are preserved under permutations we could compute the singular values of A by computing the singular values of J_m or equivalently computing the eigenvalues of the smaller real symmetric tridiagonal matrix $L_m \equiv J_m^T J_m$ where $L_m(j,j) = \beta_{2j}^2 + \beta_{2j-1}^2$ and $L_m(j,j+1) = \beta_{2j}\beta_{2j+1}$. This is just the matrix obtained earlier when we considered $A^T A$. The singular value program in the **LINPACK** Library [1979] works directly with J_m. In that program the singular values of J_m are computed using a modified QR algorithm. Golub, Luk, Overton [1981] also use J_m. In our Lanczos procedure we work directly with the real symmetric tridiagonal matrices T_{2m}. These matrices are needed for our identification test which picks out the 'spurious' eigenvalues. See Sections 4.5 - 4.6 in Chapter 4. They are also needed for computing error estimates for the computed singular values and vectors.

The Lanczos singular value procedure with no reorthogonalization which we describe below is very efficient for large matrices provided that the computation of the matrix-vector products $A^T v$ and Av requires only computer storage and arithmetic operations counts which are linear in $\ell = \max(\ell, n)$. This is true for example for sparse A. The amount of computing time required for a given matrix A depends directly upon how fast the products $A^T u$ and Av can be computed, upon the gap distribution of the singular values considered as eigenvalues of B and upon which subsets of singular values are to be computed. Generally, the denser or more clustered the desired singular values of A are as eigenvalues of B, the more difficult it is to compute them. The convergence observed is analogous to that obtained for real symmetric matrices. See Chapter 4, Section 4.7.

FORTRAN code for this procedure is included in Chapter 6 of Volume 2. Below we outline this procedure for the case $\ell \geq n$. The analog for the case $\ell \leq n$ is easily obtained.

LANCZOS SINGULAR VALUE/VECTOR PROCEDURE ($\ell \geq n$)

Step 1. Set $v_0 = 0$ and $\beta_1 = 0$ and generate a random unit vector u_1 of length n.

Step 2. For i = 1, 2, ... M generate β_j, j = 2,3,...,2M + 1 and unit vectors v_i, u_{i+1} such that

$$\beta_{2i} v_i = Au_i - \beta_{2i-1} v_{i-1}$$
$$\beta_{2i+1} u_{i+1} = A^T v_i - \beta_{2i} u_i$$

(Store the scalars β_j, $1 \leq j \leq 2M + 1$ and the last 2 Lanczos vectors v_M and u_{M+1}.)

Step 3. For some even-ordered $m \leq 2M$, compute the nonnegative eigenvalues of T_m in the intervals of interest. Determine the numerical multiplicities of each of these eigenvalues. Accept each numerically-multiple eigenvalue as a converged singular value of A. Test each simple computed eigenvalue μ of T_m to determine whether or not it is also an eigenvalue of the corresponding $\hat{T}_2$ (the matrix obtained by deleting the first row and column of T_m). If it is, reject that eigenvalue as 'spurious', otherwise accept it as good and label it is an approximation to a singular value of A. In the code in Volume 2, Chapter 6 the test for numerical multiplicity and the $\hat{T}_2$ test for spurious eigenvalues are done simultaneously with the eigenvalue computations with essentially no extra cost.

Step 4. Using inverse iteration on $(T_m - \mu I)$ compute an error estimate for each μ which is an 'isolated' (considered as an eigenvalue of T_m) approximate singular value of A.

Step 5. Test for convergence of the computed approximate singular values, using the error estimates. If convergence is observed terminate the eigenvalue procedure and proceed to the singular vector computations. Otherwise increment m and return to Step 3 to compute the eigenvalues of T_m for the larger m on those intervals of interest where convergence has not yet been observed. Depending upon the original choice of M, this may mean enlarging T_{2M}, if so go to Step 2 instead of Step 3 and enlarge T_{2M}.

Step 6. Compute left and right singular vectors for each of a user-selected subset of the computed converged singular values by computing the corresponding eigenvectors of B. This computation uses the Lanczos eigenvector procedure described in Chapter 4, Section 4.8, appropriately modified to incorporate the special structure of T_m and of the Lanczos vectors. For each computed Ritz vector $(y,x)^t$ of B, compute $\bar{y} \equiv y/\|y\|$ as an approximation to a left singular vector of A, and compute $\bar{x} \equiv x/\|x\|$ as an approximation to a right singular vector of A.

The determination (see Step 3) of the 'good' eigenvalues of the Lanczos matrices is accomplished using the same test as was used in the real symmetric case. See Chapter 4 Section 4.5 for a discussion of this test.

To demonstrate how our singular value procedure works in practice we applied this procedure to a fairly large $\ell = n = 411$ test problem A_1 which was generated in a power system study. We computed a few of the largest and a few of the smallest singular values of A_1 and some of the corresponding singular vectors. A_1 is actually a real symmetric matrix. However, for test purposes we can treat it as a nonsymmetric matrix, computing the eigenvalues as singular values and the eigenvectors as singular vectors. We also applied our real symmetric Lanczos procedure to A_1. The numerical results are summarized in Tables 5.4.1 - 5.4.4.

TABLE 5.4.1

Example A_1. Convergence of Largest Singular Values.

Order of T_m	Computed Singular Value	Computed Amingap	Error Estimate
20=.024N	2.36524	.76	4×10^{-7}
	1.60550	.04	5×10^{-2}
	1.56259	.04	4×10^{-2}
40=.049N	2.36524	.746	1×10^{-13}
	1.61887	.045	2×10^{-5}
	1.57375	.037	5×10^{-5}
	1.53648	.037	2×10^{-4}
	1.44614	.02	2×10^{-3}
	1.42350	.02	1×10^{-3}
60=.073N	1.61887		*
	1.57375		*
	1.53648		8×10^{-12}
	1.44615		6×10^{-10}
	1.42350		4×10^{-10}

The eigenvalues of A_1 are not known analytically. The real symmetric Lanczos procedure tells us that A_1 has eigenvalues which range from -2.365 to -1.46×10^{-8}, and that the gaps between successive eigenvalues vary from .746 at $\lambda_1 = -2.365$ to 1.04×10^{-5} at $\lambda_{58} = -.65344$. Most of the gaps vary between 2×10^{-3} and 10^{-4}.

In the power system application the small eigenvalues (singular values) and corresponding eigenvectors (singular vectors) are of interest. Here however, we consider the convergence of both ends of the singular value (eigenvalue) spectrum. First consider the convergence of the 6 largest singular values as m is increased. The observed convergence is summarized in Table 5.4.1. In practice one would not compute the eigenvalues of T_m for so many different values of m.

TABLE 5.4.2

Example A_1. Convergence of Smallest Singular Values.

Order T_m	Computed Singular Value	Computed Amingap	Error Estimate
412=.5N	$.2236 \times 10^{-3}$		4×10^{-2}
	$.1295 \times 10^{-1}$		2×10^{-2}
822=N	$.4159 \times 10^{-4}$		6×10^{-2}
	$.33 \times 10^{-2}$		1×10^{-2}
	$.129 \times 10^{-1}$		5×10^{-3}
	$.143 \times 10^{-1}$		5×10^{-2}
1234=1.5N	$.233 \times 10^{-6}$	.0027	2×10^{-1}
	$.27129 \times 10^{-2}$	.0007	9×10^{-4}
	$.34476 \times 10^{-2}$	.0007	4×10^{-4}
	$.12839 \times 10^{-1}$	.0003	6×10^{-5}
	$.13140 \times 10^{-1}$	.0003	1×10^{-4}
1644=2N	$.1457 \times 10^{-7}$	.0027	2×10^{-6}
	$.27128 \times 10^{-2}$	.0007	2×10^{-9}
	$.34467 \times 10^{-2}$	1.1×10^{-5}	6×10^{-8}
	$.34579 \times 10^{-2}$	1.1×10^{-5}	2×10^{-7}
	$.12838 \times 10^{-1}$	.0003	7×10^{-12}
	$.13140 \times 10^{-1}$	.0003	1×10^{-11}
1850=2.25N	$.1457 \times 10^{-7}$		3×10^{-10}
	$.27128 \times 10^{-2}$		*
	$.34467 \times 10^{-2}$		8×10^{-12}
	$.34579 \times 10^{-2}$		3×10^{-11}
	$.12838 \times 10^{-1}$		*
	$.13140 \times 10^{-1}$		*

Table 5.4.1 illustrates that the largest singular values (those in the interval (1.3, 2.5)) can be computed easily. By m = 60 = .073N, where N=822 is the order of B_1, all 6 have been computed accurately. In fact this holds true for many more of the large singular values. For example by m = 80 = .097N, the 13 largest singular values are obtained accurately. In each of the Tables, if by a particular value of m a given singular value has converged, then that singular value is not listed at any larger value of m used in that Table. Error estimates which are $< 10^{-12}$ are denoted by an asterisk. The table headers have the same meanings as they had in Chapter 4, Sections 4.7. and 4.8.

Table 5.4.2 summarizes the convergence of the 6 smallest singular values; these are in the interval (0, .02). These values are in the center of the eigenvalue spectrum of B_1. Convergence of the 6 smallest singular values is achieved by m = 2N = 1644 which is quite reasonable relative to the size of B_1.

TABLE 5.4.3

Example A_1. Convergence of Eigenvalues, Largest and Smallest in Magnitudes.

Order T_m	Computed Good Eigenvalues	Computed Amingap	Error Estimate
18=.044N	-2.36524	.764	3×10^{-9}
	-1.61827	.046	2×10^{-2}
	-1.57266	.043	2×10^{-2}
	-1.52928	.043	5×10^{-2}
	-1.4252	.104	2×10^{-2}
	-1.240	.127	2×10^{-1}
30=.073N	-2.36524	.764	2×10^{-12}
	-1.61887	.045	2×10^{-6}
	-1.57375	.037	5×10^{-6}
	-1.53648	.037	2×10^{-5}
	-1.44615	.023	3×10^{-4}
42=.102N	-2.36524		*
	-1.61887		*
	-1.57375		*
	-1.53648		3×10^{-10}
	-1.44615		2×10^{-10}
	-1.42350		8×10^{-8}
104=.25N	$-.14576 \times 10^{-7}$	.0027	2×10^{-8}
	$-.27128 \times 10^{-2}$	.0007	2×10^{-5}
	$-.344762 \times 10^{-2}$	.0007	3×10^{-5}
	$-.48854 \times 10^{-2}$	.0014	2×10^{-2}
	$-.12839 \times 10^{-1}$	.0003	4×10^{-4}
	$-.13144 \times 10^{-1}$	.0003	8×10^{-4}
206=.50N	$-.14576 \times 10^{-7}$	.0027	*
	$-.27128 \times 10^{-2}$	.0007	*
	$-.34467 \times 10^{-2}$	1.1×10^{-5}	*
	$-.34579 \times 10^{-2}$	1.1×10^{-5}	*
	$-.12839 \times 10^{-1}$	.0003	*
	$-.1314 \times 10^{-1}$	.0003	*

Now we compare the above convergence of the singular values of A_1 with the convergence of the corresponding eigenvalues of A_1 obtained when we applied our real symmetric Lanczos procedure directly to A_1. Such a comparison gives us some understanding of the penalties incurred in working with a real nonsymmetric matrix versus working with a real symmetric matrix of the same order whose eigenvalue distribution matches the singular value distribution of the nonsymmetric matrix.

We observe from Tables 5.4.1 - 5.4.3 that the large singular values converged even better than might be expected. Less than 50% more work was required to compute these quantities than what was required when they were considered as eigenvalues of the real symmetric matrix

A_1. This extra cost is understandable since B_1 has twice as many eigenvalues as A_1. Clearly the relative positions of these eigenvalues in the spectrum of B_1 are the same as in A_1, i.e., still extreme and with the same gap structure.

The amount of work required to get the smallest singular values is however much larger that that required to get the smallest eigenvalues. The smallest singular value appears at the center of the spectrum of B_1 as a pair of tiny eigenvalues $\sim \pm 10^{-8}$, instead of as an isolated tiny eigenvalue on the extreme of the spectrum of A_1. Thus, we see from Table 5.4.3 that .145 x 10^{-7} appeared as an eigenvalue of the Lanczos matrix generated by the real symmetric procedure as early as m = 104 whereas when it is considered as an eigenvalue of B_1 it did not appear accurately as an eigenvalue of a corresponding Lanczos matrix until somewhere between

TABLE 5.4.4a

Example A_1.   Inner Products Between Computed Singular Vectors.

No.	Singular Values	1	2	3	4	5	6
1	2.3652	1.0					
2	1.6189	2×10^{-12}	1.0				
3	1.5737	4×10^{-12}	7×10^{-13}	1.0			
4	1.5365	1×10^{-11}	4×10^{-13}	6×10^{-13}	1.0		
5	1.4461	5×10^{-11}	4×10^{-13}	6×10^{-13}	8×10^{-16}	1.0	
6	1.4325	2×10^{-11}	6×10^{-13}	2×10^{-13}	5×10^{-13}	2×10^{-12}	1.0
7	.01314	1×10^{-10}	2×10^{-11}	3×10^{-13}	6×10^{-11}	3×10^{-11}	8×10^{-14}
8	.01283	1×10^{-12}	8×10^{-13}	4×10^{-13}	3×10^{-13}	4×10^{-14}	1×10^{-13}
9	.003458	2×10^{-12}	3×10^{-14}	3×10^{-13}	1×10^{-13}	4×10^{-13}	4×10^{-13}
10	.003446	3×10^{-12}	3×10^{-14}	3×10^{-13}	7×10^{-13}	5×10^{-13}	2×10^{-13}
11	.002713	3×10^{-12}	7×10^{-13}	9×10^{-14}	3×10^{-13}	6×10^{-14}	5×10^{-14}
12	1.45 x 10^{-8}	4×10^{-12}	2×10^{-13}	3×10^{-13}	5×10^{-13}	8×10^{-13}	9×10^{-14}

m = 1234 and m = 1644. This clearly illustrates the effect of the transformation from A_1 to B_1 on any very small singular values. However, if we use the more reasonable comparison of the size of B_1 to the values of m used: m = 2N = 1644 and m = 2.25N = 1850, then the observed convergence is very reasonable since B_1 has N = 822 distinct nonzero eigenvalues. In fact by m = 1850 = 2.25N we actually have all of the singular values of A_1 accurately. Thus, a user could compute any portion of the spectrum that is of interest not just a few of the smallest or largest singular values.

To demonstrate the performance of the corresponding Lanczos singular vector procedure we computed the singular vectors for the 6 smallest singular values and the 6 largest singular values. The inner products of each pair of these computed singular vectors are given in Table 5.4.4. These factors are uniformly small.

TABLE 5.4.4b

Example A_1. Inner Products Between Computed Singular Vectors.

No.	Singular Values	7	8	9	10	11	12
7	.01314	1.0					
8	.01283	$2x10^{-9}$	1.0				
9	.003458	$2x10^{-10}$	$3x10^{-12}$	1.0			
10	.003446	$1x10^{-10}$	$2x10^{-13}$	$2x10^{-9}$	1.0		
11	.002713	$1x10^{-10}$	$2x10^{-12}$	$4x10^{-12}$	$2x10^{-11}$	1.0	
12	1.45×10^{-8}	$1x10^{-10}$	$2x10^{-12}$	$4x10^{-12}$	$7x10^{-12}$	$9x10^{-11}$	1.0

Table 5.4.5 lists the uniformly good error estimates obtained for each of these singular vectors. Since A_1 is actually a real symmetric matrix, the left and the right singular vectors should be identical and equal to the corresponding eigenvectors of A_1. We used this fact to

TABLE 5.4.5a

Example A_1. Error Estimates for Computed Singular Vectors

No.	Singular Value	Order T_m	Computed Amingap	Aerror	$\dfrac{Aerror}{Amingap}$
1	2.36524	28	.746	$2x10^{-10}$	$2x10^{-10}$
2	1.61887	56	.045	$1x10^{-11}$	$2x10^{-10}$
3	1.57375	61	.037	$2x10^{-12}$	$5x10^{-11}$
4	1.53648	66	.037	$1x10^{-12}$	$3x10^{-11}$
5	1.44615	72	.0226	$2x10^{-12}$	$7x10^{-11}$
6	1.42350	72	.0226	$2x10^{-12}$	$1x10^{-10}$
7	.01314	1669	$3x10^{-4}$	$3x10^{-10}$	$1x10^{-6}$
8	.01283	1769	$3x10^{-4}$	$4x10^{-12}$	$1x10^{-8}$
9	.003458	2111	$1x10^{-5}$	$4x10^{-12}$	$4x10^{-7}$
10	.003446	2081	$1x10^{-5}$	$6x10^{-12}$	$5x10^{-7}$
11	.002713	1820	$7x10^{-4}$	$7x10^{-12}$	$1x10^{-8}$
12	$.145 \times 10^{-7}$	2158	$3x10^{-3}$	$1x10^{-11}$	$4x10^{-9}$

compare the eigenvectors obtained using the real symmetric procedure with the corresponding computed 'singular vectors'. In all cases except for the smallest singular value, the norm of the difference between the eigenvector and the corresponding singular vector was less than 4×10^{-9}. For the smallest singular value $\sigma_n = 1.4x10^{-8}$, this norm was 1.4×10^{-6}. Table 5.4.5 also lists the sizes of the Lanczos matrices used and the error estimates obtained in the corresponding real symmetric Lanczos eigenvector computations.

TABLE 5.4.5b

Example A_1.   Error Estimates for Computed Eigenvectors

No.	Eigenvalue	Order T_m	Computed Amingap	Aerror	$\dfrac{\text{Aerror}}{\text{Amingap}}$
1	2.36524	20	.746	2×10^{-10}	3×10^{-10}
2	1.61887	40	.045	3×10^{-12}	7×10^{-11}
3	1.57375	43	.037	1×10^{-12}	3×10^{-11}
4	1.53648	46	.037	9×10^{-13}	2×10^{-11}
5	1.44615	50	.0226	1×10^{-12}	7×10^{-11}
6	1.42350	49	.0226	9×10^{-12}	4×10^{-10}
7	.01314	178	3×10^{-4}	2×10^{-11}	7×10^{-8}
8	.01283	194	3×10^{-4}	9×10^{-13}	3×10^{-9}
9	.003458	199	1×10^{-5}	5×10^{-11}	5×10^{-6}
10	.003446	185	1×10^{-5}	1×10^{-12}	1×10^{-7}
11	.002713	160	7×10^{-4}	5×10^{-12}	7×10^{-9}
12	$.145 \times 10^{-7}$	135	3×10^{-3}	2×10^{-12}	7×10^{-10}

Thus, the proposed procedure can be used to compute any of the singular values (and vectors) desired, not just a few large ones. However, the cost is proportional to the degree of difficulty in resolving the desired singular values. Its performance on a given matrix A will depend upon the resulting distribution of the singular values in the eigenvalue spectrum of the associated matrix B given in Eqn(5.4.1) and which singular values are desired. Tiny, dense singular values will be difficult to compute because each will appear as a $\pm$ pair of tiny eigenvalues located at the very center of the spectrum of B. The largest singular values can often be computed readily, along with many of the other singular values.

Therefore, this procedure is not always practical in the sense that it will compute the desired quantities in a reasonable amount of time. However, there are many matrices for which it can provide accurate singular values and corresponding vectors with a non-trivial but still reasonable amount of work. We emphasize the very important fact that the singular value Lanczos procedure requires a minimal amount of storage. The example which we have given does not illustrate the phenomena addressed in the following Lemma even though the smallest singular value in our example was fairly small. However, there is a problem with using this procedure to compute singular vectors corresponding to very small singular values, and Lemma 5.4.6 provides a statement of this problem.

LEMMA 5.4.6   Let A be a real rectangular ℓxn matrix. W.l.o.g we assume that $\ell \geq n$. Apply the singular value/vector Lanczos procedure described in this section to A with the starting vector $(0, u_1)^t$ where u_1 is $n \times 1$. (i) If $0 < \mu = O(\varepsilon)$ where ε is very small is an eigenvalue of some even-ordered Lanczos matrix generated, then the corresponding approximation to the right singular vector of A will be more accurate than the corresponding approximation to the left singular vector of A is. The computed unnormalized approximation to the left singular vector will be $O(\varepsilon)$. (ii) The smaller ε is, the larger the relative error in the left singular vector will be. In the limit as $\varepsilon \Rightarrow 0$, the unnormalized approximation to the left singular vector will converge to 0. If the starting vector is of the form $(v_1,0)^t$, where v_1 is $\ell \times 1$, then the left singular vector will be more accurate than the right singular vector and in the limit the unnormalized approximation to the right singular vector will converge to 0.

Proof. The left singular vector is formed by combining the even-numbered Lanczos vectors with the even-numbered components of any computed eigenvector of the Lanczos matrix corresponding to μ. However when μ is very small, the eigenvalues $\pm \mu$ are too close together, and the inverse iteration procedure on $(T_m - \mu I)$ which we use to determine an eigenvector for μ yields a vector which is a linear combination of the eigenvectors for $\pm \mu$. From Lemma 5.4.5 we have that the even-numbered components of the eigenvectors corresponding to any eigenvalue $\mu = O(\varepsilon)$ and to $-\mu$ are equal in magnitude but different in sign and of size $O(\varepsilon)$. Furthermore, the odd-numbered components are identical and of size $O(1)$.

Therefore for any such mixed eigenvector, the subvector composed of the odd-numbered components is $O(1)$, since the odd-numbered components for the two eigenvectors corresponding to $\pm \mu$ are identical and of size $O(1)$. However the corresponding subvector formed from the even-numbered components is of size $O(\varepsilon)$. Therefore the relative errors in the even-numbered components in this mixed vector are larger by a factor $O(1/\varepsilon)$ than the relative errors in the odd-numbered components. Therefore the left singular vector approximation is less accurate than the right singular vector approximation, and this difference in accuracy increases as $\varepsilon \Rightarrow 0$. We have a similar situation if we use the starting vector $(v_1,0)^t$. (ii) follows directly from (i). $\qquad \square$

CHAPTER 6

NONDEFECTIVE COMPLEX SYMMETRIC MATRICES

SECTION 6.1   INTRODUCTION

Complex symmetric matrices arise naturally in electron spin resonance (ESR) and nuclear magnetic resonance (NMR) studies, see for example Moro and Freed [1981]. The reader is referred to this reference for more details on these applications. Such matrices are typically large and sparse. In this chapter we present a Lanczos procedure for computing distinct eigenvalues and corresponding eigenvectors of large, sparse, nondefective, complex symmetric matrices. Specifically, for a given nondefective complex symmetric matrix A, we consider the problem of computing complex scalars λ and corresponding complex vectors $x \neq 0$ such that

$$Ax = \lambda x. \tag{6.1.1}$$

DEFINITION 6.1.1   The complex nxn matrix $A \equiv (a_{ij})$, $1 \leq i,j \leq n$, is complex symmetric if and only if for every i and j, $a_{ij} = a_{ji}$. It is nondefective if and only if it is diagonalizable. (See Definition 6.1.4.)

If we write a given complex symmetric matrix A as $A = B + iC$ where B and C are real matrices and $i = \sqrt{-1}$, then from the symmetry we see that B and C must be real symmetric matrices. It is also easy to prove that if λ and μ are any two distinct eigenvalues of a complex symmetric matrix A and x and y are corresponding eigenvectors of A, then the Euclidean 'inner product' of x and y is zero. That is,

$$x^T y = 0 \tag{6.1.2}$$

where the superscript T denotes transpose. It is important to observe at this point that there are nonzero complex vectors x such that $x^T x = 0$. Such vectors will be called quasi-null. For example the 2-dimensional vector $x \equiv (1,i)$ is quasi-null.

DEFINITION 6.1.2   A complex vector $x \neq 0$ is a quasi-null vector if and only if $x^T x = 0$.

DEFINITION 6.1.3   A set of complex vectors $X \equiv \{x_1,...,x_q\}$ is real orthogonal if and only if $X^T X = I_q$ where I_q denotes the qxq identity matrix.

For any nxn real orthogonal set X of vectors $X^T = X^{-1}$, so that in particular such a set of vectors must be linearly independent. We have the following characterization of quasi-null vectors which follows directly from Definition 6.1.2.

LEMMA 6.1.1   A nonzero complex n-vector x is a quasi-null vector if and only if $x_R^T x_R = x_I^T x_I$ and $x_R^T x_I = 0$, where x_R denotes the real n-vector composed of the real parts of the components of x, and x_I is the real n-vector of the imaginary parts of the components of

194

x. An equivalent statement is that the following four, real 2n-vectors are orthogonal:

$$y_1 = \begin{bmatrix} x_R \\ x_I \end{bmatrix} \quad y_2 = \begin{bmatrix} x_R \\ -x_I \end{bmatrix}, \quad y_3 = \begin{bmatrix} x_I \\ x_R \end{bmatrix}, \quad y_4 = \begin{bmatrix} -x_I \\ x_R \end{bmatrix} \quad (6.1.3)$$

DEFINITION 6.1.4 A matrix is diagonalizable if and only if there exists a nonsingular matrix X such that $X^{-1}AX = \Lambda$ where Λ is a diagonal matrix.

In Definition 6.1.4 it is trivial to demonstrate that the entries in Λ are the eigenvalues of A, that the jth column of X is a right eigenvector of the corresponding eigenvalue λ_j, and that the ith row of X^{-1} is a left eigenvector of A corresponding to λ_i. Here we use the Wilkinson [1965] definition of left-eigenvectors. That is, given an eigenvalue λ of a matrix A, a vector y is a left-eigenvector of A corresponding to λ if and only if $y \neq 0$ and $A^T y = \lambda y$. Other references, see for example Stewart [1973], give the following definition. Given an eigenvalue λ of a matrix A, a vector z is a left-eigenvector of A corresponding to λ if and only if $z \neq 0$ and $A^H z = \bar{\lambda} z$. The superscript H denotes the complex-conjugate transpose of the matrix A and the bar denotes the complex conjugate of the scalar. Obviously these two definitions are equivalent since for any y defined as above, we could define a corresponding $z \equiv \bar{y}$ and vice-versa. Throughout the discussions in this chapter we use the transpose definition. Note that any matrix with distinct eigenvalues is diagonalizable.

A literature search for algorithms specifically designed for computing eigenvalues of complex symmetric matrices uncovered only the Eberlein [1971] paper and its predecessors. The EISPACK Library [1976,1977] contains subroutines which can be used on any complex matrix, including of course any complex symmetric one. However the storage requirements for the Eberlein and for the EISPACK procedures are $O(n^2)$ where n is the order of the given matrix and the arithmetic operations counts are $O(n^3)$. Therefore, these procedures are not practical for very large matrices.

Eberlein [1971] is a Jacobi procedure. It uses a sequence of complex plane rotations P_k, $k = 1,2,....$ At each iteration k, the rotated matrix (the kth iterate)

$$A_k \equiv Q_k^{-1} A Q_k,$$

where $Q_k \equiv \prod_{j=1}^{k} P_j$, and the complex rotations P_k are chosen to maximize the decrease in the norm $\|A_k\|$ where $\|B\| \equiv \sum_{i,j} |b_{ij}|^2$ is the Frobenius norm.

Each rotation matrix $P_k \equiv (p_{m\ell})$ satisfies $P_k^T = P_k^{-1}$ and has the following form.

$$p_{MM} = p_{LL} = \cos z, \quad p_{LM} = -p_{ML} = \sin z \quad \text{and}$$
$$p_{m\ell} = \delta_{m\ell} \quad \text{otherwise,} \quad (6.1.4)$$

where $M < L$ are the indices of rotation, and $\delta_{m\ell} = 0$ if $m \neq \ell$ and $\delta_{mm} = 1$. Eberlein proves that this procedure has global convergence. That is for large N we have that

$$Q_N^{-1}AQ_N \approx \Lambda$$

where the entries of the diagonal matrix Λ are the eigenvalues of the A-matrix. Anderson and Loizou [1973] prove that if the rotation indices of each P_k are chosen properly, then in the later stages of this procedure the convergence rate is quadratic, provided that the given complex symmetric matrix is diagonalizable.

As we said earlier the Eberlein and the EISPACK procedures have storage requirements which grow quadratically with the size of the given matrix and are therefore not practical for large matrices. In Section 6.3 we describe a single-vector Lanczos procedure with no reorthogonalization for complex symmetric matrices which is well-suited for large but sparse matrices, but first we have to introduce some basic definitions and lemmas needed for our discussions.

SECTION 6.2   PROPERTIES OF COMPLEX SYMMETRIC MATRICES

Real symmetric and Hermitian matrices possess many desirable properties, see for example Chapter 1 and Stewart [1973]. A real matrix is real symmetric if and only if $A^T = A$, and a complex matrix A is Hermitian if and only if $A^H = A$. Real symmetric and Hermitian matrices are always diagonalizable. In the real symmetric case the eigenvectors form orthogonal sets of vectors. In the Hermitian case the eigenvectors form unitary sets of vectors. Both types of matrices have real eigenvalues. Real symmetric matrices also have real eigenvectors. Hermitian matrices have complex eigenvectors. Small symmetric (Hermitian) perturbations in the entries of a real symmetric (Hermitian) matrix result in small (in absolute size) perturbations in the eigenvalues of the matrix. The eigenvalue computations for such matrices are perfectly conditioned, see for example Wilkinson [1965].

Complex symmetric matrices are quite different. They do not retain any of these desirable properties. In general the eigenvalues of a complex symmetric matrix are complex as are their eigenvectors. In fact from Theorem 5, Chapter 1 of Gantmacher [1959] we know that given any set of eigenvalues with any arbitrary associated geometric and algebraic multiplicities, there exists a complex symmetric matrix with those eigenvalues and associated multiplicities. Therefore, in particular it is clear that a complex symmetric matrix need not be diagonalizable. Even when a complex symmetric matrix is diagonalizable, the eigenvectors need not form a unitary set of vectors. Small perturbations in a complex symmetric matrix may cause large perturbations in the eigenvalues.

We restrict our discussion in this chapter to nondefective complex symmetric matrices. Defective matrices are difficult to deal with even when they are small and dense. The accuracy to which a given defective eigenvalue can be computed is limited by the multiplicity of the defectiveness, see Wilkinson [1965]. For example for an eigenvalue with a defectiveness of 3 one can expect to be able to compute that eigenvalue only to within the cube root of the nominal accuracy. Nondefective matrices have properties similar to those of a matrix whose eigenvalues are all distinct.

LEMMA 6.2.1 Let A be a complex symmetric nxn matrix. (i) Let λ be any eigenvalue of A and x be a corresponding right eigenvector. Then x is also a left eigenvector of A. (ii) If A is a nondefective complex symmetric matrix, then A is diagonalizable by a real orthogonal set of eigenvectors X of A. That is,

$$X^T A X = \Lambda \text{ where } X^T X = I. \tag{6.2.1}$$

Proof. Part (i) follows directly from the fact that $A^T = A$. Part (ii) is a direct consequence of Theorem 4, Chapter 1, Gantmacher [1959] which states that if two complex symmetric matrices are similar, then they are orthogonally similar. Since A is diagonalizable, it is similar to a diagonal matrix Λ whose entries are the eigenvalues of A. Therefore there exists a real orthogonal matrix X which diagonalizes A. That is X and A satisfy Eqn(6.2.1). However, it is clear that the vectors X must then be eigenvectors of A. $\qquad\square$

Thus whenever a complex symmetric matrix A is not defective, there is a basis of eigenvectors which are linearly independent, real orthogonal and therefore not quasi-null. Thus Hermitian matrices have sets of eigenvectors X such that $X^H X = I_n$, whereas nondefective complex symmetric matrices have sets of eigenvectors such that $X^T X = I_n$. If however A is defective and λ is a defective eigenvalue of A, then the eigenvector corresponding to λ is quasi-null. This difference between the defective and the nondefective cases clearly illustrates the level of difficulty associated with the computation of the eigenvalues and eigenvectors of defective matrices. For completeness we include a simple example of a defective complex symmetric matrix.

EXAMPLE 6.2.1 Let

$$A = \begin{bmatrix} a & b & 0 \\ b & c & b \\ 0 & b & a \end{bmatrix}$$

where $a \neq c$ and $(a - c)/2 = i\sqrt{2}b$. The eigenvalues of A are $\mu, \mu,$ and ν where $\mu = (a + c)/2$ and $\nu = a$. Let

$$W = \begin{bmatrix} ib/\sqrt{2} & 0 & 1 \\ b & i/\sqrt{2} & 0 \\ ib/\sqrt{2} & 0 & -1 \end{bmatrix} \text{ then } J = W^{-1}AW = \begin{bmatrix} \mu & 1 & 0 \\ 0 & \mu & 0 \\ 0 & 0 & \nu \end{bmatrix}.$$

The Jordan matrix J clearly illustrates the fact that the eigenvalue μ is defective. The eigenvector for μ is $(ib/\sqrt{2}, b, ib/\sqrt{2})^T$ and is a quasi-null vector.

Lemma 6.2.1 gives the following spectral expansion for a nondefective complex symmetric matrix.

$$A = \sum_{i=1}^{n} \lambda_i x_i x_i^T \qquad (6.2.2)$$

where $X = \{x_1,...,x_n\}$ is such that $X^T X = I$. In Lemmas 6.2.2 and 6.2.3 we state some of the relevant results from the perturbation theory for general complex matrices. For more details see the paper by Kahan, Parlett and Jiang [1982]. The following lemma is a specialization of Theorem 4.2, Chapter 6, Stewart [1973] for general complex matrices.

LEMMA 6.2.2 Stewart [1973] Let A be a complex symmetric matrix. Let λ be a simple eigenvalue of A and let x be a corresponding eigenvector scaled so that $x^H x = 1$. Then for any matrix $B \equiv A - E$ there exists a corresponding eigenvalue μ of B which satisfies the inequality

$$|\lambda - \mu| \leq |(x^T E x)/(x^T x)| + O(\|E\|^2) \qquad (6.2.3)$$

where $O(\|E\|^2)$ indicates the given quantity is bounded by a constant C times $\|E\|^2$.

We are interested in the case when $\|E\| < \epsilon$ where ϵ is small. Typically, E would also be complex symmetric. Lemma 6.2.2 tells us that the sensitivity of any simple eigenvalue of a complex symmetric matrix to perturbations or to errors in the eigenvalue computations depends upon how nearly quasi-null the corresponding eigenvector is. The following lemma applies to a general, not necessarily complex symmetric, matrix. This lemma connects the accuracy of the eigenvalue computations to the condition number of the matrix of eigenvectors.

LEMMA 6.2.3 Bauer and Fike [1960]. Assume that B is diagonalizable where $B \equiv A - E$. That is, for some nonsingular matrix X and diagonal matrix Σ, $X^{-1} B X = \Sigma$. Then to each eigenvalue λ_i of A there is an eigenvalue σ_j of B such that

$$|\lambda_i - \sigma_j| \leq \|X\| \|X^{-1}\| \|E\| \qquad (6.2.4)$$

where $\| \ \|$ is any matrix norm such that for any diagonal matrix Λ, $\|\Lambda\| \equiv \max |\lambda_i|$.

Corresponding statements for relationships between corresponding eigenvectors of A and eigenvectors of B are weaker than the statements for eigenvalues given in Eqns(6.2.3) and (6.2.4) and incorporate the effects of the gaps between the eigenvalue whose eigenvector is being computed and the other eigenvalues of A. See for example, Chapter 6 of Stewart [1973] for details.

It is clear from the preceding discussion that the proper 'norm' for the complex symmetric case is the Euclidean norm. Unfortunately, as we said earlier, the Euclidean norm is only a quasi-norm when the vectors involved are complex since there are quasi-null complex vectors. We will however use the associated quasi-inner product in our complex symmetric Lanczos procedure. Restricting ourselves to diagonalizable complex symmetric matrices, we minimize the effects of quasinull vectors.

SECTION 6.3   LANCZOS PROCEDURE, NONDEFECTIVE MATRICES

We want to construct a storage efficient procedure for computing the distinct eigenvalues and corresponding eigenvectors of large sparse complex symmetric matrices. We restrict our discussion to diagonalizable complex symmetric matrices. We consider the two-sided Lanczos recursion given in Wilkinson [1965], specializing it to complex symmetric matrices. The complex symmetric recursion given in Lemma 6.3.1 has been used by Moro and Freed [1981].

Wilkinson [1965] presents the following two-sided recursions for a general nxn complex matrix A. Let v_1 and w_1 be randomly-generated starting vectors scaled so that $v_1^T w_1 = 1$. Then for $i = 1,2,...$ use the following recursion to define vectors v_{i+1}, w_{i+1} such that

$$\gamma_{i+1} v_{i+1} = A v_i - \alpha_i v_i - \beta_i v_{i-1}$$
$$\gamma_{i+1}^* w_{i+1} = A^T w_i - \alpha_i w_i - \beta_i^* w_{i-1}$$

(6.3.1)

where the scalars α_i, β_i, γ_i, β_i^*, and γ_i^* are chosen so that for each m, $V_m^T W_m = I$ where $W_m \equiv \{w_1,...,w_m\}$ and $V_m \equiv \{v_1,...,v_m\}$.

LEMMA 6.3.1 (Exact Arithmetic) Let A be a complex symmetric matrix. Then if we set $v_1 = w_1$ in Eqn(6.3.1), those recursions reduce to the following single-vector Lanczos recursion.

$$\beta_{i+1} v_{i+1} = A v_i - \alpha_i v_i - \beta_i v_{i-1} \equiv r_{i+1}$$

(6.3.2)

where

$$\beta_{i+1} \equiv (r_{i+1}^T r_{i+1})^{1/2}, \quad \alpha_i \equiv v_i^T (A v_i - \beta_i v_{i-1}).$$

(6.3.3)

Proof. Since $A = A^T$ and the starting vectors are equal, the two recursions are identical, reducing to

$$\gamma_{i+1} v_{i+1} = A v_i - \alpha_i v_i - \beta_i v_{i-1} \equiv r_{i+1}$$

(6.3.4)

where γ_i and β_i must be chosen so that for each i, $v_{i+1}^T v_{i+1} = 1$ and $v_{i+1}^T v_{i-1} = 0$. Therefore we have that for each i, $\gamma_i = v_i^T A v_{i-1} = v_{i-1}^T A v_i = \beta_i$, so that Eqn(6.3.4) reduces to Eqn(6.3.2). Moreover for each i, $\alpha_i = v_i^T (A v_i - \beta_i v_{i-1}) = v_i^T A v_i$.

First we must prove that for each m, $V_m^T V_m = I$. We do this by induction. By construction $v_1^T v_1 = 1$, $v_2^T v_2 = 1$, and $v_1^T v_2 = 0$. Assume that for some K>1 and all $j,i \leq K$ that for $i \neq j$, $v_i^T v_j = 0$. Note that by construction $v_i^T v_i = 1$. Moreover by construction,

$v_{K+1}^T v_K = 0$ and for $j<K$

$$v_j^T v_{K+1} \beta_{K+1} = v_j^T A v_K - \beta_K v_j^T v_{K-1}. \tag{6.3.5}$$

If we use the recursion in Eqn(6.3.2) to replace Av_j in Eqn(6.3.5), then we get that $v_j^T v_{K+1} = 0$ for all $j \leq K$. That is, for $j \leq (K-2)$ all terms vanish, and for $j=K-1$ the two terms $\beta_{j+1} v_{j+1}^T v_K$ and $\beta_K v_j^T v_{K-1}$ are equal and cancel. Thus, we obtain global real orthogonality of the Lanczos vectors.

Using the orthogonality it is trivial to demonstrate that for any i,

$$\beta_{i+1}^2 \equiv r_{i+1}^T r_{i+1}.$$

Furthermore, it is straight-forward to demonstrate that setting β_{i+1} equal to either the $\pm$ principal square root results in changes in the off-diagonal entries of the Lanczos matrices of only multiplies of ± 1 and similar changes in the Lanczos vectors. Since the eigenvalues of any irreducible tridiagonal complex symmetric matrix depend only on the products of the off-diagonal entries, we can set each β_{i+1} equal to the principal square root of the Euclidean 'norm' of the vector r_{i+1}. $\qquad \square$

The recursion in Eqns(6.3.2) - (6.3.3) generates a family of complex symmetric tridiagonal matrices T_m with diagonal entries α_i and nonzero off-diagonal entries β_{i+1}. In practice the modified Gram-Schmidt step given in the definition of α_i in Eqn(6.3.3) enhances the numerical stability of the procedure. We note that the Euclidean 'norm' $r_{i+1}^T r_{i+1}$ will vanish if r_{i+1} is a quasi-null vector. We do not have an argument which says that this cannot happen but however because of Lemma 6.1.1, we expect it to be a rare event. It is not difficult to incorporate a test into the complex symmetric Lanczos codes to check for this possibility. If this should happen the user must change the seed supplied to the random number generator and restart the Lanczos procedure using a different starting vector. In the complex symmetric case if the Lanczos matrices required are large, then typically the Lanczos matrix regeneration will take much less time than any of the associated eigenvalue computations.

The reader should contrast the complex symmetric recursion with the one obtained for the real symmetric case, see Chapter 4. Note that symbolically the recursions are identical. However, numerically they differ since in the complex symmetric case the matrix A and the Lanczos vectors v_i are complex vectors. In fact as we said earlier, complex symmetric and real symmetric matrices do not share any nice properties other than their symmetry. The following lemma tells us that the family of complex symmetric tridiagonal matrices encompasses the whole family of irreducible, nonsymmetric tridiagonal matrices. Therefore, forcing the Lanczos matrices to be complex symmetric tridiagonal does not impose any implicit restrictions on the eigenvalue distributions or eigenvectors being considered. This symmetry is very important numerically.

LEMMA 6.3.2 Let T^* be a general nonsymmetric tridiagonal matrix with no zero entries on either the sub or super diagonals. Then there exists a complex symmetric tridiagonal matrix T_{CS} which has the same eigenvalues as T^*.

Proof. Let

$$\alpha_i \equiv T^*(i,i), \quad \beta_{i+1} \equiv T^*(i,i+1), \quad \gamma_{i+1} \equiv T^*(i+1,i).$$

Define the diagonal similarity transformation $D \equiv (d_j)$ where

$$d_1 = 1, \quad \text{and} \quad (d_{j+1}/d_j) \equiv (\gamma_{j+1}/\beta_{j+1})^{1/2}.$$

Then $T_{CS} \equiv D^{-1}T^*D$ is a complex symmetric tridiagonal matrix with the same eigenvalues.

$$T_{CS}(i,i) \equiv \alpha_i, \quad \text{and} \quad T_{CS}(i,i+1) = T_{CS}(i+1,i) = (\beta_{i+1}\gamma_{i+1})^{1/2}.$$

The complex symmetric Lanczos eigenelement procedures mimic the corresponding real symmetric Lanczos procedures presented in Chapter 4. However, in the complex symmetric case Sturm sequencing is not valid and we do not have an analog of the BISEC subroutine used for the real symmetric problems. The computations are in two phases. First the desired eigenvalues are computed, and then eigenvectors of user-specified eigenvalues are computed using a separate program. There is no reorthogonalization.

COMPLEX SYMMETRIC LANCZOS EIGENVALUE PROCEDURE.

Step 1. For a user-specified choice of M, generate the Lanczos complex symmetric tridiagonal matrix T_M using the recursion in Eqns(6.3.2) - (6.3.3).

Step 2. For a user-specified size $m \leq M$ compute the eigenvalues of the tridiagonal Lanczos matrix T_m.

Step 3. Compute the eigenvalues of the corresponding complex symmetric tridiagonal matrix $\hat{T}_2$ obtained from T_m by deleting the first row and column of T_m.

Step 4. Determine the numerical multiplicities of each of the eigenvalues of T_m. Then for each simple eigenvalue of T_m use the eigenvalues of $\hat{T}_2$ to label it a 'good' or a 'spurious' eigenvalue. Note that all numerically-multiple eigenvalues are assumed to be 'good'.

Step 5. Estimate the error in each simple 'good' eigenvalue. Error estimates are computed using inverse iteration on the Lanczos matrix, T_m.

Step 6. Terminate if all these estimates indicate convergence. Otherwise, using these estimates as a guide, decide upon an increment for m, and enlarge the Lanczos matrix and repeat Steps 2-6.

To summarize the above, note that first a complex symmetric tridiagonal Lanczos matrix is generated, then eigenvalues of specified principal submatrices of this matrix are determined. An identification test is then applied to select a subset of the eigenvalues of the Lanczos matrix used as approximations to the eigenvalues of A. These are the 'good' eigenvalues. Error estimates are computed and used to check the computed 'good' eigenvalues for convergence.

There are fundamental differences between the amount of storage and the amount of computation required by the complex symmetric codes versus that required by the real symmetric codes. First, all of the complex symmetric computations are done in double precision complex arithmetic, and all of the vectors used are complex vectors. Thus, the basic storage requirements for the complex symmetric Lanczos procedure are twice those required by the real symmetric version of the Lanczos codes. Second, since a bisection subroutine cannot be used to compute the eigenvalues of the complex symmetric tridiagonal Lanczos matrices, we are forced to do a complete eigenvalue computation on each complex symmetric tridiagonal matrix which we consider. In fact our procedure requires, see Steps 2 and 3 above, two complete complex symmetric tridiagonal eigenvalue computations for each Lanczos tridiagonal matrix which we consider. Two are required because our identification test for labelling the computed eigenvalues of the Lanczos matrices as either 'good' or 'spurious' uses the eigenvalues of the corresponding tridiagonal matrix obtained from the given Lanczos matrix by deleting the first row and column of that matrix. This identification test is a two-dimensional analog of the identification test which we used in the real symmetric case. See Chapter 4, Section 4.5. In the real symmetric case, this test could be accomplished for each eigenvalue as a natural part of the eigenvalue computation at essentially no extra cost. A second tridiagonal eigenvalue computation was not required.

As stated earlier, the standard EISPACK subroutines for general complex matrices cannot easily be used on large matrices because of the large amount of computer storage which they require and the large number of arithmetic operations required. Thus, out of necessity we had to devise a new procedure for computing the eigenvalues of the complex symmetric tridiagonal Lanczos matrices which recursions (6.3.2) - (6.3.3) generate, designed to use as little computer storage as possible. Such a procedure was obtained by constructing a complex symmetric analog of the EISPACK subroutine IMTQL1 which is for real symmetric matrices. We call this modified subroutine CMTQL1 and it is discussed in the next section. Storage requirements for CMTQL1 are minimal. The only storage required is the storage in which the given complex symmetric tridiagonal matrix is stored.

Volume 2, Chapter 7 of this book contains a FORTRAN implementation of our complex symmetric Lanczos eigenvalue procedure. The main program CSLEVAL for the complex symmetric eigenvalue computations generates the user-specified Lanczos matrix. CSLEVAL then calls the subroutines COMPEV and CMTQL1 to compute the eigenvalues of the Lanczos matrix T_m specified by the user and the eigenvalues of the related complex symmetric tridiagonal matrix $\hat{T}_2$ obtained from T_m by deleting the first row and first column. COMPEV then determines the numerical multiplicities of the eigenvalues of T_m and sorts these eigenvalues into the two classes, the 'good' eigenvalues and the 'spurious' eigenvalues. Simple eigenvalues of the Lanczos matrix T_m which match eigenvalues of the associated matrix $\hat{T}_2$ are 'spurious' and not eigenvalues which we want, the other eigenvalues are called 'good'. The 'good' eigenvalues are 'accepted' as approximations to eigenvalues of the user-specified matrix A. Error estimates are computed for the isolated 'good' eigenvalues. All other 'good' eigenvalues are assumed to have converged. These error estimates are used to check the convergence of the good eigenvalues. Error estimates are computed using a complex version of the inverse iteration subroutine INVERR used in the real symmetric case. If convergence has not yet occurred and a larger Lanczos matrix has been specified by the user, the program will continue on to the larger matrix, repeating the above procedure on the larger matrix.

The convergence observed is the convergence of the eigenvalue in question as an eigenvalue of the associated Lanczos matrices as they are increased in size. That is, the error estimates tell us which eigenvalues of the Lanczos matrix have stabilized as eigenvalues of these matrices for all larger orders. (See Chapter 2). These stabilized eigenvalues are taken as approximations to eigenvalues of A. How accurately such converged eigenvalues approximate the desired eigenvalues of the given matrix A will depend upon the numerical condition of the associated eigenspace of the given A-matrix for the particular eigenvalue in question.

These codes can be used to compute either a very few or very many of the distinct eigenvalues of a nondefective, complex symmetric matrix. As the documentation provided for CSLEVAL in Volume 2 indicates, the A-multiplicity of a given computed eigenvalue can be obtained only with additional computation, and the modifications required to do this additional computation are not included in the versions of the FORTRAN codes in Volume 2. This is analogous to the real symmetric case. There is no reorthogonalization of the Lanczos vectors at any stage in any of the computations.

Once convergence is observed, the user can select a subset of the 'converged' eigenvalues for which corresponding eigenvectors are to be computed. The main program CSLEVEC for computing eigenvectors of complex symmetric matrices, is then used to compute these desired eigenvectors. The basic steps in the eigenvector computations are given below and again parallel the basic steps required by the corresponding real symmetric eigenvector procedure. Volume 2, Chapter 7 contains a FORTRAN implementation of CSLEVEC. Please see Volume 2 for the details of the implementation and for the associated program documentation.

COMPLEX SYMMETRIC LANCZOS EIGENVECTOR PROCEDURE.

Step 1.   Read in the eigenvalues for which eigenvectors are to be computed.

Step 2.   Read in the Lanczos matrix T_M generated for the eigenvalue computations and enlarge if necessary.

Step 3.   Determine appropriate size Lanczos matrices $T_{m(\mu)}$ for each eigenvalue μ supplied, simultaneously computing the associated Lanczos matrix eigenvector using inverse iteration.

Step 4.   Compute the associated Ritz vectors as the required Lanczos vectors are regenerated. These Ritz vectors are taken as the desired approximate eigenvectors of the given matrix A.

The accuracy achievable for these computations is noticeably less than that achievable in the real symmetric case. This is to be expected from the perturbation analysis for the complex symmetric case. Thus, in the FORTRAN codes provided in Volume 2 we reduced the anticipated accuracy of the computed eigenvalues and used larger tolerances in our multiplicity and spuriousness tests than what was used in the real symmetric cases. These larger tolerances decrease the resolution capabilities of these codes. However, these tolerances appear to be realistic.

SECTION 6.4   QL ALGORITHM, COMPLEX SYMMETRIC TRIDIAGONAL MATRICES

The complex symmetric Lanczos procedure transforms a given general complex symmetric matrix into a family of complex symmetric tridiagonal matrices, T_m, $m = 1,2,...$. The eigenvalues of the given matrix A are computed by computing eigenvalues of one or more of these tridiagonal matrices and then selecting subsets of these eigenvalues as approximations to eigenvalues of the given matrix A. If only a few large extreme eigenvalues of the given matrix are required, then it may be possible to compute these eigenvalues using small Lanczos matrices. In this situation it is perfectly acceptable to use an EISPACK subroutine which is actually designed for a general complex matrix to do these tridiagonal computations. However, in general we cannot necessarily compute the desired eigenvalues by restricting our considerations to relatively small Lanczos matrices. To be able to perform many practical eigenvalue computations we will need to be able to work with fairly large Lanczos matrices. Note that this may be true even if the desired eigenvalues are on the extremes of the spectrum of A. In this section we present a storage efficient procedure for computing the eigenvalues of any complex symmetric tridiagonal matrix. The only storage used is that which is required to define the complex symmetric tridiagonal matrix.

The IMTQL1 subroutine in the EISPACK library [1976,1977] is designed specifically for computing the eigenvalues of real symmetric tridiagonal matrices. It is a very clever implementation of the basic QL method, see Martin and Wilkinson [1968]. Each iteration step of the basic QL method consists of a factorization of a shifted tridiagonal matrix

$$T - \omega I = QL \text{ where } Q^T Q = I, \tag{6.4.1}$$

and L is a lower triangular matrix, followed by the computation of the next iterate as

$$T_+ = LQ + \omega I. \tag{6.4.2}$$

Note that

$$T_+ = Q^T T Q, \tag{6.4.3}$$

is orthogonally similar to the original matrix, and therefore has the same eigenvalues as T. In the limit T_+ becomes a diagonal matrix. The shift parameter ω is chosen to accelerate the convergence and at each iteration is the eigenvalue of the current uppermost 2x2 principal minor which is closest to the uppermost diagonal element of that minor. Each matrix Q is a sequence of plane rotations and each iterate T_+ is real symmetric and tridiagonal.

The IMTQL1 implementation of the QL method does not use explicit factorizations as given in Eqn(6.4.1) and (6.4.2). Instead on each iteration IMTQL1 simply constructs a sequence of plane rotations P_{m-1}, P_{m-2}, ..., P_1 such that the (m-1,m) element of the matrix $P_{m-1}(T - \omega I)$ is zero, and the remaining rotations are used to simply chase the additional nonzero quantity introduced by the first rotation P_{m-1} up and out of the matrix. This preserves the real symmetric tridiagonal structure.

These rotations are applied in bottom-up order. That is, within a given iteration, starting with the matrix T that exists at the beginning of that iteration, IMTQL1 forms sequentially the matrices

$$T^k \equiv P_k T^{k+1} P_k^T, \quad k = (m - 1),....,1 \tag{6.4.4}$$

where $T^m \equiv T$ and $T^1 \equiv T_+$, the next iterate.

Each plane rotation $P_k \equiv (p_{ij})$ satisfies $P_k^T = P_k^{-1}$ and has the following form. $P_k \equiv (p_{ij})$

$$p_{kk} = p_{k+1,k+1} = \cos x, \quad p_{k+1,k} = -p_{k,k+1} = \sin x$$
$$p_{ij} = \delta_{ij} \text{ otherwise,} \tag{6.4.5}$$

where the angle x is selected based upon the entries in the associated tridiagonal matrix, and the parameter ω.

The equivalence of repeatedly doing the factorizations in Eqn(6.4.1) followed by the reverse recombination of the factors as in Eqn(6.4.2) with using an outer sequence of sequences of plane rotations as defined in the IMTQL1 implementation in EISPACK, follows from the fact that each iterate T_+ is also real symmetric and tridiagonal and the following theorem which is given in Martin and Wilkinson [1968].

THEOREM 6.4.1 Let T be a real symmetric matrix. Let Q be a real orthogonal matrix, that is $Q^T = Q^{-1}$. Define $T_+ = Q^T T Q$ and suppose that T_+ is also tridiagonal with all positive entries on the superdiagonal. Then the matrix T_+ and the columns of Q are uniquely determined from only the last column of Q.

Thus one needs only to choose the first rotation to be the same rotation which one would use if the factorization in Eqn(6.4.1) were being carried out explicitly and then combine that with a construction for the rest of that iteration which preserves the real symmetry and tridiagonal properties of the original matrix. If this is done, as it is done in IMTQL1, then the tridiagonal iterates which are generated will be identical to those (at least theoretically) which would be generated by doing the explicit factorizations. Note that Theorem 6.4.1 assumed that all of the nonzero off-diagonal entries of T_+ were positive. However, it is not necessary to assume this in the sense that all that is really needed is that all of the super-diagonal entries are nonzero. In this more general situation the diagonal entries of T_+ are uniquely determined. However, the off-diagonal entries are determined only to within a multiple of ± 1. Since the eigenvalues of a real symmetric tridiagonal matrix depend only on the squares of the superdiagonal entries (and of course on the diagonal entries), all possible matrices would have the same eigenvalues so in that sense the matrix T_+ is uniquely determined. Similarly, each column in the Q matrix is determined to within a factor of ± 1. A proof of convergence for the QL method for real symmetric matrices is given in Wilkinson [1968].

In our procedure CMTQL1 for computing the eigenvalues of complex symmetric tridiagonal matrices we have combined the idea of complex plane rotations used in Eberlein [1971] together with complex symmetric analogs of the ideas used in IMTQL1 to produce a storage efficient procedure specifically designed for complex symmetric tridiagonal matrices. CMTQL1 is based upon the following direct complex symmetric analog of the implicit QL procedure for real symmetric matrices.

At each iteration we have a complex symmetric tridiagonal matrix T and as in Eqn(6.4.1) we factor a shifted T,

$$T - \omega I = QL \text{ where } Q^T Q = I \tag{6.4.6}$$

and L is lower triangular. Symbolically this is identical to the real symmetric case. However contrary to the real symmetric case, our Q in Eqn(6.4.6) is a complex matrix. Thus, the orthogonality imposed in Eqn(6.4.6) is only a real orthogonality. Normally in developing procedures for complex matrices, unitary matrices are used. However, as we discussed in

Section 6.2 the natural transformation matrices for complex symmetric matrices are the real orthogonal not the unitary matrices. We then form the next iterate as

$$T_+ = LQ + \omega I. \tag{6.4.7}$$

Clearly, $T_+ = Q^T TQ$, so that T_+ has the same eigenvalues as T. As before the parameter ω introduced at each iteration is used to accelerate the convergence. The selection procedure for ω is the same as that used in IMTQL1.

On each iteration the matrix Q is obtained as a sequence of complex plane rotations. In particular $Q \equiv \prod\limits_{k=1}^{m-1} P_{m-k}$ where each rotation $P_k \equiv (p_{ij})$ must satisfy the following conditions.

$$p_{k+1,k+1} = -p_{k,k} = c_k, \quad p_{k+1,k} = p_{k,k+1} = s_k,$$
$$p_{i,j} = \delta_{ij}, \text{ otherwise, } \text{ and } c_k^2 + s_k^2 = 1. \tag{6.4.8}$$

The scalars c_k and s_k are complex-valued and each rotation matrix P_k is a complex symmetric and real orthogonal matrix. That is for each k, $P_k^T = P_k = P_k^{-1}$ and therefore the corresponding Q matrix is real orthogonal, $Q^T = Q^{-1}$.

Mimicking IMTQL1, we choose not to explicitly implement the factorizations and reverse recombinations of factorizations defined in Eqns(6.4.6) and (6.4.7). The argument that this is legitimate parallels that for the real symmetric case. The argument that the real symmetric structure was preserved was strictly structural, and it therefore applies directly to the complex symmetric case. Moreover, the proof of Theorem 6.4.1 also extends readily to the complex symmetric case. We can therefore state the following theorems.

THEOREM 6.4.2 The matrix T_+ in Eqn(6.4.7) is complex symmetric and tridiagonal.

THEOREM 6.4.3 Let T be an mxm complex symmetric matrix. Let Q be a real orthogonal matrix, that is $Q^T = Q^{-1}$. Define $T_+ \equiv Q^T TQ$. If T_+ is complex symmetric and tridiagonal with nonzero super-diagonal entries, then the diagonal entries of T_+ are uniquely determined, the superdiagonal entries are determined to within a factor of ± 1, and the columns of Q are determined to within factors of ± 1, from only the last column of Q.

Proof. For $k = 1,...,m$, let $d_k^+ \equiv T_+(k,k)$, and $e_{k+1}^+ \equiv T_+(k+1,k)$, where $e_{m+1}^+ \equiv 0$. Define $q_{m+1} \equiv 0$ and assume that the last column of Q, q_m, is given. Since $TQ = QT_+$, we obtain the following recursion for the vector q_{k-1} and the corresponding coefficients in the T_+ matrix.

$$e_k^+ q_{k-1} = Tq_k - d_k^+ q_k - e_{k+1}^+ q_{k+1} \equiv u_{k-1}. \tag{6.4.9}$$

Using the orthogonality of q_{m-1} and q_m, we get for k=m that

$$d_m^+ \equiv q_m^T Tq_m \text{ and } (e_m^+)^2 \equiv u_{m-1}^T u_{m-1}.$$

That is, the diagonal entry of T_+ and the square of the super-diagonal entry are uniquely determined. Thus, e_m^+ is uniquely defined up to a multiple of ± 1 and likewise q_{m-1} is uniquely defined up to a multiple of ± 1. Using induction on k and the recursion in Eqn(6.4.9), it is straight-forward to demonstrate that the ± 1 multiplies do not affect the sizes of any of the quantities, and that therefore the matrix T_+ and the columns of Q are uniquely determined up to factors of ± 1. We can further state that because the matrices involved are complex symmetric tridiagonal, the eigenvalues of all of the possible matrices which could be generated by picking different ± 1 combinations are identical. $\qquad\qquad\square$

Thus at each iteration, CMTQL1 executes a sequence of two-sided complex plane rotations. Specifically we have at a given iteration that for some ℓ and j, the starting matrix

$$T^j \equiv T_{\ell,j},\qquad\qquad(6.4.10)$$

and the internal iteration matrices

$$T^k = P_k T^{k+1} P_k$$

for $k = j-1, j-2,...,\ell$, are generated with the next iterate $T_+ \equiv T_{\ell,j}^+ \equiv T^\ell$. In Eqn(6.4.10) $T_{\ell,j}$ denotes a principal submatrix of T consisting of rows and columns $\ell, \ell + 1,...,j$ of the current iterate T. Initially, $\ell = 1$ and $j = m$. However as convergence occurs the top portion diagonalizes and perhaps the bottom portion of the T-matrix splits off.

For a given ℓ and j, the first transformation P_{j-1} is determined by the following conditions.

$$-c_{j-1}e_j^j + s_{j-1}(d_j^j - \omega) = 0,$$
$$c_{j-1}^2 + s_{j-1}^2 = 1 \qquad\qquad(6.4.11)$$

where ω is the eigenvalue of the uppermost 2x2 principal submatrix of $T_{\ell,j}$ which is closest to the diagonal entry d_ℓ^j. Thus, we obtain

$$c_{j-1} = (d_j^j - \omega)/\tau,\quad s_{j-1} = e_j^j/\tau$$
$$\tau = [(e_j^j)^2 + (d_j^j - \omega)^2]^{1/2} \qquad\qquad(6.4.12)$$

Note that since the quantities used in the computation of the denominator τ in Eqn(6.4.12) are complex-valued, τ could vanish without either e_j^j or $(d_j^j - \omega)$ vanishing. CMTQL1 contains the following check for this eventuality where a and b denote the two particular complex numbers being combined. CMTQL1 terminates if this occurs.

$$| (a^2 + b^2)^{1/2} | \le 10*\text{MACHEP}*(|a|^2 + |b|^2)^{1/2}. \qquad\qquad(6.4.13)$$

Since $T^{j-1} \equiv P_{j-1}T^j P_{j-1}$, it follows that the (j-2,j)-th entry of T_j is different from zero. Subsequent rotations P_k are designed to zero out this and succeeding terms introduced as this new nonzero term is chased out of the matrix to preserve the complex symmetric tridiagonal structure. These two-sided complex rotations are applied sequentially until we reach $T_+ \equiv T^\ell$. After each iteration the resulting tridiagonal matrix is checked for possible splitting into submatrices. An off-diagonal entry e_{i+1}^+ is said to be negligible if and only if

$$|e_{i+1}^+| \leq \text{MACHEP}*(|d_i^+| + |d_{i+1}^+|)$$

where d_i^+ and d_{i+1}^+ are used to denote the corresponding diagonal entries in the particular iterate under consideration. The matrix diagonalizes from the top. The plane rotations are applied from the bottom up.

A FORTRAN listing of CMTQL1 is provided in Chapter 7 of Volume 2 of this book. The interested reader may look there for the details of the implementation. The question of the convergence of this method has been answered in the affirmative, if we assume that each of the plane rotations required is well-defined. That is the proof makes the assumption that the exception given in Eqn(6.4.13) is never encountered. CMTQL1 has storage requirements which are O(m) where m is the order of the matrix, and arithmetic operation counts which are $O(m^2)$.

CHAPTER 7

BLOCK LANCZOS PROCEDURES, REAL SYMMETRIC MATRICES

SECTION 7.1   INTRODUCTION

The single vector Lanczos procedure described in Chapter 4 can be used to compute either a few extreme or many eigenvalues of any given real symmetric matrix A. However, that procedure does not directly determine the A-multiplicities of the computed 'good' eigenvalues. It also does not directly determine a complete basis for the invariant subspace corresponding to any such multiple eigenvalue. Additional computation is required. See Section 4.8. Here we consider an alternative approach which allows the direct determination of A-multiplicities and of a basis for a corresponding invariant subspace. This approach uses a block variant of the basic Lanczos recursion together with iteration. Such an approach however allows only the computation of a few of the extreme eigenvalues (and a basis for the corresponding invariant subspace.) The eigenvalue and the eigenvector approximations are computed simultaneously. This is in contrast to the single-vector Lanczos procedure discussed in Chapter 4 which can be used to compute eigenvalues in any portion of the spectrum, and which computes the eigenvalue and the eigenvector approximations separately.

We consider a block analog of the Lanczos recursions given in Eqns(4.3.1) - (4.3.2). Define $B_1 \equiv 0$, and $Q_0 \equiv 0$. Define Q_1 as an nxq matrix whose columns are orthonormalized, randomly-generated vectors. For i=2,...,s define Lanczos blocks Q_i where

$$Q_{i+1}B_{i+1} = P_i \equiv AQ_i - Q_iA_i - Q_{i-1}B_i^T, \qquad (7.1.1)$$

$$A_i \equiv Q_i^T(AQ_i - Q_{i-1}B_i^T) \text{ and } Q_{i+1}B_{i+1} \equiv P_i \qquad (7.1.2)$$

and the matrix B_{i+1} is obtained by applying a modified Gram-Schmidt orthonormalization to the columns of P_i. The dimension q must be greater than or equal to the number of eigenvalues wanted. We use the symbol q in two ways, as the number of vectors in Q_1 and as a column of some Q_j. It should be clear from the context which is intended. Using a straight-forward block analog of the proof of Theorem 2.2.1 for the single-vector recursion, we obtain the following Lemma which tells us that the blocks Q_j, j=1,2,...,s form an orthonormal basis for the Krylov subspace

$$\mathscr{K}^s(Q_1,A) \equiv \text{sp}\{Q_1,AQ_1,...,A^{s-1}Q_1\} \qquad (7.1.3)$$

corresponding to the first block.

LEMMA 7.1.1 (Exact Arithmetic) Given a real symmetric matrix A of order n and an orthonormal starting block Q_1 of size nxq , use the block Lanczos recursion in Eqns(7.1.1) and (7.1.2) to generate blocks $Q_2,...,Q_s$ where qs<n. These blocks form an orthonormal basis for

the Krylov space $\mathscr{K}^s(Q_1,A)$. Furthermore,

$$Q_j^T A Q_i = 0 \quad \text{for} \quad i < (j-1), \tag{7.1.4}$$

so that the associated Lanczos matrices generated by this recursion are block tridiagonal.

Proof. By induction. We have by construction that $Q_i^T Q_j = 0$ for $1 \le i < j \le 3$. Therefore, using Eqn(7.1.1), we have that

$$Q_3^T A Q_1 = Q_3^T Q_2 B_2 + Q_3^T Q_1 A_1 = 0.$$

Assume that for some $J \ge 3$, $Q_i^T Q_J = 0$ for $i < J$ and that $Q_i^T A Q_J = 0$ for $i < J - 1$. We prove that these relationships hold for $J+1$. By construction,

$$Q_i^T Q_{J+1} B_{J+1} = Q_i^T A Q_J - Q_i^T Q_J A_J - Q_i^T Q_{J-1} B_J^T. \tag{7.1.5}$$

But by the induction hypothesis, the right-hand side of this equation equals 0 for any $i < J - 1$. It is zero for $i = J$ because in exact arithmetic $A_j = Q_j^T A Q_j$.

It is theoretically possible that some vector in the block

$$P_J \equiv A Q_J - Q_J A_J - Q_{J-1} B_J^T \equiv Q_{J+1} B_{J+1} \tag{7.1.6}$$

may vanish identically, in which case the number of vectors in Q_{J+1} would be less than the number in the preceding block Q_J. In this situation the matrix B_{j+1} would not be square but it would have full rank, and therefore the matrix $B_{J+1} B_{J+1}^T$ would be invertible. Thus, we can conclude from Eqn(7.1.5) that $Q_i^T Q_{J+1} = 0$ for all $i \le J$.

Now consider the conjugacy relationship Eqn(7.1.4) for $j=J+1$. By construction we have that

$$Q_i^T A Q_{J+1} = B_{i+1}^T Q_{i+1}^T Q_{J+1} + A_i Q_i^T Q_{J+1} + B_i Q_{i-1}^T Q_{J+1}.$$

But since the blocks are pairwise orthogonal, the right-hand side of this equation must vanish for any $i < J$. Therefore, we have immediately that the projection of A onto the Krylov space $\mathscr{K}^s(Q_1,A)$ is represented by the block tridiagonal matrix $T_s \equiv (Q_j^T A Q_i)$, $1 \le j,i \le s$. If we set $\overline{Q}_s \equiv \{Q_1,...,Q_s\}$ then $T_s = (\overline{Q}_s)^T A \overline{Q}_s$. $\qquad \square$

If we rewrite Eqn(7.1.1) in matrix form, then after s Q-blocks have been generated we obtain

$$A\overline{Q}_s = \overline{Q}_s T_s + Q_{s+1} B_{s+1} E_{s+1}^T \quad \text{where} \quad \overline{Q}_s \equiv \{Q_1,...,Q_s\} \tag{7.1.7}$$

and $E_{s+1}^T \equiv (0,...,0,I_q)$ is a qxqs matrix. Here we have assumed that every block generated is nxq. In practice this need not be the case. We will say more about this possibility in Section 7.4. The corresponding small projection matrix T_s is a real symmetric, block tridiagonal matrix of order m = qs.

$$
T_s \equiv \begin{bmatrix} A_1 & B_2^T & & \\ B_2 & A_2 & & \\ & & \ddots & B_s^T \\ & & B_s & A_s \end{bmatrix} \tag{7.1.8}
$$

Typically, we will have that m = qs << n, the order of the given matrix A . We approximate eigenvalues of A by eigenvalues of the T_s matrices. Notice the difference in the notation. The order of T_s is m = qs not s. If we attempted to use both indices s and m, the notation would become even more horrendous than it is.

There is no uniquely defined way to construct a practical block Lanczos procedure from these recursions. The literature contains two basically different types of block Lanczos procedures, iterative and noniterative. For iterative procedures see for example, Cullum and Donath [1974] and Golub and Underwood [1975,1977]. For noniterative procedures see for example, Lewis [1977], Ruhe [1979], and Scott [1979]. Noniterative procedures mimic the single-vector Lanczos procedures. Chains of blocks are generated, the length of the chain depends upon what one is trying to compute and upon the amount of storage available. The Lanczos blocks may or may not be reorthogonalized as they are generated. Ritz vectors may or may not be computed simultaneously with the eigenvalues. In each case subsets of the eigenvalues of the block tridiagonal matrices T_s generated are used as approximations to eigenvalues of A. Iterative procedures, on each iteration k, use the block recursion to generate a small projection matrix. First the relevant eigenvalues and eigenvectors of these small projection matrices are computed, and then the corresponding Ritz vectors are computed and used as updated approximations to the desired eigenvectors. If on iteration k convergence has not yet been achieved, another iteration is carried out using these updated eigenvector approximations as the starting block. The iterations continue until convergence is achieved.

In this chapter we focus on the Cullum and Donath block iterative procedure [1974] and on a hybrid single-vector block procedure which we have devised. We address the question of computing a few of the extreme eigenvalues and a basis for the corresponding invariant subspace. Iterative block Lanczos procedures require relatively large amounts of storage and do not seem to be suitable for computing large numbers of eigenvalues and eigenvectors of large matrices. Moreover, as we will see in the discussion in Section 7.3, they are not directly capable of computing other than the extremes of the spectrum. Sometimes it is feasible to compute other portions of the spectrum by working with the inverse of a shifted matrix $(A - \sigma I)$ for some shift σ, using a sparse matrix factorization. In this case the block procedure can compute eigenvalues of A closest to σ. A FORTRAN implementation of the hybrid algorithm is given in Chapter 8 of Volume 2 of this book.

The basic iterative block Lanczos procedure on which the Cullum and Donath [1974] and the Golub and Underwood [1975,1977] procedures are both based is stated below. Here we assume that the user wants a few of the algebraically-largest eigenvalues of the given matrix. A few of the algebraically-smallest eigenvalues can be obtained by using -A instead of A. In Section 7.3 we will see that we can actually use such a procedure to simultaneously compute a few eigenvalues from both ends of the spectrum of the given matrix.

BASIC ITERATIVE BLOCK LANCZOS PROCEDURE

Step 1.   For k=1 select a starting block, Q_1^k of size nxq where q is at least as large as the number of eigenvalues desired and the columns of Q_1^k are orthonormal.

Step 2.   Given Q_1^k, compute $P_1^k \equiv AQ_1^k - Q_1^k A_1^k$ where $A_1^k \equiv (Q_1^k)^T AQ_1^k$ and use the norms of the columns of P_1^k to test for convergence. If convergence has occurred terminate. Otherwise, go to Step 3.

Step 3.   Use the recursion in Eqns(7.1.1) and (7.1.2) to generate a sequence of blocks Q_j^k, j = 2,...,s satisfying that recursion. Use the small matrices A_j^k, B_{j+1}^k generated to define the real symmetric block tridiagonal Lanczos matrix, T_s^k.

Step 4.   Compute the q algebraically-largest eigenvalues of T_s^k and a corresponding set W_1^k of eigenvectors.

Step 5.   Compute the updated approximation to the eigenvectors of A as

$$Q_1^{k+1} \equiv \overline{Q}_s^k W_1^k \text{ where } \overline{Q}_s^k \equiv \{Q_1^k,... ,Q_s^k\},$$

increment k to k+1 and then go to Step 2.

We used Q_i to denote a subblock, and $\overline{Q}_s$ to denote a set of s Q-subblocks. Typically in the literature, we find that the letter Q is reserved for matrices with orthonormal columns. In contrast to our single-vector Lanczos procedures, we will attempt to maintain the near-orthogonality of all of the blocks of Lanczos vectors generated within each iteration of our block Lanczos procedures. However, as in the single vector case, this cannot be accomplished by a straightforward implementation of the block recursions. The numerical problems will be discussed in detail in Section 7.4 along with the remedies which were incorporated in Cullum and Donath [1974].

In the next two sections we provide an optimization interpretation for the basic iterative block Lanczos procedure. In order to do this we first look at an iterative version of the single-vector Lanczos procedure which generates (s+1) Lanczos vectors on each iteration, and show that such a procedure is equivalent to using an s-step conjugate gradient procedure to maximize the Rayleigh quotient function associated with the given matrix A. (See Eqn(7.2.1).) We then develop a block analog of this correspondence between an iterative Lanczos procedure and an s-step conjugate gradient procedure for the Rayleigh quotient function.

SECTION 7.2   ITERATIVE SINGLE-VECTOR, OPTIMIZATION INTERPRETATION

Each iteration of the basic iterative block Lanczos procedure defined in Section 7.1 consists of an inner and of an outer loop computation. In each inner loop eigenvalues and eigenvectors of a small projection matrix T_s^k are computed. In each outer loop the block recursion in Eqns(7.1.1) - (7.1.2) is applied to A using the current approximation Q_1^{k+1} to the desired eigenspace as the first block, to generate a new projection matrix T_s^{k+1}. Karush [1951] proposed such an inner and outer loop single-vector Lanczos algorithm for computing the algebraically-largest eigenvalue, but using a slightly different version of the single-vector Lanczos recursions in Eqn(4.3.1). We can view the Karush algorithm as an iterative block Lanczos procedure with only one vector in each of the blocks. In the next few paragraphs we demonstrate that the Karush algorithm is nothing more than a clever implementation of a s-step conjugate gradient procedure for maximizing the Rayleigh quotient function r(x) defined in Eqn(7.2.1).

$$r(x) \equiv [x^T A x]/[x^T x] \quad \text{for} \quad x \neq 0. \tag{7.2.1}$$

OPTIMIZATION INTERPRETATION It is easy to demonstrate that the values of the Rayleigh quotient function in Eqn(7.2.1) at the 'points' x at which its gradient vanishes are eigenvalues of A. Furthermore, the corresponding 'points' or x-values are eigenvectors of A. In particular, the global-maximum of this function is the algebraically-largest eigenvalue of A, and any 'point' at which this global maximum is attained is an eigenvector corresponding to the algebraically largest eigenvalue of A.

Hestenes and Karush [1951] applied the method of steepest ascent to r(x) to obtain the algebraically-largest eigenvalue of A At each iteration k the steepest ascent procedure selects the current direction of steepest increase in the given function f as the direction of movement and starting at the current iterate moves along that direction until the maximum of f along that line is achieved. The direction of steepest ascent at any iterate x_k is the ray through x_k whose direction is the gradient of f at x_k. We denote the gradient of r at x by $g_r(x)$ and the gradient of r at x_k simply by g_k. The next iterate is the point on the line at which the maximum value is attained. Hestenes and Karush [1951] proved that for the Rayleigh quotient function, the values $r(x_k)$ and the iterates x_k generated in this manner must converge respectively, to the algebraically-largest eigenvalue of A and to a corresponding eigenvector, as the number of iterations $k \to \infty$.

It is not immediately apparent how the iterative, single-vector Lanczos algorithm in Karush [1951] is related to the steepest ascent procedure for r(x). We want to use a similar analogy in our discussions of our iterative block Lanczos procedures to justify the use of such procedures. Therefore we consider the relationship between these two methods in detail.

LEMMA 7.2.1 (Exact Arithmetic) (i) For any $\sigma \neq 0$, the function $r_\sigma(x) \equiv r(\sigma x)$ has the same maximum value as the function r(x) and any maximizing point for r(x) is a maximizing point for $r_\sigma(x)$. (ii) Maximizing r(x) on a line $\mathscr{L} \equiv \{x \mid x = x_k + \omega g_k, \ \omega \neq 0\}$ is equivalent to maximizing r(x) on the corresponding 2-dimensional subspace $\mathscr{K}^2 = sp\{x_k, g_k\} \equiv sp\{x_k, Ax_k\}$. (iii) The value $r(x_{k+1})$ obtained by applying steepest ascent to r(x) at x_k is the same as the maximum value of r(x) on the corresponding subspace $sp\{x_k, g_k\}$. (iv) The maximum of r(x) on this subspace can be obtained by computing the algebraically-largest eigenvalue of the projection of A onto this subspace.

Proof. (i) follows directly from the fact (see Eqn(7.2.1)) that r(x) is a homogeneous function of degree zero. That is, for any scalar $\sigma \neq 0$ and any $x \neq 0$, that $r(\sigma x) = r(x)$. To prove (ii) consider the following. Clearly, $\mathscr{L} \subseteq \mathscr{K}^2$. Therefore, if the maximum of r(x) on $\mathscr{K}_2$ occurs at a 'point' $x^* = \theta x_k + \omega g_k$ where $\theta \neq 0$, then the corresponding point $\hat{x} \equiv x^*/\theta \in \mathscr{L}$ and $r(\hat{x}) = r(x^*)$. Therefore, all we need to show is that θ cannot be 0. If θ were 0, then since $r(\omega g_k) = r(g_k)$ for any $\omega \neq 0$, it is sufficient to consider the possibility that $x^* = g_k$. But we have that $x_k^T g_k = 0$ so that $g_k^T Ax_k = (g_k^T g_k)(x_k^T x_k) > 0$. But then, as can be verified by direct substitution, for large enough ω we have the contradiction that $r(g_k) < r(x_k + \omega g_k)$. Therefore, $x^* \neq g_k$. (iii) is simply a restatement of (ii). (iv) follows from the fact that for any $x = x_k \theta + g_k \omega$ we have that

$$x^T Ax = [x_k\theta + g_k\omega]^T A[x_k\theta + g_k\omega] = [\theta, \omega] A_p \begin{bmatrix} \theta \\ \omega \end{bmatrix} \text{ where } A_p \equiv \begin{bmatrix} x_k^T \\ g_k^T \end{bmatrix} A[x_k, g_k] \quad .$$

Therefore, as we vary θ and ω, we are computing Rayleigh quotients for the projection matrix A_p which is simply the projection of A onto $sp\{x_k, g_k\}$. $\square$

The asymptotic rate of convergence of the steepest ascent procedure depends upon the stiffness of the second derivative matrix at the maximizing point. A straight-forward computation of the second derivative of r(x) yields

$$H_r(x) = \frac{2[A - r(x)I]}{[x^T x]} - \frac{4[x(g_r(x))^T + (g_r(x))x^T]}{[x^T x]^2} \tag{7.2.2}$$

At any stationary point $g_r(x) = 0$ so that at such points $H_r(x) = 2[A - r(x)I]/[x^T x]$. The rate of convergence of steepest ascent for a function which at its maximizing point has a second derivative matrix which is negative definite, is linear with the coefficient of convergence no greater than

$$\frac{[\mu_{max} - \mu_{min}]}{[\mu_{max} + \mu_{min}]}^2$$

where μ_{max} and μ_{min} denote respectively the largest and smallest magnitudes of the eigenvalues of the second derivative matrix at the maximizing point. See for example, Luenberger [1973, p.154]. This result can be extended to include functions with semidefinite second derivative matrices. In this case the smallest (in magnitude) nonzero eigenvalue is used in computing the coefficient of convergence.

The eigenvalues of $H_r(x^*)$ are $\{ (\lambda_n - \lambda_1), \dots, (\lambda_2 - \lambda_1), 0 \}$ where $\lambda_1 \geq , \dots, \geq \lambda_n$ are the eigenvalues of A. The spread of these eigenvalues is $|\lambda_1 - \lambda_n|$ and the gap between the algebraically-largest and the next largest eigenvalue of A is just $|\lambda_1 - \lambda_2|$. If we let z_1 denote the desired eigenvector of A, then we have that for $k=2,3,\dots$

$$\| x_{k+1} - z_1 \|^2 \leq Kc^k \| x_1 - z_1 \|^2 \quad \text{where}$$
$$K \equiv [spread/gap] \quad \text{and} \quad |c| \leq [(spread-gap)/(spread+gap)]^2 \tag{7.2.3}$$

Therefore the larger the spread and the smaller the gap, the slower the convergence.

Lemma 7.2.1 states that a Karush procedure [1951] which uses two steps of the Lanczos recursion on each iteration is identical to applying the steepest ascent procedure to r(x). Below we show that this analogy extends to using an s-step conjugate gradient procedure on r(x) and using s+1 steps of the Lanczos recursion on each iteration in a Karush procedure.

In the discussions in this chapter we will always use the letter k to denote the iteration number. If the procedure being discussed has only one internal step on each iteration then we will use the iteration label k as a subscript on the relevant variables involved in the iterations. If, however, the internal step of a given procedure consists of more than one step then we use the iteration label k as a superscript on the variables involved in the procedure and will use some other subscript variable to denote the number of the internal step being considered on iteration k. For example, when we discuss the basic conjugate gradient procedure we use the iteration number as a subscript. However, when we discuss the s-step conjugate gradient procedure we will use the iteration number as a superscript and use some other variable to label the s conjugate gradient steps within each iteration.

We need to review the definition of the conjugate gradient optimization procedure for the special case when the given function is quadratic. Let

$$f(x) = (x^T Bx)/2 + d^T x + e \tag{7.2.4}$$

be a quadratic function of n parameters where B is a negative definite, real symmetric nxn matrix. We denote the gradient of f at x_k by g_k.

CONJUGATE GRADIENT PROCEDURE FOR A QUADRATIC FUNCTION

Step 1. Let $f(x)$ be as defined in Eqn(7.2.4). Select a starting vector x_1 and set the starting direction $p_1 = g_1$.

Step 2. Maximize f on the line $\{x \mid x = x_k + \omega p_k, \ \omega \neq 0\}$ and set x_{k+1} equal to the maximizing point of f on this line. Test for convergence. If convergence has occurred terminate. Otherwise, go to Step 3.

Step 3. Update the direction of movement $p_{k+1} = g_{k+1} + \gamma_k p_k$ where the scalar $\gamma_k = -[g_{k+1}^T B p_k / p_k^T B p_k]$ is chosen such that the new direction p_{k+1} is B-conjugate to the previous direction. Go to Step 2.

Therefore, the conjugate gradient procedure for a quadratic function consists of the successive generation of directions that are B-conjugate and the successive maximization of the given function along these directions as they are generated. Step 3 is sufficient to guarantee (in exact arithmetic) that the directions p_k, $k = 1,2,..,s$. are B-conjugate. From the definitions we see that these directions are simply specific combinations of the gradients of the iterates generated, and thus we can view this as a procedure for conjugating those gradients. Note that for any k, $sp\{g_1,...,g_k\} = sp\{p_1,...,p_k\}$.

For any negative definite quadratic function $f(x)$ this successive maximization over B-conjugate 1-dimensional subspaces can be shown to be equivalent to maximizing f over the entire s-dimensional subspace $sp\{ g_1, ..., g_s \}$. See for example Luenberger [1973]. Therefore in exact arithmetic, the conjugate gradient procedure terminates at the maximum of f in at most n steps.

In practice when the number of parameters is large, even if f is quadratic, it is desirable and even necessary to modify the straightforward conjugate gradient procedure so that it is restarted after each s-steps, where s is specified by the user. That is after each s-steps, the conjugate gradient procedure reverts to one steepest ascent step. This restarts the conjugate gradient recursions with the current gradient.

S-STEP CONJUGATE GRADIENT PROCEDURE FOR A QUADRATIC FUNCTION

Step 1. Let f be a quadratic function given by Eqn(7.2.4). Specify the number of steps s for each conjugate gradient cycle. Set $k=j=1$ and select a starting vector, $x_1 \equiv x_1^1$ and set the starting direction $p_1^1 = g_1$.

Step 2. Maximize f on the line $\{x \mid x = x_j^k + \omega p_j^k, \ \omega \neq 0\}$ and set x_{j+1}^k equal to the maximizing point of f on this line. Test for convergence. If convergence has occurred terminate. Otherwise, go to Step 3.

Step 3. If $j<s$ set $j=j+1$ and update the direction of movement $p_{j+1}^k = g_{j+1}^k + \gamma_j^k p_j^k$ where γ_j^k is chosen to make p_{j+1}^k B-conjugate to p_j^k. If $j = s$ then set $j=1$ and $x_{k+1} = x_1^{k+1} = x_{s+1}^k$ and $p_1^{k+1} = g(x_{k+1})$, $j = 1$ and then go to Step 2.

Within each iteration k a set $\{p_j^k, j = 1,...,s\}$ of B-conjugate directions is generated. As each new direction is generated the given function is maximized over the corresponding 1-dimensional subspace $\{x \mid x = x_j^k + \omega p_j^k, \omega \neq 0\}$. The following theorem, which can be found for example in Luenberger [1973, page 180], states that if a given function has a second derivative with a few large eigenvalues, then the detrimental effect of these few eigenvalues upon the rate of convergence observed for steepest ascent can be eliminated by using s-step conjugate gradients with a large enough value of s.

THEOREM 7.2.1 (s-Step Conjugate Gradient Method) Define f by Eqn(7.2.4). Let $\{\mu_n \leq ... \leq \mu_1 < 0\}$ be the eigenvalues of B. Then the iterates generated by the method of s-step conjugate gradients when it is applied to f and it is restarted every s steps, satisfy

$$| E(x_{k+1}) | \leq [(\mu_1 - \mu_{n-s+1})/(\mu_1 + \mu_{n-s+1})]^2 | E(x_k) | \qquad (7.2.5)$$

where

$$E(x_{k+1}) \equiv [(x_{k+1} - x^*)^T B(x_{k+1} - x^*)] / 2$$

and x^* is the maximizing point of f.

The extension of the conjugate gradient procedure to nonquadratic functions rests upon the assumption that near its maximizing point, the given function can be approximated well by a negative definite quadratic function. (We need only semi-definiteness). If this is the case, then once an iterate has been obtained that is close to the maximizing point of f, we might expect the subsequent terminal convergence to mimic that expected for a quadratic function. This happens however only if we restart the conjugate gradient procedure once a suitable neighborhood of the maximizing point has been reached. Typically, as we said earlier some type of restarting strategy is essential. For a general function we have the following definition of a s-step conjugate gradient procedure. If $s=1$ then this reduces to the standard conjugate gradient procedure.

S-STEP CONJUGATE GRADIENT PROCEDURE FOR A GENERAL FUNCTION

Step 1. Specify the number of steps s for each conjugate gradient cycle. Set $k=j=1$ and select a starting vector $x_1 = x_1^1$ and set the starting direction $p_1 = g(x_1)$.

Step 2. Maximize f on the line $\{x \mid x = x_j^k + \omega p_j^k, \omega \neq 0\}$ and set x_{j+1}^k equal to the maximizing point of f on this line. Test for convergence. If convergence has occurred terminate. Otherwise, go to Step 3.

Step 3. If $j<s$ set $j=j+1$ and update the direction of movement $p^k_{j+1} = g^k_{j+1} + \gamma^k_j p^k_j$ where

$$\gamma^k_j = -[(g^k_{j+1})^T (\Delta g)^k_{j+1}/(p^k_j)^T (\Delta g)^k_{j+1}] \quad \text{with} \quad (\Delta g)^k_{j+1} \equiv g(x^k_{j+1}) - g(x^k_j).$$

If $j=s$ then set $j=1$ and

$$x_{k+1} \equiv x^k_{s+1} \equiv x^{k+1}_1 \quad \text{and} \quad p^{k+1}_1 = g(x_{k+1}) \quad \text{and go to Step 2.}$$

If we apply this s-step procedure to the Rayleigh quotient function $r(x)$, we obtain the following lemma.

LEMMA 7.2.2 (Exact Arithmetic) If $\{g^k_1, g^k_2, ..., g^k_s\}$ are the gradients obtained by applying one step of the s-step conjugate gradient method to the Rayleigh quotient function $r(x)$ starting at an iterate x_k with $g^k_1 \equiv g(x_k)$, then

$$\{x \mid x \in x_k + \text{sp}\{g^k_1, ..., g^k_s\}\} \subseteq \mathcal{K}^{s+1}(x_k, A)$$

and the maximum of $r(x)$ over this subspace equals the maximum of $r(x)$ over the entire subspace, $\mathcal{K}^{s+1}(x_k, A)$.

Proof. For simplicity we ignore the k superscript. By construction we have that for any $\sigma \neq 0$, $(x_k + g_1\sigma) \in \mathcal{K}^2(x_k, A)$. Moreover, since $r(x)$ is homogeneous of degree zero we have that for any $\gamma \neq 0$, $r(x_k\gamma + g_1\sigma) = r(x_k + g_1\sigma\gamma^{-1})$. Now assume that the desired containment is true for some j. We know that

$$g_{j+1} = [Ax_{j+1} - \mu(x_{j+1})x_{j+1}] / [x^T_{j+1}x_{j+1}]$$

and that each direction of movement is some combination of the gradients so that $x_{j+1} \subseteq \text{sp}\{x_1, g_1, ..., g_j\}$. Therefore by induction $x_{j+1} \in \mathcal{K}^{j+1}(x_1, A)$ and therefore g_{j+1} is in the Krylov subspace $\mathcal{K}^{j+2}(x_1, A)$. The homogeneity argument that the two maxima must be identical that was used in Lemma 7.2.1 for the case $j=2$ is equally applicable for any j. □

Therefore, we have that applying s-step conjugate gradients to the Rayleigh quotient function corresponds to successively maximizing $r(x)$ on the Krylov subspaces, $\mathcal{K}^{s+1}(x_k, A)$, $k = 1, 2, ...$. This maximization can be achieved using an iterated, truncated Lanczos recursion. On each iteration the Lanczos recursion is used to generate an orthonormal basis for the Krylov subspace generated by x_k and the maximum of $r(x)$ on this subspace is simply the largest eigenvalue of A constrained to this subspace. This restriction of A is given by the small Lanczos tridiagonal matrices generated. Thus, to perform s-steps of conjugate gradients on $r(x)$, we simply compute the relevant eigenvalues and eigenvectors of the small Lanczos tridiagonal matrices and then using the Lanczos vectors transform these eigenvectors into Ritz vectors in n-space. This is precisely Karush's algorithm [1951], so we have the following lemma. The Lanczos procedure has the advantage that at each stage the maximizations over the relevant subspaces are carried out exactly by simply computing eigenvalues and eigenvectors of small projection matrices.

LEMMA 7.2.3   The Karush [1951] procedure which uses s+1 steps of the Lanczos recursion on each iteration, is equivalent to applying an s-step conjugate gradient algorithm to r(x).

We can now use this relationship to get an estimate of the rate of convergence for the Karush algorithm.   Eqn(7.2.2) tells us that near any stationary point of r(x), $H_r(x)$ is essentially a shifted A-matrix.   We get an estimate of the rate of convergence by using the eigenvalues of this matrix.   Theorem 7.2.1 states that if we use s-step conjugate gradients on a quadratic function then the 'spread' in Eqn(7.2.3) is replaced by the 'effective spread', $\lambda_1 - \lambda_{n-s+1}$, and the coefficient of convergence c in Eqn(7.2.3) is no greater than

$$[(\lambda_{n-s+1} - \lambda_1) - (\lambda_2 - \lambda_1)]^2 / [(\lambda_{n-s+1} - \lambda_1) + (\lambda_2 - \lambda_1)]^2 \qquad (7.2.6)$$

K is still $|\lambda_n - \lambda_1| / |\lambda_2 - \lambda_1|$.   As before, the λ_j, $1 \leq j \leq n$, are the eigenvalues of A and z_1 is the eigenvector of A corresponding to λ_1.   Thus, the coefficient of convergence can sometimes be significantly reduced by increasing the number of steps used within each iteration of s-step conjugate gradients.

Obviously, this procedure could also be used to compute the algebraically-smallest eigenvalue, simply by minimizing r(x) rather than maximizing it.   However it is designed to compute only the algebraically-largest (or the algebraically-smallest) eigenvalue of A and a corresponding eigenvector.   We want a procedure which can be used to compute several of the algebraically-largest eigenvalues and a basis for the corresponding invariant subspace.   In Section 7.3 we demonstrate that the basic iterative, block Lanczos procedure will do this.   The arguments in the next section are constructive.   We extend the optimization analogy to the block case and then use this analogy to make some comments about the expected rate of convergence of the iterative block procedures.

SECTION 7.3   ITERATIVE BLOCK, OPTIMIZATION INTERPRETATION

As we saw in the preceding section, the use of an iterative, single vector Lanczos algorithm which uses s+1 steps on each iteration is equivalent to the application of a s-step conjugate gradient procedure to the Rayleigh quotient function r(x) in Eqn(7.2.1).   In this section we use a block analog of this equivalence to justify the use of an analogous iterative block Lanczos procedure.   The literature contains alternative justifications, see for example Golub and Underwood [1977], which parallel the Kaniel [1966] arguments for the single-vector procedures.   Those arguments use direct estimates for how well the eigenvalues and eigenvectors of the projection of A onto the particular Krylov space obtained on one iteration can be expected to approximate the actual eigenvalues and eigenvectors of A.

We choose to view an iterative block Lanczos procedure as a block generalization of the s-step conjugate gradient procedure to a 'block' version R(X) of the Rayleigh quotient function r(x). This allows us to estimate the overall rate of convergence of the iterative procedure by heuristically extrapolating the estimates obtained for the scalar case to the block case. To do this we need the following theorem which is a combination of results from Fan [1949] and Cullum, Donath and Wolfe [1975]. This theorem identifies the relevant block generalization of r(x). In particular it states that for any real symmetric matrix A and for any q, the q algebraically largest eigenvalues of A and a corresponding set of eigenvectors can be obtained by maximizing a sum of q Rayleigh quotients of A subject to orthonormality constraints on the set of vectors used in the Rayleigh quotients.

THEOREM 7.3.1 For any real symmetric matrix A of order n, and any $q < n$ define the block Rayleigh quotient function

$$R(X) \equiv \sum_{i=1}^{q} x_i^T A x_i \quad \text{where } X = \{x_1,...,x_q\} \tag{7.3.1}$$

and each $x_i \in E^n$. Then

$$\max_{\mathscr{S}} R(X) = \sum_{i=1}^{q} \lambda_i(A), \tag{7.3.2}$$

where $\mathscr{S} \equiv \{X \mid X^T X = I_q\}$. Furthermore, this maximum is attained only on sets X such that (i) X spans the eigenspace of A corresponding to the eigenvalues $\lambda_i(A) > \lambda_q(A)$, and (ii) X is contained in the eigenspace of A corresponding to $\lambda_i(A)$, $1 \le i \le q$.

In Eqn(7.3.2) as elsewhere $\lambda_i(A)$ denotes the ith eigenvalue of A with the ordering $\lambda_n(A) \le \lambda_{n-1}(A) \le ... \le \lambda_1(A)$. We simplify the notation by using λ_j instead of $\lambda_j(A)$. The constraint set $\mathscr{S}$ is simply the statement that the set X of n-vectors x_i, $1 \le i \le q$, used must be pairwise orthonormal. Theorem 7.3.1 tells us that we can obtain the quantities we want by maximizing R(X) over $\mathscr{S}$.

Let tr[B] denote the trace of B. We have that

$$R(X) = \text{tr } [X^T A X] \tag{7.3.3}$$

since the trace of a matrix is just the sum of its diagonal elements. We know moreover, that the trace of a matrix equals the sum of its eigenvalues. Therefore, we have that given any nxq block X and any orthogonal qxq matrix Γ that

$$R(X) = R(X\Gamma) \tag{7.3.4}$$

since the matrices $X^T A X$ and $\Gamma^T X^T A X \Gamma$ have the same eigenvalues. In particular we have that for any such Γ with $R_\Gamma \equiv R(X\Gamma)$ that

$$\max_{\mathscr{S}} R_\Gamma(X) = \max_{\mathscr{S}} R(X). \tag{7.3.5}$$

Obviously, the maximizing block is not unique. Any orthogonal basis for a corresponding invariant subspace will maximize R(X).

How can we maximize R(X)? It is a function of the qn parameters which define the components of the matrix X. Therefore, we could consider it as a function of qn parameters subject to constraints corresponding to the orthonormality conditions, and attempt to apply the s-step conjugate gradient procedure directly in that context. It is however, advantageous to preserve the block structure. We work directly with the matrix X and define a block analog of the s-step conjugate gradient procedure.

What is an appropriate block analog? Several different versions of block conjugate gradient algorithms for solving systems of linear equations have been proposed, see for example O'Leary [1980]. She considered the question of using a block version of the conjugate gradient procedure to speed up the computation of either (i) several solutions of a given matrix equation or (ii) the solution of a system of equations where the given matrix has several widely separated eigenvalues at one extreme of the spectrum. The formulas which we introduce are somewhat different from hers. These formulas reduce to the corresponding scalar algorithms for the Rayleigh quotient function r(x) when blocks containing only one vector are used.

We first define a block analog of steepest ascent for R(X). We will then extend this to a s-step block conjugate gradient procedure. For any given change ΔX in the parameter matrix X, we have the following change in R(X).

$$\Delta R \equiv R(X + \Delta X) - R(X) = 2\text{tr}\,[(\Delta X)^T AX + o(\Delta X)],$$

where $\|o(\Delta X)\| / \|\Delta X\| \to 0$ as $\|\Delta X\| \to 0$. This follows directly from the facts that the trace of a matrix is a linear function of its matrix argument and that a matrix and its transpose have the same trace. We need the following definition.

DEFINITION 7.3.1 The Frobenius norm of an nxq matrix $B \equiv (b_{ij})$ is defined as follows:

$$\| B \|_F^2 \equiv \text{tr}\,[B^T B]. \tag{7.3.6}$$

The associated inner product of two nxq matrices B_1 and B_2 is given by

$$< B_1, B_2 >_F \equiv \text{tr}\,[B_1^T B_2]. \tag{7.3.7}$$

Using this definition we can now write the first order change in R(X) as

$$\delta R(X) \equiv 2\text{tr}\,[(\Delta X)^T AX] = 2< \Delta X, AX >_F. \tag{7.3.8}$$

Thus w.r.t the matrix inner product space defined by the Frobenius norm, the gradient of R(X) is just AX. We must however constrain ΔX such that $(X + \Delta X) \in \mathscr{S}$. We use the following Lemma.

LEMMA 7.3.1 Let $A_X \equiv X^T AX$ for $X \in \mathscr{S}$ and A_X diagonal. Then for any ΔX such that

$X + \Delta X \in \mathscr{S}$,

$$\text{tr } [(\Delta X)^T X A_X] = o(\|\Delta X\|). \tag{7.3.9}$$

Proof. By hypothesis $(\Delta X)^T X = -X^T \Delta X + o(\Delta X)$. Therefore, to first order the matrix $(\Delta X)^T X$ is skew symmetric and its diagonal entries are $o(\|\Delta X\|)$. But A_X is diagonal so that Eqn(7.3.9) is satisfied. $\qquad\square$

LEMMA 7.3.2 The first order change $\delta R(X)$ in $R(X)$ along the constraint set $\mathscr{S}$ satisfies

$$\delta R(X) = 2 \langle \Delta X, G(X) \rangle_F \text{ where } G(X) \equiv AX - XA_X. \tag{7.3.10}$$

Proof. From Eqn(7.3.4) we have that for any orthogonal matrix Γ, $R_\Gamma \equiv R(X\Gamma) \equiv R(X)$. Therefore, in any maximization scheme which we use on $R(X)$ we can assume for theoretical purposes and w.l.o.g. that each iterate X that we generate is chosen such that A_X is a diagonal matrix. (In practice in our block Lanczos procedure we will choose each block iterate Q_1^k such that the corresponding A_1^k is diagonal.) Using this fact and Lemma 7.3.1 we get that for any such X the first order change in $R(X)$ along the constraint $\mathscr{S}$ is given by Eqn(7.3.10). $\qquad\square$

Thus, w.r.t. the Frobenius inner product the projected gradient of $R(X)$ on $\mathscr{S}$ is, to within the scalar 2 which we will choose not to write explicitly each time, the nxq matrix $G(X)$ defined in Eqn(7.3.10). Note that this is just the block matrix which we would have obtained by formally projecting the gradient AX onto the constraint $X^T X = I$.

LEMMA 7.3.3 W.r.t the Frobenius space, the subspace of steepest ascent for the block Rayleigh quotient function $R(X)$ on $\mathscr{S}$ is spanned by the vectors

$$\Delta X = G(X)/\|G(X)\|_F. \tag{7.3.11}$$

Proof. The proof is trivial. To determine the 'direction' of steepest ascent we must determine ΔX which maximizes

$$\langle \Delta X, G(X) \rangle_F \text{ over all } \Delta X \text{ such that } \|\Delta X\|_F = 1.$$

By the triangle inequality we have that for any such ΔX, $\langle \Delta X, G(X) \rangle_F \leq \|G(X)\|_F$. But we achieve this if we set $\Delta X = G(X)/\|G(X)\|_F$. $\qquad\square$

Lemma 7.3.3. tells us that the 'direction' of steepest ascent is just the subspace defined by the projected gradient. We use this subspace to define a block analog of the steepest ascent procedure for the block Rayleigh quotient function $R(X)$.

BLOCK STEEPEST ASCENT PROCEDURE FOR R(X)

Step 1. Let R(X) be the function defined in Eqn(7.3.1) for some q. Set k=1 and select a starting block X_1 of size nxq such that $X_1^T X_1 = I_q$.

Step 2. Given k, maximize R(X) on the intersection of the constraint set $\mathscr{S} = \{X \mid X^T X = I_q\}$ with the set of all matrices

$$\{ X \mid X = X_k \Phi_k + G(X_k)\Omega_k \} \qquad (7.3.12)$$

where Φ_k, Ω_k are qxq matrices and Φ_k is invertible. Set the updated block iterate X_{k+1} equal to the block which maximizes R on this intersection. Test for convergence. If convergence has occurred terminate. Otherwise, set k=k+1 and repeat Step 2.

This procedure reduces to ordinary steepest ascent when q = 1. That is, when R(X) = r(x). How would we implement such a procedure?. We use the correspondence developed in Section 7.2 between the steepest ascent procedure for r(x) and the iterative, single-vector Lanczos procedure to tell us what to do here. It is easy to prove the following lemma.

LEMMA 7.3.4 Restricting the matrix X to the set of matrices defined by Eqn(7.3.12) is equivalent to restricting A to the subspace $\text{sp}\{X_k, G(X_k)\}$ and then maximizing the associated block Rayleigh quotient function $R_p(W) \equiv \text{tr}(W^T A_p^k W)$ where W is any 2qxq matrix with orthonormal columns and A_p^k represents the projection of A onto the above subspace.

We then have the following Corollary as a direct consequence of Lemma 7.3.4 and Theorem 7.3.1.

COROLLARY 7.3.4 The maximum value of R(X) on the subspaces $\text{sp}\{X_k, G(X_k)\}$ defined by Eqn(7.3.12) and a basis for the corresponding eigenspace can be obtained by computing the extreme eigenvalues and corresponding eigenvectors of the small projection matrices A_p^k.

LEMMA 7.3.5 If corresponding to each iteration of the block steepest ascent procedure, we set $Q_1^k \equiv X_k$ and then use the basic block Lanczos recursion given in Eqns(7.1.1) - (7.1.2) to generate Q_2^k, then

$$\text{sp}\{Q_1^k, Q_2^k\} \equiv \text{sp}\{X_k, G(X_k)\}.$$

In fact the 2nd block Q_2^k is just an orthonormalized version of $G(X_k)$ and the block tridiagonal Lanczos matrix T_2^k is just the desired projection matrix A_p^k of A.

Thus, the block steepest ascent procedure for R(X) is just the sequential computation of eigenelements of small projection matrices. This computation can be implemented as an iterative block Lanczos procedure which uses two blocks on each iteration.

If we wish to compute q' eigenvalues, then we let the initial starting block Q_1^1 be a nxq block of orthonormalized vectors where $q' \leq q$. We use the block Lanczos recursion in an iterative manner to compute on each iteration k an orthonormal basis for the subspaces $sp\{Q_1^k, G(Q_1^k)\}$. The corresponding restriction of A to the subspace $sp\{Q_1^k, Q_2^k\}$ is given by the 2qx2q matrix

$$T_2^k \equiv (\overline{Q}_2^k)^T A \overline{Q}_2^k \text{ where } \overline{Q}_2^k \equiv \{Q_1^k, Q_2^k\} .$$

We compute the q algebraically largest eigenvalues μ_j^k, $1 \leq j \leq q$, of T_2^k, as well as a corresponding orthonormal basis W_1^k for an associated eigenspace of T_2^k. Then $\overline{Q}_2^k W_1^k$ maximizes $R(X)$ on this subspace. Of particular interest in this computation is the fact that $\|Q_2^k B_2^k\| \equiv \|AQ_1^k - Q_1^k A_1^k\|$ is just the norm of the error in $sp\{Q_1^k\}$ being an invariant subspace of A. We will say more about this in Section 7.4. Invariance would mean that $AQ_1^k \subseteq sp\{Q_1^k\}$.

The preceding arguments demonstrate that we can view a basic iterative, block Lanczos procedure which uses two blocks on each iteration as a steepest ascent procedure for the simultaneous maximization of q Rayleigh quotients subject to orthonormality constraints. However, as in the scalar case, the convergence rate can be improved by using not just two blocks on each iteration but using s blocks. We therefore, look at a block analog of s-step conjugate gradients for $R(X)$. We need the following lemma.

LEMMA 7.3.6 Given the block Rayleigh quotient function $R(X)$ defined in Eqn(7.3.1) and the constraint set $\mathscr{S} \equiv \{X \mid X^T X = I_q\}$, we have that if $X \epsilon \mathscr{S}$ then for any s, the Krylov subspace,

$$\mathscr{K}^s(X,A) = \hat{\mathscr{K}}^s(X,A) \equiv sp\{X, G(X), ..., A^{s-2}G(X).\} \tag{7.3.13}$$

Proof. The proof is by induction. By Eqn(7.3.10) we have that $\mathscr{K}^2 \equiv \hat{\mathscr{K}}^2$. Assume that for some j, $\mathscr{K}^j \equiv \hat{\mathscr{K}}^j$. Clearly, $\hat{\mathscr{K}}^{j+1} \subseteq \mathscr{K}^{j+1}$. Conversely, for any matrix $W \epsilon \mathscr{K}^{j+1}$, we have by definition that for some Θ and Φ

$$W = W_1 \Theta + A^j X \Phi \text{ where } W_1 \epsilon \mathscr{K}^j.$$

But by Eqn(7.3.10) there exists a matrix Ω with $A^j X \Phi = A^{j-1} G(X) \Phi + A^{j-1} X \Omega$. But by definition, $A^{j-1} G(X) \epsilon \hat{\mathscr{K}}^{j+1}$ and by induction $A^{j-1} X \Omega \epsilon \hat{\mathscr{K}}^{j+1}$. □

Given the block analog of steepest ascent we can define a corresponding block analog of the s-step conjugate gradient procedure as follows. This reduces to the s-step conjugate gradient algorithm for $r(x)$ if $q=1$. We have used P_i in two ways. The directions P_j^k defined below are not related to the P_i defined in Eqn(7.1.2).

S-STEP BLOCK CONJUGATE GRADIENT PROCEDURE FOR R(X)

Step 1. Specify the number of blocks s for each conjugate gradient cycle. Select q, set $k=j=1$, and select a starting block $X_1 \equiv X_1^1$ of size nxq such that $X_1^T X_1 = I_q$. Set the starting 'direction' $P_1^1 = G(X_1)$.

Step 2. Maximize R(X) on the intersection of the set of all matrices

$$\{ X \mid X = X_j^k \Phi_j^k + P_j^k \Omega_j^k \text{ where } \Phi_j^k, \Omega_j^k \text{ are qxq matrices } \} \quad (7.3.14)$$

with the constraint set $\mathscr{P}$ and set the updated block iterate X_j^{k+1} equal to the maximizing block of R on this intersection. Test for convergence. If convergence has occurred terminate. Otherwise, go to Step 3.

Step 3. If $j<s$ set $j=j+1$ and update the block direction of movement $P_{j+1}^k = G_{j+1}^k + P_j^k \Gamma_j^k$ where

$$\Gamma_j^k = -[(\Delta G_{j+1}^k)^T P_{j+1}^k]^{-1} [(\Delta G_{j+1}^k)^T G_{j+1}^k] \quad (7.3.15)$$

where $G_{j+1}^k \equiv G(X_{j+1}^k)$ and $(\Delta G_{j+1}^k) \equiv G(X_{j+1}^k) - G(X_j^k)$. If $j=s$ set $j=1$ and $X_{k+1} \equiv X_1^{k+1} \equiv X_{s+1}^k$ and go to Step 2.

This procedure is well-defined only as long as the required matrix inverses exist. If at some iteration a singular matrix is encountered, then the size of the Q-subblocks must be reduced. In the next section we will see how this is done in practice. We now relate this s-step block conjugate gradient procedure for R(X) to the iterative block Lanczos procedure which uses $s+1$ blocks on each internal iteration. The following Lemma is easily established.

LEMMA 7.3.7 On each iteration k, the block 'directions' P_j^k and the iterates X_{j+1}^k for $j=1,2,...,s$ generated by the block analog of s-step conjugate gradients for the function R(X) belong to the subspaces $\mathscr{K}^{j+1}(X_k, A)$.

Proof. On iteration k, the s-step conjugate gradient cycle starts at $X_1^k \equiv X_k$. By construction the initial block direction $P_1^k \equiv G(X_k)$. The proof is by induction on $j = 1,2,...,s$. The result follows for P_1^k and X_2^k directly from Lemma 7.3.5. Therefore assume that the result has been established for all $j<J$, and show that it is valid for $j=J$. However, by induction and the fact that $G(X_J^k) = AX_J^k - X_J^k A_J^k$ and $P_J^k \equiv G(X_J^k) + P_{J-1}^k \Gamma_{J-1}^k$, we have that $P_J^k \in \mathscr{K}^{J+1}(X_k, A)$. Therefore, $X_{J+1}^k = X_J^k \Phi_J^k + P_J^k \Omega_J^k$ is also in $\mathscr{K}^{J+1}(X_k, A)$. $\qquad \square$

Thus, from Lemma 7.3.7 we have that within each iteration of the s-step block conjugate gradient procedure for R(X) that the iterates X_j^k, $j = 1,...,s + 1$, are all in the associated Krylov spaces $\mathscr{K}^j(X_k, A)$. But by an easy extension of Lemma 7.3.5 to $s>1$ and the definition of G(X), we have that the orthonormal bases generated by the block Lanczos recursion in Eqns.(7.1.1) and (7.1.2) are bases for these subspaces, $\hat{\mathscr{K}}^j(X_k, A) = \mathscr{K}^j(X_k, A)$. Furthermore, by Theorem 7.3.1 the maximization of R(X) on these subspaces can be accomplished by computing the eigenvalues and eigenvectors of the projection matrices of the given matrix A on these subspaces. But these matrices are just the Lanczos matrices T_s^k. Thus, the basic iterative block Lanczos procedure which uses s+1 blocks on each internal iteration is an implementation of the general s-step block conjugate gradient procedure for maximizing the block Rayleigh quotient function R(X).

The predicted convergence rates which we obtain by simply extrapolating the convergence estimates for the scalar s-step conjugate gradient procedure to the block setting appear to be achieved in practice. In particular, using a starting block of size nxq and s+1 blocks on each iteration, numerical tests indicate that the convergence rate of the basic iterative block Lanczos procedure depends upon the effective spread $|\lambda_{n-s+1} - \lambda_q|$ and upon the effective gaps $|\lambda_j - \lambda_{q+1}|$ between the first eigenvalue λ_{q+1} not being approximated by the block procedure and the eigenvalues λ_j, $1 \leq j \leq q$, which are being approximated. Thus, increasing the number of blocks which are used on each iteration will increase the rate of convergence of the block procedure if in so doing we improve the effective spread of the matrix. Increasing the number of vectors in the first block will improve the local gaps $|\lambda_j - \lambda_{q+1}|$. However, increasing the size of the first block may result in an accompanying decrease in the number of blocks that can be generated on any given iteration because in practice the amount of storage available is limited.

As we mentioned earlier, alternative estimates of convergence can also be obtained by estimating how well the eigenvalues of one of the small projection matrices T_s generated by the block Lanczos recursions approximate eigenvalues of A. However, this type of estimate applies only to each iteration of an iterative block procedure. Saad [1980] provides a variety of estimates of this type. Below we restate a theorem from Golub and Underwood [1977].

THEOREM 7.3.2 (Golub and Underwood [1975,1977]) Let $\lambda_n \leq \ldots \leq \lambda_1$ be the eigenvalues of real symmetric matrix A . Let $z_1, \ldots, z_n$ be corresponding orthonormalized eigenvectors of A. Apply the block Lanczos recursion to A generating s+1 blocks and let $\mu_{q(s+1)}^k \leq \ldots \leq \mu_1^k$ be the eigenvalues of the small block tridiagonal matrix T_{s+1}^k generated on one iteration k. Let

$$Y \equiv \begin{bmatrix} Y_1 \\ Y_2 \end{bmatrix} \equiv Z^T Q_1^k.$$

be the nxq matrix of the projections of the starting block Q_1^k on the eigenvectors of A, where Y_1 is the qxq matrix composed of the corresponding projections on the desired eigenvectors of A. Assume that σ_{min}, the smallest singular value of Y_1, is greater than 0. Then for i=1,2,...,q

$$\lambda_i \geq \mu_i^k \geq \lambda_i - \varepsilon_i^2 \qquad (7.3.16)$$

where

$$\varepsilon_i^2 \equiv [\,(\lambda_i - \lambda_n)\,\tan^2\theta\,]\,/[\,\mathscr{T}_s^2([1 + \gamma_i]/[1 - \gamma_i])\,] \tag{7.3.17}$$

and $\theta = \arccos[\sigma_{\min}]$, $\gamma_i = [\lambda_i - \lambda_{q+1}]/[\lambda_i - \lambda_n]$, and $\mathscr{T}_s$ is the s-th Chebyshev polynomial of the first kind.

This type of estimate illustrates the importance of the effective local gaps $|\lambda_i - \lambda_{q+1}|$, but does not illustrate the potential positive effect of the outer loop iteration of an iterative block Lanczos procedure on reducing the overall effective spread and thereby improving the convergence rate.

In Lemma 7.1.1 we demonstrated that (in exact arithmetic) the blocks generated within one iteration of an iterative block Lanczos procedure are orthonormal and that in fact the corresponding Lanczos matrices were block tridiagonal. In the next section we will see how this type of orthogonality relates to a practical implementation of an iterative block Lanczos procedure.

COMPUTING BOTH ENDS OF THE SPECTRUM SIMULTANEOUSLY

Before proceeding with a discussion of the practical implementation of a block Lanczos procedure, we digress to consider the possibility of using a block procedure to simultaneously compute extreme eigenvalues from both ends of the spectrum of the given matrix. We define the following generalization of the block Rayleigh quotient function $R(X)$ which was defined in Eqn(7.3.1).

$$\overline{R}(X,Y) \equiv \sum_{i=1}^{q} x_i^T A x_i - \sum_{j=1}^{t} y_j^T A y_j, \tag{7.3.18}$$

where $X = \{x_1, \ldots, x_q\}$, $Y = \{y_1, \ldots, y_t\}$ and each x_i and $y_j \in E^n$. A straightforward generalization of Theorem 7.3.1 yields the following theorem.

THEOREM 7.3.3 For any real symmetric matrix A of order n and any q, t > 0 with q+t < n,

$$\operatorname*{Max}_{\overline{\mathscr{P}}} \overline{R}(X,Y) = \sum_{i=1}^{q} \lambda_i(A) - \sum_{j=n}^{n-t+1} \lambda_j(A), \tag{7.3.19}$$

where

$$\overline{\mathscr{P}} \equiv \{(X,Y): X^T Y = 0, X^T X = I_q, Y^T Y = I_t\} \tag{7.3.20}$$

Furthermore, this maximum is attained only on sets (X,Y) such that (i) X spans the eigenspace of A corresponding to the eigenvalues $\lambda_i > \lambda_q$, and the span of X is contained in the eigenspace of A corresponding to λ_i, $1 \le i \le q$. (ii) Y spans the eigenspace of A corresponding to the eigenvalues $\lambda_j < \lambda_{n-t+1}$, and the span of Y is contained in the eigenspace of A corresponding to λ_j, $n - t + 1 \leq j \leq n$.

Theorem 7.3.3 tells us that if we maximize $\overline{R}(X,Y)$ subject to the orthonormality constraints $\mathscr{P}$, then any globally maximizing set of vectors (X^*,Y^*) must span a subspace that contains the invariant subspaces corresponding to all $\lambda_i > \lambda_q$ and to all $\lambda_j < \lambda_{n-t+1}$, and contains r orthonormal eigenvectors corresponding to λ_q and u orthonormal eigenvectors corresponding to λ_{n-t+1} where r and u are respectively the interior multiplicities of the border eigenvalues λ_q and λ_{n-t+1}. In particular r is chosen such that $\lambda_{q-r} > \lambda_{q-r+1} = \ldots = \lambda_q$, and the interior multiplicity u is chosen similarly. Therefore, we can compute simultaneously a few of the algebraically-largest and a few of the algebraically-smallest eigenvalues of A and a corresponding eigenspace by 'simply' performing the maximization of $\overline{R}(X,Y)$.

How can we maximize $\overline{R}(X,Y)$? Observe that

$$\overline{R}(X,Y) = \text{tr}[X^T AX] - \text{tr}[Y^T AY] = R(X) - R(Y). \tag{7.3.21}$$

Because the derivative is a linear operator, we have that the gradient of $\overline{R}(X,Y)$, which we will denote by $\overline{G}(X,Y)$, satisfies

$$\overline{G}(X,Y) = (AX,-AY). \tag{7.3.22}$$

If we incorporate the constraint $\mathscr{P}$ by projecting $\overline{G}(X,Y)$ onto this constraint, we obtain the projected gradient

$$\overline{G}_p(X,Y) = (AX, -AY) - (X, Y)[(X, Y)^T A(X, -Y)]. \tag{7.3.23}$$

Here we have assumed that $(X,Y) \in \mathscr{P}$. Rewriting we obtain

$$\overline{G}_P(X,Y) = ([AX - XA_X - YA_{XY}^T], [-AY + YA_Y + XA_{XY}]) \tag{7.3.24}$$

where $A_X \equiv X^T AX$, $A_Y \equiv Y^T AY$, and $A_{XY} \equiv X^T AY$.

First consider an analog of projected steepest ascent for $\overline{R}(X,Y)$. On each iteration k of steepest ascent we must maximize $\overline{R}(X,Y)$ on the subspaces $\text{sp}\{(X_k, Y_k), \overline{G}_p(X_k, Y_k)\}$. These optimizations can be performed as follows. First compute an orthonormal basis Q^k for the given subspace. Then compute the maximum of the following restriction of $\overline{R}$ to $\text{sp}\{Q^k\}$.

$$\overline{R}_p(\Theta,\Phi) \equiv \text{tr}[\Theta^T A_p^k \Theta] - \text{tr}[\Phi^T A_p^k \Phi] \tag{7.3.25}$$

where $A_p^k \equiv (Q^k)^T AQ^k$, and Θ and Φ must satisfy the constraints $\Theta^T \Phi = 0$, $\Theta^T \Theta = I_q$ and $\Phi^T \Phi = I_t$. Define $R_p^1(\Theta) \equiv \text{tr}[\Theta^T A_p^k \Theta]$ and

$R_p^2(\Phi) \equiv tr[\Phi^T(-A_p^k)\Phi]$. Then clearly,

$$\max_{(\Theta,\Phi)^T(\Theta,\Phi)=I_{q+t}} \overline{R}_p(\Theta,\Phi) \leq \max_{\Theta^T\Theta=I_q} R_p^1(\Theta) + \max_{\Phi^T\Phi=I_t} R_p^2(\Phi).$$

But, by Theorem 7.3.1 we have that

$$\max_{\Theta^T\Theta=I_q} R_p^1(\Theta) = \sum_{i=1}^{q} \lambda_i(A_p^k) \quad \text{and} \quad \max_{\Phi^T\Phi=I_t} R_p^2(\Phi) = - \sum_{j=n}^{n-t+1} \lambda_j(A_p^k).$$

Furthermore, these maxima are achieved on vectors Θ_k and Φ_k that span the appropriate eigenspaces of A_p^k. If the order of A_p^k is larger than q+t, then in fact $\Theta_k^T\Phi_k = 0$, so that the corresponding Ritz blocks $X_k \equiv Q^k\Theta_k$ and $Y_k \equiv Q^k\Phi_k$ belong to the constraint set $\overline{\mathscr{P}}$. Hence we can maximize $\overline{R}(\Theta,\Phi)$ on the given subspaces by independently maximizing the functions R_p^1 and R_p^2 and observing that the resulting maximizing quantities actually satisfy all of the constraints. Equivalently, this says that we can maximize $\overline{R}_p$ by computing the q algebraically largest and the t algebraically-smallest eigenvalues and corresponding eigenvectors of the small projection matrices A_p^k, for $k = 1,2,\ldots$.

Thus, given orthonormal bases for the subspaces associated with steepest ascent we can maximize $\overline{R}(X,Y)$ on these subspaces by computing the eigenvalues and eigenvectors of the corresponding projection matrices. Moreover, if we compute the iterates in this fashion we can assume w.l.o.g that the iterates that we generate satisfy the constraints. In fact we can assume that if $Z_k \equiv (X_k,Y_k)$ then the matrix $Z_k^TAZ_k$ is diagonal. Thus the question of applying steepest ascent to $\overline{R}(X,Y)$ reduces to the question of generating the required orthonormal bases for the associated subspaces.

But consider the block Rayleigh quotient function $R(X)$ defined in Eqn(7.3.1). Replace X by Z. Assume that Z^TAZ is a diagonal matrix. Then we have that the gradient of $R(Z)$ is $G(Z) = AZ - ZA_Z$ where $A_Z = Z^TAZ$. If we apply steepest ascent to $R(Z)$ then we must maximize R on the subspaces $sp\{Z_k,G(Z_k)\}$. The iterative block Lanczos procedure which uses two blocks on each iteration yields orthonormal bases for these subspaces. But observe that for such Z_k, because of the way in which they were constructed, we have that $sp\{Z_k,G(Z_k)\} = sp\{Z_k,\overline{G}_p(Z_k)\}$. Therefore, the iterative block Lanczos procedure which uses two blocks on each iteration can be used to generate the orthonormal bases for the subspaces required on each iteration of steepest ascent when it is applied to $\overline{R}(X,Y)$. We restate these comments as a Lemma.

LEMMA 7.3.8 Applying projected steepest ascent to the generalized block Rayleigh quotient function $\overline{R}(X,Y)$ defined in Eqn(7.3.18) is equivalent to using the basic iterative block Lanczos algorithm with two blocks generated on each iteration and at each iteration choosing the updated iterate (X_{k+1},Y_{k+1}) such that X_{k+1} is the block of Ritz vectors corresponding to the q algebraically-largest eigenvalues of the corresponding projection matrix $T_2^k = A_P^k$, and Y_{k+1} is the block of Ritz vectors corresponding to the t algebraically-smallest eigenvalues of this matrix.

Lemma 7.3.4 can be extended easily to the function $\overline{R}(X,Y)$ and to s-dimensional subspaces, so that we can generalize the above relationship to an s+1 step block Lanczos procedure for $\overline{R}(X,Y)$. Therefore, we have the following iterative Block Lanczos procedure for computing simultaneously a few of the algebraically-largest and a few of the algebraically-smallest eigenvalues of a given real symmetric matrix A. Numerical tests indicate that in fact this is an appropriate mechanism for computing both ends of the spectrum simultaneously.

BASIC BLOCK PROCEDURE, COMPUTING BOTH EXTREMES SIMULTANEOUSLY

Step 1. Steps 1,2,3 and 5 are identical to Steps 1,2,3 and 5 in the basic iterative block Lanczos procedure defined in Section 7.1 except that the starting block must of size nxj where $j \geq q + t$.

Step 4. On each iteration k, compute the q algebraically-largest eigenvalues and the t algebraically-smallest eigenvalues of the small projection matrix T_s^k generated and corresponding sets W_1^k and W_2^k of eigenvectors. The remaining j-q-t vectors needed for completing the new first block on each iteration can be computed from either end of the spectrum of T_s^k. A strategy for deciding which end of the spectrum to use must be incorporated.

The optimization interpretation of an iterative block Lanczos procedure which was presented in this section yields a simple explanation for why initial iterations of an iterative block Lanczos procedure often yield good approximations to eigenvalues at both extremes of the spectrum of the original matrix. On the initial iteration, the space being generated is identical to the space which would be generated for getting the same number of either the largest or the smallest eigenvalues. Since the iterative block Lanczos procedure works with subspaces spanned by the gradients, we can simultaneously maximize the generalized block Rayleigh quotient in the X variables and minimize it in the Y variables. In practice however we find that if both ends of the spectrum are desired, it is often quicker to compute the ends of the spectrum separately, rather than simultaneously. See Cullum [1978] for examples.

If the basic iterative block Lanczos procedure is implemented as it is stated in this section, it will be numerically-unstable. Block Lanczos procedures suffer from the same types of numerical difficulties as the single-vector procedures do. In the next two sections we discuss practical computer implementations.

SECTION 7.4   ITERATIVE BLOCK, A PRACTICAL IMPLEMENTATION

The arguments substantiating the use of an iterative block Lanczos procedure all assumed that within each iteration the Lanczos blocks which were generated were orthogonal. However in practice, as in the single-vector Lanczos procedure, this orthogonality is not preserved if the block Lanczos procedure is implemented as it is stated in Section 7.1. Such losses may cause a simple eigenvalue of A to appear in the eigenvalue approximations with a multiplicity greater than one and may introduce extra eigenvalues not directly related to the original matrix. This is directly analogous to what happens in the single-vector procedure. Examples of this behavior are given in Cullum and Donath [1974b]. The basic procedure must be modified for practical applications. We describe two possible modifications. In this section we consider the modifications incorporated in the Cullum and Donath procedure [1974]. In Section 7.5 we consider a hybrid algorithm. FORTRAN code for this hybrid algorithm is contained in Chapter 8 of Volume 2 of this book.

COMPUTATION OF SECOND Q-BLOCK   The unnormalized second block computed on each iteration k, $P_1^k \equiv AQ_1^k - Q_1^k A_1^k$ is, as we saw in the last section, essentially the projected gradient of the generalized block Rayleigh quotient function R(X) given in Eqn(7.3.1). If Q_1^k is a maximizing block of R then $P_1^k \equiv 0$. Therefore, $\| P_1^k \| \to 0$ as we maximize R(X). Thus, as convergence commences several leading digits in AQ_1^k and $Q_1^k A_1^k$ become identical and these digits cancel during the computation of P_1^k. If enough digits cancel, the roundoff error in what remains will adversely affect the orthogonality of P_1^k to the first block Q_1^k. At issue also is the question of allowing roundoff errors to propagate meaningless quantities throughout the block generation. Initially attempts were made to minimize these cancellation difficulties by modifying the quantities used in the computation of the blocks P_1^k so that these quantities could be computed more accurately. These modifications used a mixture of single and double precision arithmetic.

Kahan [1972] made the following suggestions and these were implemented in the original version of our block Lanczos code. Round the vectors in the current first block Q_1^k to single precision to obtain a matrix W and also round the current approximations to the eigenvalues of A and store them in a diagonal matrix Γ. Then if the key quantity $S \equiv AW - W\Gamma$, which we expect to have very small components when convergence occurs, is computed in double precision and the given matrix is in single precision then essentially every digit in S will be meaningful. The related matrix $Z \equiv W^T S$ should also be computed in double precision. (For simplicity we drop the superscripts and subscripts.) In order to compute the second block we have to correct for the fact that W is not orthogonal. Therefore, define $H \equiv W^T W = I + \Delta$. Then the correct formula is $P_1 = AW - WH^{-1}(W^T AW)$. We expect Δ to be small so we can approximate $H^{-1} \cong (1 - \Delta + \Delta^2)$ and substitute that into the expression for P_1. We then obtain the following expression which uses S, which is known precisely and where the parentheses are important.

$$P_1 = S - W(Z - (\Delta - \Delta^2)Z).$$

Moreover, compute $W^T AW = (Z + \Delta\Gamma) + \Gamma$. The second block was then computed from P_1 by applying modified Gram-Schmidt with pivoting for size at each stage to this block.

As anticipated these changes, along with appropriate other changes in the program, did result in improvements in the orthogonality between the first and the second block. However, on the tests run, the orthogonality benefits were not reflected in significantly enough improved convergence to justify the required increases in program complexity and number of arithmetic operations. The modifications discussed above also triggered other modifications which further complicated the program. See Cullum and Donath [1974b] for details. Rounding the approximate eigenvectors to single precision also limited the achievable accuracy. After much debate we finally removed these mixed precision computations from our code. Perhaps if double and extended precision had been used we would have seen greater improvements. However, that would have been very expensive both in terms of the storage requirements and in terms of the computation time.

The numerical problems which we encounter because of these cancellation errors seem to be adequately handled by incorporating the combination of a procedure for not allowing 'converged' eigenvectors to generate descendant Lanczos vectors, together with the reorthogonalization of new Lanczos blocks as they are generated w.r.t. any such 'converged' eigenvectors. This alternative approach requires no extra arrays, no mixture of single and double precision variables and arithmetic and yields a much simpler program. These modifications are described below.

DROPPING 'ZERO' LANCZOS VECTORS. As convergence occurs, from some iteration k on, the norms of one or more of the vectors in the unnormalized second block P_1^k become of the same order of magnitude as the error incurred in computing those norms. If this happens for some eigenvalue μ_j^k, the ancestor q_j^k in the first block of the vector p_j^k in P_1^k cannot be allowed to generate any descendants. The notation here may be somewhat confusing. We use lower case letters to denote the columns in the blocks Q_1^k and P_1^k. The discussion is confined to dropping vectors in the second block. Although it is theoretically possible that vectors could be dropped from any block, in practice we see this happening only in the second block. Not allowing some $q_j^k \in Q_1^k$ to generate descendants is equivalent to assuming that its descendant $p_j^k \equiv 0$, which of course it is not. Whenever a Lanczos vector is 'dropped' from P_1^k on some iteration k, the second Q-block on that iteration will contain fewer vectors than the corresponding first block. The size of the first block is not changed.

This systematic reduction in the size of the second and succeeding blocks as 'convergence' of the vectors in Q_1^k occurs can have very desirable effects upon the rate of convergence of the remaining eigenvector approximations. Consider the following simple example.

EXAMPLE 7.4.1. Let A be the diagonal matrix of order 211 whose entries are 0., -1., 100., -100.1, -100.2, -100.5, -101.5, -103.0, -104.5, -106.0, -107.5, -107.55, ... , -117.5 where ... represents eigenvalues which are equally spaced .05 units apart. We computed the four algebraically-largest eigenvalues of A using q=4, and Lanczos matrices no larger than 20.

Using the block procedure which allows vectors to be dropped from the second Q-block, took 5 iterations and 108 matrix-vector multiplies. Five subblocks were used on iterations 1 and 2, and nine subblocks were used on iterations 3 through 5. The errors in the computed eigenvalues were 3×10^{-21}, 0., 2×10^{-12}, and 2×10^{-11}. The norms of the errors in the corresponding eigenvectors were 5×10^{-12}, 9×10^{-12}, $< 10^{-6}$, and 2×10^{-6}. Running the same procedure except that no vectors were allowed to be dropped from the second block, took 10 iterations and 208 matrix-vector multiplies. The errors in the computed eigenvalues were 10^{-31}, 0., 2×10^{-14}, and 2×10^{-10}. The norms of the errors in the corresponding eigenvectors were 10^{-16}, 10^{-16}, $< 10^{-6}$, and 3×10^{-5}.

Example 7.4.1 illustrates the benefit which can be gained by allowing the size of the second subblock to decrease once convergence is observed. 'Dropping' vectors, however, causes other problems which must be addressed. Subblocks which are generated after a vector has been dropped from the second block will not be orthogonal to the ancestor of that vector in the first block. This orthogonality is normally propagated by the descendants of that vector. If we drop a vector from the second block, the equation $Q_2^k B_2^k = AQ_1^k - Q_1^k A_1^k$ is no longer satisfied. To see what is happening consider the following simple example.

EXAMPLE 7.4.2 Let $Q_1 = \{q_1, q_2\}$ and let $P_1 = \{p_1, p_2\}$. Suppose that $\| p_2 \| > \| p_1 \|$ then $q_3 \equiv p_2 / \| p_2 \|$. Then compute $p'_1 = p_1 - (p_1^T q_3) q_3$. Then if $\| p'_1 \| < \varepsilon_{tol}$, where ε_{tol} is the 'zero' or convergence tolerance, we have that $\dim(Q_2) = 1$, $B_2 = (p_1^T q_3, \| p_2 \|)$, $Q_2 = \{q_3\}$ and

$$q_3 B_2 = (p_1 - p'_1, p_2) = AQ_1 - Q_1 A_1 - (-p'_1, 0). \tag{7.4.1}$$

Furthermore, $Q_1^T Q_3 B_3 = -p'_1{}^T Q_2 \neq 0$ and if we compute $Q_1^T Q_4$, we find that any error in orthogonality has been amplified by A_1.

Thus to correct for these losses in orthogonality we must explicitly reorthogonalize subsequently-generated blocks w.r.t any vectors in the first block that are not generating descendants. If when there are fewer vectors in the second block than there are in the first block, we were to use the block recursions in Eqns(7.1.1) - (7.1.2) without any alteration, the resulting losses in orthogonality in the blocks Q_j^k, $j \geq 3$ would accumulate rapidly, causing the algorithm to fail. The following lemma demonstrates, at least theoretically, that this limited reorthogonalization within a given iteration of block Lanczos procedure is sufficient to maintain the orthogonality of the Lanczos subblocks.

Because the unnormalized second block is just the projected gradient of the generalized Rayleigh quotient function we expect that any ancestor of a vector 'dropped' from the 2nd block will be in the eigenspace of A being approximated. This question is addressed in Lemma 7.4.2. We can without any loss of generality, assuming otherwise just makes the proof more complicated, assume that any dropping of vectors occurs only in the second block.

LEMMA 7.4.1 (Exact Arithmetic) Given an orthonormal starting block Q_1 define Q_2, P_1 and B_2 as in Eqns(7.1.1) - (7.1.2) except that within the modified Gram-Schmidt orthonormalization of the second block we drop any vector whose norm is not greater than a specified tolerance ε_{tol}. For $j \geq 2$ define P_j as in Eqn(7.1.1) but define Q_{j+1} by

$$Q_{j+1}B_{j+1} = P_j - \tilde{Q}_1((\tilde{Q}_1)^T P_j)$$

where $\tilde{Q}_1$ denotes those vectors in Q_1 that were not allowed to generate descendants. Similarly, define $\hat{Q}_1$ as that portion of the first block that is generating descendants. Assume that each block Q_j for $j \geq 2$ contains the same number of vectors. Then, at least theoretically, the blocks generated are orthonormal,

$$Q_j^T Q_i = 0 \text{ for } i<j, \ j = 2,... \tag{7.4.2}$$

Moreover, they are almost block A-conjugate in the sense that

$$Q_j^T A Q_i = 0 \text{ for } 1<i<j-1. \tag{7.4.3}$$

For $i = 1$ and $j = 3,...$ we have that

$$Q_j^T A \hat{Q}_1 = 0 \text{ and } Q_j^T A \tilde{Q}_1 = O(\varepsilon_{tol}) \tag{7.4.4}$$

where ε_{tol} is the tolerance used in dropping vectors.

Lemma 7.4.1 tells us that even with the reorthogonalization we will still effectively have block tridiagonal matrices. The sizes of the reorthogonalization terms are controlled by the size of the 'zero' tolerance ε_{tol}. For the purposes of the proof we will assume that the first block $Q_1 = (\tilde{Q}_1, \hat{Q}_1)$. That is, that both subsets of vectors, those not generating descendants and those that are, are contiguous. This need not happen in practice, but it often does. In any case, there is no loss of generality in assuming that it does happen.

Proof. By induction. By construction $Q_1^T Q_2 = 0$ and $Q_2^T Q_3 = 0$. Moreover, it is easy to show that $Q_1^T Q_3 = 0$. Also by construction we have that

$$Q_2 B_2 = A Q_1 - Q_1 A_1 - (P'_1, 0) \tag{7.4.5}$$

where P'_1 denotes the residual left after performing the modified Gram-Schmidt procedure with pivoting and discarding vectors with norms less than ε_{tol} generated during the modified Gram-Schmidt process. Therefore, by construction the maximum of the norms of the columns of P'_1 is less than ε_{tol}. Then we have directly from Eqn(7.4.5) that

$$Q_3^T A \widetilde{Q}_1 = Q_3^T P'_1 \quad \text{and} \quad Q_3^T A \widehat{Q}_1 = 0. \tag{7.4.6}$$

By construction we have that $Q_j^T \widetilde{Q}_1 = 0$. Furthermore,

$$\widehat{Q}_1^T Q_4 B_4 = -\widehat{Q}_1^T Q_3 A_3 + \widehat{Q}_1^T A Q_3 - \widehat{Q}_1^T Q_2 B_3^T - \widehat{Q}_1^T \widetilde{Q}_1 (\widetilde{Q}_1)^T P_3 = 0.$$

This statement and Eqn(7.4.6) combine to yield $Q_2^T A Q_4 = 0$. Thus, Eqns(7.4.2) and (7.4.3) are satisfied for $j \le 4$. Now assume that they are valid for some $J > 4$ and show that they must then be valid for $J+1$. Note that theoretically, $A_J = Q_J^T A Q_J$ although in practice Eqn(7.1.3) is used. We have that

$$Q_i^T Q_{J+1} B_{J+1} = Q_i^T A Q_J - Q_i^T Q_J A_J - Q_i^T Q_{J-1} B_J^T - Q_i^T \widetilde{Q}_1 ((\widetilde{Q}_1)^T P_J).$$

By assumption B_{J+1} is square and invertible. By induction

$$Q_i^T A Q_J = 0 \quad \text{for} \quad 1 < i < J - 1 \quad \text{and} \quad Q_i^T Q_J = 0 \quad \text{for} \quad i < J.$$

Therefore, for $i < J - 1$, $Q_i^T Q_{J+1} = 0$. Verification for $i = J - 1, J$ is straightforward. Conjugacy for $1 < i < J$ now follows immediately since for $j \ge 2$,

$$Q_i^T A Q_{J+1} = A_i Q_i^T Q_{J+1} + B_{i+1}^T Q_{i+1}^T Q_{J+1} + B_i^T Q_{i-1}^T Q_{J+1} + (P_i^T \widetilde{Q}_1)(\widetilde{Q}_1)^T Q_{J+1}.$$

The proof will be completed if we note that from Eqn(7.4.5) that for $j \ge 3$

$$Q_j^T A \widehat{Q}_1 = Q_j^T (Q_1 \widehat{A}_1) + Q_j^T (Q_2 \widehat{B}_2) = 0 \quad \text{and}$$
$$Q_j^T A \widetilde{Q}_1 = Q_j^T (Q_1 \widetilde{A}_1) + Q_j^T (Q_2 \widetilde{B}_2) + Q_j^T P'_1 \cong O(\varepsilon_{tol}).$$

Here for example $Q_1 \widetilde{A}_1$ denotes the columns of $Q_1 A_1$ whose indices are the same as the indices of $\widetilde{Q}_1$. $\quad\square$

Thus, we see that if the number of vectors in the second block is less than the number in the first block it is sufficient, at least in exact arithmetic, to reorthogonalize the Lanczos vectors w.r.t. those vectors that are not generating descendants. We will argue in Lemma 7.4.2 that those vectors dropped from the second block correspond to converged eigenvectors of A. Note that due to the specific nature of the loss in orthogonality we expect that the theoretical result obtained in Lemma 7.4.1 will remain valid in practice and in fact that is what we observe in practice. This modification requires only the use of those vectors in the first block that are not generating descendants. There is no increase in the amount of storage required relative to the case where all the blocks that are generated have the same size.

When we allow the second Lanczos block to decrease in size as eigenvector approximations converge, the storage required for each of the subsequently-generated blocks Q_j, $j \geq 2$ decreases and thus often the total number of blocks that can be generated on a given iteration within the storage supplied by the user is larger than when we force all the blocks to be of the same size. For example if we have 4 vectors in the first block and 2 have converged and we have room for 16 vectors, then we can generate 7 blocks, one of size 4 and 6 of size 2 each instead of 4 blocks each of size 4. This increase in the number of blocks generated combined with the removal of quantities that are numerically no longer meaningful can greatly enhance the terminal convergence of the iterative block procedure. Thus the user does not specify the number of blocks which are to be used on each iteration. Instead the user specifies the maximum storage that is available for the block generation on each iteration, which is consistent with the computational time and computer storage limitations. The user also specifies the number of vectors in the first block. The procedure then varies the number of blocks generated on a given iteration depending upon the amount of storage available and upon the current state of the convergence of the eigenvectors being approximated. Lemma 7.4.2 will demonstrate that the vectors dropped correspond to 'converged' Ritz vectors and that therefore these reorthogonalizations are w.r.t such vectors. Thus, we are performing an implicit deflation of the matrix w.r.t. to these vectors.

An alternative approach to our implicit deflation is explicit deflation either directly on the matrix A or w.r.t the vectors generated by the Lanczos procedure. Explicit deflation of the matrix, that is replacing A by $A - \lambda xx^T$ where λ is the computed eigenvalue and x is the corresponding computed eigenvector, is not recommended. Explicit vector deflation has been used successfully, see Golub and Underwood [1977] for example. In this approach on each iteration the size of the starting block is reduced by the number of converged Ritz vectors and the starting block is made orthogonal to these converged vectors.

Since in implicit vector deflation the 'converged' eigenvectors are still included directly in the computations at each stage, the use of implicit deflation circumvents the question of whether or not the eigenvectors have converged sufficiently to merit explicit deflation from the computations. Also the accuracy of these vectors can continue to improve as the iterations are continued to compute the remaining eigenvectors which have not yet converged. The penalty which we pay for not allowing explicit deflation is that the converged eigenvectors still occupy space and perhaps because of that fewer blocks can be generated on each iteration. The major gain achieved by this approach is the improved accuracy. There is essentially no decrease in the accuracy of the computed eigenvectors as we increase the number of such vectors to be computed.

A modified Gram-Schmidt procedure with pivoting for size is used for determining which vectors in the block P_1^k are negligible.

LOSSES OF ORTHOGONALITY   If we correct for 'dropped' vectors as described above, then our numerical experience indicates that the subsequent losses in orthogonality between the blocks generated caused by the roundoff error increase slowly as we increase the number of blocks. Numerical experience also indicates that the rate of this increase depends upon the dominance of the largest eigenvalues. The greater the dominance, the more rapid this decay may be. If at some stage these losses have accumulated sufficiently, then the assumption that the block tridiagonal Lanczos matrices are the projections of the given matrix A on the subspaces $\overline{Q}^k \equiv \{Q_1^k,...,Q_s^k\}$ will be false. In such a situation the q largest (or smallest) extreme eigenvalues of these matrices may not be accurate reflections of the q largest (or smallest) extreme eigenvalues of the given matrix A. Since we do not want to specify a priori the number of blocks generated at any particular iteration, we must have a test for losses in orthogonality and for terminating the block generation if there is a problem with losses in orthogonality. If at some iteration k, a block Q_j^k violated this test, then the matrix T_{j-1}^k would be used in the eigenelement computations on that iteration.

Therefore, we ask the question, What do we need to guarantee that $(\overline{Q}_s^k)^T A \overline{Q}_s^k$ is a 'good' approximation to A on $sp\{Q_1^k,...,Q_s^k\}$? Clearly, a very weak test such as $\cos\theta < .001$ where θ is the maximum angle between any two of the vectors in $\overline{Q}_s^k$ is more than sufficient to guarantee the near-orthogonality of these vectors since $arccos(.001) = 89^\circ\ 59'$. However such a test requires the computation of all of the inner products $(Q_j^k)^T Q_i^k$ for all $1 \leq i < j$. We do not want to do that much computation so we ask the question, is some subset of these quantities sufficient? We have the following heuristic argument.

We drop the superscript k from $\overline{Q}$ and from Q^* to simplify the notation. On iteration k, let $Q^* \equiv \{Q_2^k,...,Q_s^k\}$ and $\overline{Q} \equiv \{Q_1^k,Q^*\}$ where s is the number of blocks generated on that iteration. The subspace $sp\{Q_1^k\}$ is an invariant subspace of A if and only if

$$T_c \equiv \overline{Q}^T A \overline{Q} = \begin{bmatrix} A_1 & | & 0 \\ & | & \\ -\ -\ -\ & | & -\ -\ -\ - \\ & | & \\ 0 & | & Q^{*T} A Q^* \end{bmatrix}. \tag{7.4.7}$$

That is, Q_1^k is an invariant subspace of A if and only if the projection matrix decomposes as in Eqn(7.4.7). An equivalent statement is that for all $j>1$, the blocks $(Q_1^k)^T A Q_j^k = 0$. But from

Eqn(7.4.5) we have that

$$(Q_1^k)^T A Q_j^k = (B_2^k)^T (Q_2^k)^T Q_j^k + A_1^k (Q_1^k)^T Q_j^k + (P'_1)^T Q_j^k. \qquad (7.4.8)$$

If we have convergence, then as we iterate $\|P_1^k\|$ and $\|B_2^k\| \to O(\varepsilon_{tol})$. This happens because $\|B_2^k\| \leq [\|P_1^k\| + O(\varepsilon_{tol})]$ and P_1^k is just the projected gradient of the generalized Rayleigh quotient function. Moreover, the last term in Eqn(7.4.8) is also $O(\varepsilon_{tol})$. (See Example 7.4.2.) Therefore, the dominant term in Eqn(7.4.8) is $A_1^k (Q_1^k)^T Q_j^k$. The 'inner products' $(Q_1^k)^T Q_j^k$ are amplified by the Rayleigh quotient matrix A_1^k. It is clear that these inner products play a critical role in the performance of the algorithm. Therefore, we base our orthogonality test upon these factors. Since we are reorthogonalizing with respect to those vectors in the first block that are not generating descendants, this orthogonality test is only concerned with those vectors $\hat{Q}_1^k$ in Q_1^k which are still generating descendants.

How do we choose an appropriate tolerance for the orthogonality test? It is appropriate to use an unscaled tolerance to check the orthogonality factors. The only problem we may encounter in doing this is if the matrix A is very stiff and the eigenvalues we are computing are very small, then our test may not be sufficient. Thus, as each new vector is generated we compute its inner products with each vector in the first block which is still generating descendants and test to see if each of these inner products is less than some prespecified tolerance, ε_{orth} which we typically choose to be 10^{-5} or 10^{-6}. If this test is violated by some vector in some block Q_j, then block generation on that iteration is terminated and the first j-1 blocks are used in the small eigenelement computation.

CONVERGENCE We expect the desired convergence to occur since our procedure is an implementation of a maximization procedure for the block function R(X). We must however have a test for recognizing such convergence numerically. This procedure operates in 2 phases. In Phase 1 convergence is monitored by checking the norms of the vectors in the P_1^k blocks, k=1,2,.... Using an estimate of the roundoff error incurred in computing the squares of the norms of the residuals, $\| Ax - \lambda x \|^2$ where $\| x \| = 1$, we obtain the following convergence tolerance. This tolerance is also used to monitor the sizes of the vectors in the P_1^k blocks during the modified Gram-Schimdt orthonormalizations with pivoting for size.

$$\varepsilon_{conv} \equiv 2*(n + 1000)*n_{nz}*a_{aver}^2*\varepsilon_{mach} \qquad (7.4.9)$$

where n is the order of the given matrix A, n_{nz} is the average number of nonzeros in any given row or column of A, a_{aver} is the average size of the nonzero entries in A, and ε_{mach} is the machine epsilon. This tolerance can typically be expected to yield eigenvectors that are accurate to 5 or 6 digits.

In many cases the accuracy obtained in phase 1 is sufficient. However, in some cases it is not sufficient and moreover, it is often possible to significantly improve the accuracy of the basis for the computed eigenspace by invoking phase 2 of this procedure. The convergence tolerance for phase 2 is computed by using the eigenvalues computed in the first phase and a user-supplied relative tolerance. In particular,

$$\varepsilon_{conv2} \equiv \text{reltol}^2 * [\max(1, \lambda_{max}^2)] \qquad (7.4.10)$$

where reltol is the user-supplied relative tolerance and λ_{max} is the magnitude of that computed eigenvalue which is largest in magnitude.

It is very possible that the Phase 2 convergence criterion may not be achievable so this portion of our iterative block Lanczos procedure is designed to terminate automatically if the procedure cannot achieve the convergence criterion which has been specified. If at a certain point in the procedure, numerical difficulties appear, we expect them to be identifiable via losses in orthogonality of the Lanczos vectors. Therefore in phase 2, recurring losses in orthogonality trigger termination. Also the user has to specify the maximum number of matrix vector multiplies allowed for each of the phases. The convergence test which we use rests upon the following theorem from Davis and Kahan [1970].

THEOREM 7.4.1 (Davis and Kahan [1970]) Let Q_1 be an orthonormal set of vectors and let $A_1 = Q_1^T A Q_1$. Then the matrix of angles Θ required to rotate the subspace $sp\{Q_1\}$ into an invariant subspace of A satisfies the inequality

$$\| \sin(2\Theta) \| \leq 2 \| P_1 \| / [\text{gap}] \qquad (7.4.11)$$

where 'gap' represents the smallest gap between an eigenvalue which is being approximated and any eigenvalue which is not being approximated. $P_1 \equiv A Q_1 - Q_1 A_1$, is the associated residual.

As we said earlier the convergence rate which we have observed in practice is dependent upon these effective gaps between the eigenvalues which are being approximated and those which are not being approximated. However, truly multiple eigenvalues do not adversely affect the convergence. If for example the two algebraically-largest eigenvalues are being computed and $\lambda_1 > \lambda_2 = \lambda_3 > \lambda_4$ and only two vectors are used in the first subblock, the effective gap for convergence purposes is $\lambda_2 - \lambda_4$. Densely packed eigenvalues can however, if they are not all included in the first block, cause very very slow convergence. If the effective gap can be increased significantly by increasing the number of vectors in the first block, a significant improvement in performance may be observed. However, increases in block size which yield only 'small' improvements in the relevant gap and/or which are accompanied by a decrease in the number of blocks generated may actually decrease the rate of convergence. As a general rule it is usually better to make the size of the first block equal to the number of eigenvalues desired and thereby maximize the number of blocks that can be generated on each iteration.

If we are checking convergence on all of Q_1^k, then from Theorem 7.4.1 we have that it is sufficient to check the size of $\|P_1^k\|$. However, we need not require that the entire block converges and the question becomes, is checking the sizes of the norms of some of the vectors in P_1^k a necessary and sufficient test for convergence of the associated Ritz vectors? For generally chosen blocks Q_1^k this test would not be sufficient. However, the iterates Q_1^k are chosen in a very special way and if the Lanczos matrices T_s^k are good approximations to the matrices $(Q^k)^T A Q^k$, then this test is necessary and sufficient. We have the following Lemma.

LEMMA 7.4.2 Let Q_1^k be the first blocks which are obtained by an iterative block Lanczos procedure which uses Eqns(7.1.1) - (7.1.2) and on each iteration forms the next starting block by computing the q algebraically-largest eigenvalues of the Lanczos matrix T_s^k and corresponding orthonormal eigenvectors W_1^k and setting $Q_1^{k+1} \equiv Q^k W_1^k$. Then if the Lanczos blocks are sufficiently orthonormal, that is

$$T_s^k \approx (\overline{Q}^k)^T A \overline{Q}^k, \tag{7.4.12}$$

and T_s^k is numerically-diagonalizable, then we have that the convergence or the lack of convergence of the corresponding Ritz vectors can be determined by checking the sizes of the norms of the columns in P_1^k. (We say a matrix is numerically-diagonalizable if its individual eigenvectors can be computed accurately.)

Proof. To simplify the notation we drop the k superscript. Suppose that we have convergence on some subset of the Ritz vectors. That is, $A\widetilde{Q}_1 = \widetilde{Q}_1 \widetilde{R}_1$ for some subset of Q_1 where $\widetilde{R}_1 \equiv (\widetilde{Q}_1)^T A \widetilde{Q}_1$. First, clearly, $\widetilde{P}_1 = A\widetilde{Q}_1 - \widetilde{Q}_1 \widetilde{A}_1$. Without loss of generality we assume that $\widetilde{Q}_1$ consists of the first q_1 columns of Q_1. That is, $Q_1 = (\widetilde{Q}_1, \hat{Q}_1)$. Then clearly,

$$\widetilde{A}_1 = \begin{bmatrix} \widetilde{R}_1 \\ \hat{Q}_1^T A \widetilde{Q}_1 \end{bmatrix}$$

and $\hat{Q}_1^T A \widetilde{Q}_1 = \hat{Q}_1^T \widetilde{Q}_1 \widetilde{R}_1 \approx 0$ since the vectors in Q_1 are orthonormal (theoretically). Therefore,

$$\widetilde{P}_1 \approx A\widetilde{Q}_1 - \widetilde{Q}_1 \widetilde{R}_1 = 0.$$

Now consider the converse. By construction we have that the first block of Lanczos vectors is just a set of Ritz vectors $Q_1^{k+1} = Q^k W_1^k$ where W_1^k is an orthonormalized set of eigenvectors of the Lanczos matrix. Thus, by Eqn(7.4.12) we have that $A_1^k = W_1^T Q^{kT} A Q^k W_1 \cong \Sigma_1$ where Σ_1 is a diagonal matrix containing the corresponding approximations to the eigenvalues of A. But, since A_1^k is diagonal, it is clearly sufficient to test convergence by looking at the norms of the individual columns of P_1^k. $\qquad \square$

Cullum and Donath [1974b] contains many numerical examples, both examples where the block Lanczos approach works very well and examples where it does not work very well. Those examples clearly illustrated the tradeoff which exists between the number of vectors allowed in the first block and the number of blocks which can be generated on a given iteration. As we said above the preferred choice seems to be to maximize the number of blocks that are generated on each iteration. Eigenvalues that have decent gaps $(\lambda_w - \lambda_{q+1})$ between it and the closest eigenvalue of A which is not being approximated, can be readily computed by this procedure.

OPERATION COUNTS   An iterative block Lanczos procedure is expensive. Basically, 6 types of computations are required. Inner products are used to compute the block entries in the Lanczos matrices, A_j and B_{j+1}. Matrix-vector multiplies are needed to compute the new Lanczos blocks along with scalar-vector addition, subtraction, multiplication and division. A small eigenelement computation must be done on each iteration. New Ritz vectors must be computed on each iteration. Some reorthogonalization is required once some of the Ritz vectors have converged. The following Table summarizes the operation counts. Here we assume that all the blocks have the same size nxq. Operation counts are given only for multiplications. In many cases, for example in the inner products an equal number of additions (or subtractions) are required.

TABLE 7.4.1   Operation Counts for the iterative block Lanczos Procedure
n = Order of A. n_z = Average number of nonzeros in any row of A.   nxq = Size of 1st block.

Type of Computation	Arithmetic Counts
Modified Gram-Schmidt, B_{j+1}	$s * n * q^2$
Matrix-vector Multiplies	$q * s * n_z$
Small Eigenelement Computations	$O((q*s)^2)$
Compute Ritz Vectors	$n * s * q^2$
Reorthogonalize w.r.t 'Converged' Vectors	$(s-2)* q* n$
Compute the A_j	$n* s* q* (q+1)/2$

This Table clearly indicates that the block Lanczos procedure is not cheap. The pivoting for size which is done on all blocks almost doubles the cost of the modified Gram-Schmidt computations due to the extra norm computations needed to determine which of the remaining vectors has maximal norm. Practical experience indicates, however, that this test needs only to be applied to the second block and that would reduce the cost of this part of the computation considerably. In addition to the computations listed in the table we also have to orthogonalize any starting vectors provided by the user and the program also reorthogonalizes each new starting block at the beginning of each iteration.

Considering the cost of this procedure it is important to ask questions about speeding up the convergence. In the next section we look at a hybrid algorithm which performs as well as the procedure discussed in this section but outperforms it on hard problems. First however, we describe an application for this type of procedure.

AN APPLICATION The original motivation for our development of an iterative block Lanczos procedure came from graph partitioning studies, Cullum, Donath, and Wolfe [1975], Donath and Hoffman [1973], Barnes [1982]. We are given an undirected graph $G = \{V,E\}$, and are asked to partition the vertices V into q subgroups $\{V_1,...,V_q\}$ of size $m_1,...,m_q$, respectively, so that the number of the edges in E which connect vertices in different subgroups is minimal. For example let G_1 be a graph with 4 vertices $V = \{1,2,3,4\}$ and the 4 edges $E = \{(1,2),(2,3),(2,4),(3,4)\}$. Schematically, we have the following picture.

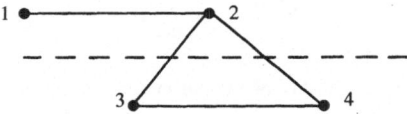

FIGURE 7.4.1 Graph G_1.

There are 3 partitions of G_1 into 2 groups of equal size; two of which result in 3 external connections between the resulting two groups. The optimal partition as indicated in Figure 7.4.1 by the dotted line, is $V_1 = \{1,2\}$ and $V_2 = \{3,4\}$. This partition has two external edges. For very small graphs such as this one it may be possible to determine the optimal partition by an exhaustive search through all of the possible partitions. However, the number of possible partitions becomes excessively large very rapidly as the number of vertices is increased.

Partitioning problems arise in the layout and placement of circuits on silicon chips in a computer. An essential objective in placement is to position the circuits on the chips in such a way that the resulting configuration is wirable, i.e. so that all of the circuits can be connected as specified by the design using the wiring patterns which have been laid out a priori on the chip. The required interconnections between the different circuits can be specified by a connection matrix A of 0's and 1's. The circuits are ordered and numbered from 1 to n, where the number of circuits n can range from several hundred to several thousand. The entry a_{ij} in the connection matrix is 1 if circuit i is connected to circuit j and 0 otherwise.

Heuristic partitioning algorithms which use eigenvectors of modified connection matrices have been developed. For such partitioning algorithms we must first compute some of the eigenvectors of these 'connection' matrices. In practice the interconnections in the circuit graphs are usually not described in terms of simple nodes and edges but in terms of small sets of nodes called 'nets'. A given node may appear in several different nets. Each such net contains a terminal corresponding to that node. Nets are really sets of terminals, rather than sets of nodes. The terminals within a given net must be connected by 'wires'. Typically, a net may contain 2 to 5 nodes, and a given node may be in 2 to 5 nets. The individual terminals are not directly represented in the matrix formulation of the interconnections. Rather the cumulative effects of the nets on each node are recorded as entries in the matrix. Some nets are connected to external nodes and the cumulative effects of such a net are measured differently from those of an internal net.

The matrices generated do not have any apparent regularities in pattern. However, they are very sparse. An example of this type of matrix can be constructed as follows Donath and Hoffman [1973]. Denote the matrix by $A = (a_{ij})$. If nodes i and j do not jointly belong to any net, set $a_{ij} = 0$. Otherwise, set

$$a_{ij} = \sum_{m \epsilon M(i,j)} f_m$$

where m is a net, $|m|$ is the cardinality of the net m, and $M(i,j) = \{ m \mid i \epsilon m$ and $j \epsilon m\}$. The summands $f_m = 4\delta/|m|^2$ if $|m|$ is even, and $4\delta/(|m|^2 - 1)$ if $|m|$ is odd. $\delta = 1$ for any net with no external connections and $1/2$ for any net with external connections.

In this particular application if we want to arrange a given number of circuits into q subsets we use at least q of the algebraically-smallest eigenvalues and corresponding eigenvectors of a modified connection matrix. Typically, the qth smallest eigenvalue was nearly multiple. Several examples are provided in Cullum and Donath [1974b]. For this particular application the ability of the procedure to reduce the block sizes when convergence was observed was critical to the overall convergence of the iterative block Lanczos procedure.

SECTION 7.5   A HYBRID LANCZOS PROCEDURE

The block Lanczos procedure described in Section 7.4 works well on many matrices. The success of that procedure however, depends upon the selection of an appropriate size for the first Q-block and that enough storage be available so that a reasonable number of blocks can be generated on each iteration. Since the user seldom has a priori information on the gaps between the eigenvalues, the best strategy in practice seems to be to set the size of the first block equal to the number of eigenvalues the user wants to compute. This strategy maximizes the number of Q-blocks which can be generated on each iteration and yields good results for many problems. There are however difficult problems for which this strategy is not sufficient to achieve a reasonable rate of convergence. In this section we propose a hybrid block-single-vector Lanczos procedure which is designed so that if for a given matrix the observed convergence rate is too slow, the procedure can automatically increase the number of vectors in the first block without incurring a corresponding significant decrease in the number of blocks which can be generated on subsequent iterations.

In Section 7.3 we demonstrated that we can compute the eigenelements which we want by applying a maximization procedure to the block Rayleigh quotient function R(X) defined in Eqn(7.3.1). We also showed that the iterative block Lanczos procedure is one way of carrying out such a maximization. Therefore, we ask ourselves if there is some other maximization procedure which we could use as the basis for developing a different Lanczos procedure which would allow for increases in the size of the first block.

We discover how to construct such a procedure by invoking some ideas from optimization theory and then examining these ideas in the context of wanting to maximize the block function R(X). The arguments are all heuristic. First we observe that the function R(X) is separable. A function is separable if and only if it is the sum of functions f_i, $1 \leq i \leq t$, whose arguments form a partition of the overall set of parameters. In R(X) this partitioning coincides with the column subdivision within the matrix X. We know from optimization theory that the maximization of any concave separable function (with no constraints) can be achieved by sequentially maximizing it over each of the subsets of the arguments in such a partitioning. We have however, a constraint set $\mathscr{S}$ which is not 'separable', so we cannot directly apply the above statement to R(X). In spite of this, we might hope that some modified form of sequential optimization along simple subspaces which are defined by the individual columns in X, might be legitimate for R(X).

A second observation is that we know that there are sequential but separable optimization procedures which are suitable for nonseparable functions. An example of this type of procedure is the Gauss-Southwell procedure, see for example Luenberger [1973, Chapter 7]. On each iteration of that procedure, the coordinate direction corresponding to the maximal component of the gradient vector of the function at the current iterate is used as the direction of movement.

A third observation is that in the preceding section we saw that the 'line' searches which were required by the s-step block conjugate gradient procedure for maximizing the Rayleigh quotient function R(X) over $\mathscr{S}$, could be carried out exactly (at least in exact arithmetic). In the corresponding iterative block Lanczos procedure these searches were accomplished by simply computing eigenvalues and eigenvectors of the Lanczos matrices generated on each iteration. We want to be able to use this same type of construction in our hybrid method so that in contrast to the inexact 'line' searches which are typically used in optimization procedures, we will again be doing 'exact' searches. This type of construction generates feasible iterates even though the constraint set is nonlinear, as long as we maintain the orthogonality of the Lanczos Q-blocks.

Thus, we have three ideas: (1) Use directions of search which are determined by using only one of the subsets of parameters defined by the columns of the parameter block X; (2) At each iteration choose this subset by determining the 'maximum component' of the block gradient of R(X) constrained to $\mathscr{S}$; (3) Use small eigenelement computations to perform 'exact' line searches. Applying these three ideas to R(X), we obtain the following hybrid procedure.

HYBRID LANCZOS PROCEDURE

Step 1. Let A be a real symmetric nxn matrix. Select q and $q' < q \ll n$ and s such that $qs \ll n$. Specify the number kmax of vectors available for the block generation on each iteration. Define a starting block $Q_1 \equiv Q_1^l$ of size nxq such that $Q_1^T Q_1 = I_q$. Set k=1.

Step 2. On each iteration k, use the block recursion in Eqn(7.1.1) - (7.1.2) to compute $P_1^k \equiv AQ_1^k - Q_1^k A_1^k$ and then compute the norms of the columns of P_1^k. If all of the norms of all of the relevant columns of P_1^k are 'small', then terminate. Otherwise select the column of maximum norm from P_1^k and normalize it to obtain a second Q-block consisting of one vector. Compute $B_2^k \equiv b_2$ by computing the projections of each of the remaining columns in P_1^k on Q_2^k. Monitor the choice of column and the changes in the size of the corresponding residual as we iterate. If convergence is too 'slow', change q to q+1 and increase the size of the first block by one vector.

Step 3. Continue with the block recursion generating 'blocks' of size nx1 and reorthogonalizing each block as it is generated w.r.t. all of the vectors in the first block Q_1^k which are not being allowed to generate descendants.

Step 4. Compute the relevant eigenvalues and eigenvectors of the Lanczos matrix generated. Compute the associated Ritz vectors and use them as the starting block of vectors for the next iteration. Set k=k+1 and go to Step 2.

On each iteration of the above procedure the Lanczos matrix which is generated has the following form. (To simplify the notation the iteration superscript has been dropped.)

$$(7.5.1) \qquad T_s \equiv \begin{bmatrix} A_1 & b_2^T & & & \tilde{Q}_1^T A\bar{Q} \\ b_2 & \alpha_2 & \beta_3 & & \\ & \beta_3 & \alpha_3 & \beta_4 & \\ \bar{Q}^T A\tilde{Q}_1 & & \beta_4 & \alpha_4 & \\ & & & & \ddots \\ & & & & & \beta_t \\ & & & & \beta_t & \alpha_t \end{bmatrix}$$

where t = kmax - q. $\tilde{Q}_1^T A\bar{Q}$ is the (q-1)x(kmax-q-1) submatrix consisting of the reorthogonalization terms. $\bar{Q} \equiv \{Q_3, Q_4, ..., Q_t\}$ and $\tilde{Q}_1$ is that portion of the first block which is not generating descendants. The picture in Eqn(7.5.1) is not exactly correct. The reorthogonalization terms are as defined. However, they do not appear in all of the first q rows. They only appear in those rows corresponding to the indices of the columns in $\tilde{Q}_1$. In the block procedure described in the preceding section we did not include the reorthogonalization terms in the Lanczos matrices because each of them corresponded to a converged Ritz vector and consequently was 'small'. Here however these quantities are not small and must be included.

Observe that in particular this approach guarantees that we are always reorthogonalizing w.r.t any 'converged' Ritz vectors since such vectors will have 'small' components in the corresponding projected block gradient. Here however we do not have to worry about what is small; we do not need a tolerance for dropping vectors. The Lanczos matrices generated on each iteration are real symmetric matrices which are tridiagonal below the first row block and to the right of the first column block. On each iteration the procedure uses a full-sized first block but all of the other blocks contain only 1 vector. The reorthogonalizations are only w.r.t. vectors in the first Q-block.

Considered as an optimization procedure, on each iteration this hybrid method computes the projected block gradient of the block Rayleigh quotient function, selects the 'component' of that gradient which has the largest norm, and then optimizes R(X) on the kmax-dimensional subspace generated by the propagation of that 'direction' through the Lanczos recursions. This subspace is $sp\{Q_1, Aq_{max}, ..., A^t q_{max}\}$ where q_{max} denotes the vector in the first block whose corresponding 'component' in the gradient block had the maximal norm on that iteration.

This procedure becomes interesting only if we can demonstrate that it outperforms the iterative block Lanczos procedure described in Section 7.4. We give several examples. In all of the test problems, this hybrid procedure always did as well as the block procedure in Section 7.4 and on problems which are difficult for iterative block Lanczos, this hybrid procedure did significantly better. Therefore, it is the procedure which is included in Chapter 8 of Volume 2 of this book. Several subroutines from the EISPACK Library [1976,1977] are required by these codes.

Diagonal test matrices with various eigenvalue distributions were generated. Tests were also run on the graph partitioning matrices mentioned in Section 7.4. In the remainder of the discussion in this section we use BLOCK to denote the iterative block Lanczos procedure described in Section 7.4, and we use HBLOCK to denote the hybrid Lanczos procedure described in this section. Example 7.5.1 is an example of a problem which is difficult for BLOCK if only a few subblocks can be generated on each iteration. The difficulty is caused by the cluster of eigenvalues at the big end of the spectrum. Example 7.5.2 is a simple illustration of the ability of the HBLOCK procedure to automatically increase the size of the first block if the convergence is not proceeding as rapidly as it might. Examples 7.5.3 and 7.5.4 are graph partitioning matrices. For Example 7.5.3 the nominal costs of the two procedures, in terms of the matrix-vector multiplies required, were essentially the same. However, for Example 7.5.4 a significant decrease in cost was achieved by using HBLOCK. Both of these procedures operate in two phases. The first phase obtains eigenvalues which are accurate to approximately 10 or more digits and eigenvectors which are typically accurate to 5 or 6 digits. Phase two can be used to obtain more accurate approximations. In many applications the accuracy achieved on phase 1 is sufficient. The numbers given below all correspond to phase 1 convergence.

EXAMPLE 7.5.1 Define a diagonal matrix DIAG405 of order n=405 with eigenvalues 0, -.1, -.1, -.1, -.1; eigenvalues spaced uniformly from -.6 to -20 with a spacing of -.05; and a cluster at -1000 of 11 eigenvalues spaced -.2 units apart. We computed the 4 algebraically-largest eigenvalues of A, using q=4, and Lanczos matrices of size ≤ 24.

This is an example of a problem which is not easy for the iterative block Lanczos procedure because the storage available is not sufficient to be able to generate enough blocks to overcome the effect of the large spread caused by the cluster of eigenvalues at -1000. The hybrid method HBLOCK used 18 iterations with 440 matrix-vector multiplies to compute the desired quantities. The squares of the residual error norms for HBLOCK were 9×10^{-12}, 8×10^{-13}, 1×10^{-12}, and 6×10^{-12}, using a convergence tolerance of 2×10^{-11}.

The corresponding BLOCK procedure used 36 iterations with 866 matrix-vector multiplies. On iterations 1-25 of BLOCK, 6 blocks were generated. On iterations 26-28, 7 blocks were generated. On iteration 29, 11 blocks were generated, and on iteration 30-36, 21 blocks were generated. The HBLOCK procedure used the 1st 'component' of the projected gradient on iterations 8,12,14,16 and 18. It used the 3rd 'component' on iterations 1 and 4. It did not use the 2nd component at all. The 4th component was used on the remaining iterations. The same convergence tolerance was used and the squares of the residual error norms were 5×10^{-12}, 9×10^{-13}, 1×10^{-12}, and 1×10^{-12}.

EXAMPLE 7.5.2 Define a diagonal matrix DIAG305 of size n=305 with eigenvalues -10^{-5}, -10^{-7}, -10, -1000, -10000, and with the remaining eigenvalues distributed evenly between -3.1 to -.11 with a spacing of .01. First we computed the two algebraically-largest eigenvalues -10^{-5} and -10^{-7}, using q=2. Then we computed just the algebraically-largest eigenvalue, using q=1. In both cases the Lanczos matrices were of size ≤ 24.

Using HBLOCK with two vectors in the first block, required 9 iterations and 221 matrix-vector multiplies to compute the first two eigenvalues and a basis for the corresponding eigenspace. The 1st component of the gradient was used on iterations 6 and 8. The 2nd component was used on the other iterations. 23 blocks were generated on each iteration. Then we looked at the question of computing only the algebraically-largest eigenvalue and setting q=1. In this case 12 iterations and 192 matrix-vector multiplies were required. On iteration 7 the procedure increased the size of the first block to two. From iteration 7 through 11 it used the 2nd component of the gradient, returning to the first component on the final iteration 12.

EXAMPLE 7.5.3 We considered a matrix CASE671 which arose in our graph partitioning studies. See Section 7.4. This matrix has order 671 and its 7 algebraically-largest eigenvalues are 2.9 x 10^{-8}, -.04248, -.085175, -.112227, -.126271, -.143557, and -.185361. Its smallest eigenvalue is -5.6. We computed the 4 algebraically-largest eigenvalues using q=4 and Lanczos matrices of size ≤ 20.

HBLOCK required 19 iterations, using 388 matrix-vector multiplies. The residuals obtained were 3 x 10^{-17}, 3 x 10^{-12}, 5 x 10^{-13}, and 2 x 10^{-13}. Iterations 9 and 19 used the 1st component of the gradient. Iterations 2, 5, and 13 used the 2nd component. Iterations 4, 7, 11 and 16 used the 3rd component. All other iterations used the 4th component. BLOCK required 20 iterations, using 404 matrix-vector multiplies. The residuals obtained were 2 x 10^{-12}, 3 x 10^{-12}, 10^{-12}, and 3 x 10^{-12}. On iterations 1-10, 5 blocks were generated. On iterations 11-14, 6 blocks were generated. On iterations 15-16, 9 blocks were generated and on iterations 17-20, 17 blocks were generated. If instead 6 eigenvectors were requested, q=6, and the maximum size Lanczos matrix allowed was increased to 30, then HBLOCK required 379 matrix-vector multiplies whereas BLOCK required 510 matrix-vector multiplies.

EXAMPLE 7.5.4 We considered a matrix CASE1239 which arose in our graph partitioning studies. See Section 7.4. This matrix has order 1239 and its 7 algebraically-largest eigenvalues are 1.27 x 10^{-8}, -.006027, -.0111228, -.016797, -.024541, -.038242, and -.0463706. We computed the 5 algebraically-largest eigenvalues using q=5 and Lanczos matrices of size ≤ 30.

HBLOCK required 22 iterations and used 670 matrix-vector multiplies. The residuals obtained were 3 x 10^{-14}, 3 x 10^{-13}, 2 x 10^{-13}, 5 x 10^{-14}, and 3 x 10^{-13}. Iterations 12 and 20 used the 1st component of the gradient. Iterations 8 and 17 used the 2nd component. Iterations 2,7,13 and 18 used the 3rd component. Iterations 5,11,16 and 21 used the 4th component. All other iterations used the 5th component. BLOCK required 37 iterations and used 1109 matrix-vector multiplies. The residuals obtained were 2 x 10^{-12}, 4 x 10^{-13}, 2 x 10^{-12}, 10^{-12}, and 4 x 10^{-13}. On iterations 1-23, 6 blocks were generated. On iterations 24-31, 7 blocks were generated. On iterations 32-34, 13 blocks were generated and on iterations 35-37, 26 blocks were generated. When 6 eigenvectors were requested, q=6, and the maximum size of the Lanczos matrices was 30, HBLOCK required 379 matrix-vector multiplies whereas BLOCK required 510 matrix-vector multiplies.

It is interesting to compare the results obtained for Example 7.5.4 with those obtained using the single-vector Lanczos procedures described in Chapter 4, Sections 4.7 and 4.8. For this particular problem the single-vector procedure required 410 matrix-vector to compute the first 7 eigenvalues. Eigenvectors for the first 5 eigenvalues were computed using a similar number of matrix-vector multiplies. The overall cost for the first 5 eigenvalues and eigenvectors computation was approximately 700 matrix-vector multiplies. However, time-wise the single-vector procedure was more than 1/3 faster than the HBLOCK procedure and used less than 1/4 as much storage.

We complete this section with a summary of the differences between the iterative block Lanczos procedures discussed in this chapter and the single-vector Lanczos procedures discussed in Chapter 4.

ITERATIVE BLOCK LANCZOS PROCEDURES

1. Procedure can determine the A-multiplicities of computed eigenvalues directly. However, procedure can compute only a few extreme eigenvalues of the given matrix along with corresponding eigenvectors.

2. User does not have to know a priori the intervals which contain the eigenvalues of A. Procedure requires only Ax's.

3. Procedure is easily restarted using the current best approximations to the desired eigenvectors.

4. Procedure is costly both in terms of storage requirements and in terms of operations counts. Ritz vectors are computed on every iteration. Orthogonality is maintained within each block. There is reorthogonalization of each Lanczos vector w.r.t certain vectors in the first Q-block. Small eigenvalue and eigenvector computations are required on each iteration. The Lanczos matrices are not tridiagonal and there is no cheap method for computing only that portion of the eigenspace required for the Ritz vector computations.

SINGLE-VECTOR LANCZOS PROCEDURES, NO REORTHOGONALIZATION

1. Any part of the spectrum can be computed. However, A-multiplicities of computed eigenvalues cannot be determined without additional computations.

2. User has to provide the program with subintervals to be used in the eigenvalues computations. Procedure requires only Ax's.

3. Procedure is easily rerun using a larger Lanczos matrix. However, there is no way to use a good approximation to the desired eigenelements as a starting 'vector' in a larger single-vector Lanczos computation.

4. Eigenvalue Procedure is very cheap both in terms of the amount of storage required and reasonably cheap in terms of the arithmetic operations counts. There is no reorthogonalization of any vectors and the Ritz vector computations are performed only once. The Lanczos matrices are real symmetric and tridiagonal so they are computationally-efficient w.r.t. both storage requirements and w.r.t. operation counts required for computing eigenvalues and eigenvectors. In particular any portion of the spectrum of any Lanczos matrix can be computed without computing any other portion.

Our experience has been that the single-vector Lanczos procedures are cheaper and provide more accurate eigenelement approximations than the iterative block Lanczos procedures. The single-vector procedures however are not as automatic as the block procedures. Overall we would recommend that the single-vector procedures be used whenever this is possible.

REFERENCES

Abo-Hamd, M., and Utku, S. (1978). Analytical study of suspension bridge flutter. *Proc. ASCE, J. Engrg. Mech. Div.* **104**, 537-550.

Abramatic, J. F., and Lee, S. U. (1980). Singular value decomposition of 2-D impulse responses. *Internat. Conf. Acoust. Speech Signal,* 749-752.

Allen, H. C. Jr., and Cross, P. C. (1963). Molecular Vib-Rotors. The Theory on Intrepretation of High Resolution Infrared Spectra. Wiley, New York.

Anderson, N., and Karasalo, I. (1975). On computing bounds for the least singular value of a triangular matrix. *BIT* **15**, 1-4.

Anderson, P. J., Liozou, G. (1973). On the quadratic convergence of an algorithm which diagonalizes a complex symmetric matrix. *J. Inst. Math. Appl.* **12**, 261-271.

Anderson, P. J., Liozou, G. (1975). A Jacobi type method for complex symmetric martices. *Numer. Math.* **25**, 347-363.

Anderson, P. W. (1958). Absence of diffusion in certain random lattices. *Phys. Rev.* **109**, 1492-1505.

Anderssen, R. S., and Bloomfield, P. (1974). A time series approach to numerical differentiation. *Technometrics* **16**, 69-75.

Anderssen, R. S., and Bloomfield, P. (1974b). Numerical differentiation procedures for non-exact data. *Numer. Math.* **22**, 157-182.

Andrews, H. C., and Patterson, C. L. (1976). Singular value decomposition and digital image processing. *IEEE Trans. ASSP-***24**, 26-53.

Andrews, H. C., and Hunt, B. R. (1977). Digital Image Restoration. Prentice-Hall, Englewood Cliffs, New Jersey.

Argyris, J. H., and Bronlund, O. E. (1975). The natural factor formulation of the stiffness for the matrix displacement method. *Comput. Methods Appl. Mech. Engr.* **5**, 97-119.

Avramovic, B., Kokotovic, P. V., Winkelman, J. R., and Chow, J. H. (1980). Area decomposition for electromechanical models of power systems. *Automatica - J. IFAC* **16**, 637-648.

Avramovic, B., Kokotovic, P. V. Winkelman, J. R., and Chow, J. H. (1980b). Coherency based decomposition and aggregation. *IFAC Congress* Kyoto, Japan.

Backus, G., and Gilbert, F. (1970). Uniqueness in the inversion of inaccurate gross earth data. *Philos. Trans. Roy. Soc. London Ser. A* **266**, 123-192.

Banfield, C. F. (1978). Singular value decomposition in multivariate analysis. In [Jacobs (1978)] 137-149.

Barakat, R. and Buder, T. E. (1979). Remote sensing of crosswind profiles using the correlation slope method. *J. Opt. Soc. Amer.* **69**, 1604-1608.

Barnes, E. (1982). An algorithm for partitioning the nodes of a graph. *SIAM J. Algebraic Discrete Methods* **3**, 541-550.

Barth, W., Martin, R. S., and Wilkinson, J. H. (1967). Calculation of the eigenvalues of a symmetric tridiagonal matrix by the method of bisection. *Numer. Math.* **9**, 386-393.

Bathe, K.-J., and Wilson, E. L. (1976). Numerical Methods in Finite Element Analysis. Prentice Hall, Englewood Cliffs, New Jersey.

Bathe, K.-J. (1977). Convergence of subspace iteration. In *Formulations and Computational Algorithms in Finite Element Analysis*. Bathe, K.-J., Oden, J. T., and Wunderlich, W. (Eds.). MIT Press, Cambridge, Massachusetts, 575-598.

Bathe, K.-J. and Ramaswamy, S. (1980). An accelerated subspace iteration method. *Comput. Methods Appl. Mech. Engrg.* **23**, 313-331.

Bauer, F. L. and Fike, C. T. (1960). Norms and exclusion theorems. *Numer. Math.* **2**, 137-141.

Björck, Å, and Golub, G. H. (1973). Numerical methods for computing angles between subspaces. *Math. Comp.* **27**, 579-594.

Björck, Å (1976). Methods for sparse linear least squares problems. In [Bunch 1976], 177-199.

Björck, Å, and Duff, I. S. (1980). A direct method for the solution of sparse linear least squares problems. *Linear Algebra Appl.* **34**, 43-67.

Björck, Å, Plemmons, R. J., and Schneider, H. (Eds.) (1980). Large Scale Matrix Problems. North Holland Publ. Co., New York.

Bowdler, H., Martin, R. S., Reinsch, C., and Wilkinson, J. H. (1968). The QR and QL algorithms for symmetric matrices. *Numer. Math.* **11**, 293-306.

Bunch, J. R., and Rose, D. E. (Eds.) (1976). Sparse Matrix Computations. Academic Press, New York.

Bunch, J. R., and Kaufman, L. (1977). Some stable methods for calculating inertia and solving symmetric linear systems. *Math. Comp.* **31**, 163-179.

Butscher, W., and Kammer, W. E. (1976). Modification of Davidson's method for the calculation of eigenvalues and eigenvectors of large real symmetric matrices: "Root-homing procedure". *J. Comput. Phys.* **20**, 313-325.

Cederbaum, H. S., Haller, E., and Domcke, W. (1980). Effective single-mode Hamiltonian for the calculation of multi-mode Jahn-Teller band shapes. *Solid State Comm.* **35**, 879-881.

Chan, S. P., Fledman, H., and Parlett, B. N. (1977). A program for computing the condition numbers of matrix eigenvalues without computing eigenvectors. *ACM TOMS* **3**, 186-203.

Chan, T. F. (1982). An improved algorithm for computing the singular value decomposition. *ACM TOMS* **8**, 72-83.

Chow, J. H., Cullum, J. and Willoughby, R. A. (1984). A sparsity-based technique for identifying slow-coherent areas in large power systems. *IEEE Trans. PAS-* **103**, 463-473.

Chowdhury, P. C. (1976). Truncated Lanczos algorithm for the partial solution of the symmetric eigenproblem. *Comput. & Structures* **6**, 439-446.

Cline, A. K., Golub, G. H., and Platzman, G. W. (1976). Calculation of normal modes of oceans using a Lanczos method. In [Bunch 1976], 409-426.

Cline, A. K., Moler, C. B., Stewart, G. W., and Wilkinson, J. H. (1979). An estimate for the condition number of a matrix. *SIAM J. Numer. Anal.* **16**, 368-375.

Cline, A. K. (1982). Counter examples to matrix condition number estimators. *SIAM Meeting*, Stanford.

Clint, M., and Jennings, A. (1970). The evaluation of eigenvalues and eigenvectors of real symmetric matrices by simultaneous iterations. *Comput. J.* **13**, 76-80.

Clint, M., and Jennings, A. (1971). A simultaneous iteration method for the unsymmetric eigenvalue problem. *J. Inst. Math. Appl.* **8**, 111-121.

Corr, R. B., and Jennings, A. (1976). A simultaneous iteration algorithm for symmetric eigenvalue problems. *Internat. J. Numer. Methods Engrg.* **10**, 647-663.

Craven, B. D. (1969). Complex symmetric matrices. *J. Austral. Math. Soc.* **10**, 341-354.

Cullum, J., and Donath, W. E. (1974). A block Lanczos algorithm for computing the q algebraically largest eigenvalues and a corresponding eigenspace for large, sparse symmetric matrices. In *Proc. 1974 IEEE Conference on Decision and Control*, IEEE Press, New York, 505-509.

Cullum, J., and Donath, W. E. (1974b). A block generalization of the symmetric s-step Lanczos algorithm. IBM T. J. Watson Research Center, RC 4845. (May 1974).

Cullum, J., Donath, W. E., and Wolfe, P. (1975). The minimization of certain nondifferentiable sums of eigenvalues of symmetric matrices. *Mathematical Programming Study* **3**, 35-55.

Cullum, J. (1978). The simultaneous computation of a few of the algebraically largest and smallest eigenvalues of a large, symmetric, sparse matrix. *BIT* **18**, 265-275.

Cullum, J. (1979). The effective choice of the smoothing norm in regularization. *Math. Comp.* **33**, 149-170.

Cullum, J., and Willoughby, R. A. (1979). Lanczos and the computation in specified intervals of the spectrum of large sparse real symmetric matrices. In [Duff 1979], 220-255.

Cullum, J., and Willoughby, R. A. (1979b). Fast modal analysis of large, sparse but unstructured symmetric matrices. In *Proc. 18th IEEE Conf. Decision Contr.*, IEEE Press, New York, 45-53.

Cullum, J., (1980). Ill-posed deconvolutions: regularization and singular value decompositions. In *19th Proc. IEEE Conf. Decision Contr.*, IEEE Press, New York, 29-35.

Cullum, J., and Willoughby, R. A. (1980). The Lanczos phenomenon - an interpretation based upon conjugate gradient optimization. *Linear Algebra Appl.* **29**, 63-90.

Cullum, J., and Willoughby, R. A. (1980b). Computing eigenvectors (and eigenvalues) of large, symmetric matrices using Lanczos tridiagonalization. In *Lecture Notes in Mathematics* **773**, 46-63. Watson, G. A. (Ed.) Springer, New York.

Cullum, J., and Willoughby, R. A. (1980c). Computing eigenvalues of large symmetric matrices-an implementation of a Lanczos algorithm with no reorthogonalization. Part II. Computer programs. IBM T. J. Watson Research Center, RC 8298.

Cullum, J., and Willoughby, R. A. (1981). Computing eigenvalues of very large symmetric matrices-an implementation of a Lanczos algorithm with no reorthogonalization. *J. Comput. Phys.* **44**, 329-358.

Cullum, J., Willoughby, R. A., and Lake, M. (1983). A Lanczos algorithm for computing singular values and vectors of large matrices. *SIAM J. Sci. Statist. Comput.* **4**, 197-215.

Cullum, J., and Willoughby, R. A. (1984). Lanczos Algorithms for Large Symmetric Matrices: Volume 2, Users Guide. Birkhauser, Boston.

Daniel, J. W., Gragg, W. B., Kaufman, L., and Stewart, G. W. (1976). Reorthogonalization and stable algorithms for updating the Gram-Schmidt QR factorization. *Math. Comp.* **30**, 772-795.

Davidson, E. R. (1975). The iterative calculation of a few of the lowest eigenvalues and corresponding eigenvectors of large real symmetric matrices. *J. Comput. Phys.* **17**, 87-94.

Davidson, E. R. (1983). Matrix eigenvector methods. In *Methods in Computational Molecular Physics.* Diercksen, G. H. F. and Wilson, S. (Eds.), Reidel, Boston, 95-113.

Davis, C., and Kahan, W. (1970). The rotation of eigenvectors by a perturbation. III. *SIAM J. Numer. Anal.* **7**, 1-46.

Dean, P. (1967). Atomic vibrations in solids. *J. Inst. Math. Appl.* **3**, 98-165.

Dean, P. (1972). The vibrational properties of disordered systems: numerical studies. *Rev. Modern Phys.* **44**, 127-168.

Dehesa, J. S. (1978). The Lanczos method and asymptotical level density of a physical system. *Lett. Nuovo Cimento* **23**, 301-305.

Dehesa, J. S. (1981). Lanczos method of tridiagonalization. Jacobi matrices and physics. *J. Comput. Appl. Math.* **7**, 249-259.

Deift, P., Nanda, T., and Tomei, C. (1983). Ordinary differential equations and the symmetric eigenvalue problem. *SIAM J. Numer. Anal.* **20**, 1-22.

Donath, W. E. (1970). Stochastic model of the computer logic design process. IBM T. J. Watson Research Center, RC 3136.

Donath, W. E., and Hoffman, A. J. (1973). Lower bounds for partitioning of graphs. *IBM J. Res. Develop.* **17**, 420-425.

Donath, W. E. (1974). Equivalence of memory to "random logic". *IBM J. Res. Develop.* **18**, 401-407.

Dongarra, J. J., Moler, C. B., and Wilkinson, J. H. (1983). Improving the accuracy of computed eigenvalues and eigenvectors. *SIAM J. Numer. Anal.* **20**, 23-45.

Doyle, J. C., and Stein, G. (1981). Multivariable feedback design: Concepts for classical/modern synthesis. *IEEE Trans.* AC-**26**, 4-16.

Dubrulle, A. A. (1970). A short note on the implicit QL algorithm for symmetric tridiagonal matrices. *Numer. Math.* **15**, 450.

Duff, I. S. (1977). A survey of sparse matrix research. *IEEE Proc.* **65**, 500-535.

Duff, I. S., and Stewart, G. W. (1979). Sparse Matrix Proceedings 1978, SIAM Press, Philadelphia.

Duff, I. S. (Ed.) (1981). Sparse Matrices and Their Uses. *IMA Numerical Analysis Group Conference,* Academic Press, New York.

Duff, I. S. (1982). Research directions in sparse matrix computations. *Harwell Report AERE-R10547.*

Dyson, F. J. (1953). The dynamics of a disordered linear chain. *Phys. Rev.* **92**, 1331-1338.

Eberlein, P. J. (1971). On the diagonalization of complex symmetric matrices. *J. Inst. Math. Anal.* **7**, 377-383.

Eckart, C., and Young, G. (1936). The approximation of one matrix by another of lower rank. *Psychometrika* **1**, 211-218.

Edwards, J. T., and Thouless, D. J. (1972). Numerical studies of localization in disordered systems. *J. Phys. C* **5**, 807-820.

Edwards, J. T., Licciardello, D. C., and Thouless, D. J. (1979). Use of the Lanczos method for finding complete sets of eigenvalues of large sparse symmetric matrices. *J. Inst. Math. Appl.* **23**, 277-283.

Eisenstat, S. C., Schultz, M. H., and Sherman, A. H. (1981). Algorithms and data structures for sparse symmetric Gaussian elimination. *SIAM J. Sci. Statist. Comput.* **2**, 225-237.

EISPACK Guide (1976). Matrix Eigensystem Routines - EISPACK Guide. Smith, B. T., Boyle, J. M., Garbow, B. S., Ikebe, Y., Klema, V. C., and Moler, C. B. *Lecture Notes in Computer Science* **6**, Second Edition, Springer, New York.

EISPACK Guide (1977). Matrix Eigensystem Routines - EISPACK Guide Extension. Garbow, B. S., Boyle, J. M., Dongarra, J. J., and Moler, C. B. *Lecture Notes in Computer Science* **51**, Second Edition, Springer, New York.

Ekstrom, M. P., and Rhoads, R. L. (1974). On the application of eigenvector expansion for numerical deconvolution, *J. Comput. Phys.* **14**, 319-340.

Englman, R. (1972). The Jahn-Teller Effect in Molecules and Crystals. Wiley-Interscience, New York.

Ericsson, T., and Ruhe, A. (1980). The spectral transformation Lanczos method for the numerical solution of large sparse generalized symmetric eigenvalue problems. *Math. Comp.* **35**, 1251-1268.

Ericsson, T. (1982). Some properties of the eigenvectors of a symmetric tridiagonal matrix. *Report UMINF-99.82,* U. Umeä.

Ericsson, T. and Ruhe, A. (1982). STLM-a software package for the spectral transformation Lanczos algorithm. *Report UMINF-101.82,* U. Umeä.

Ericsson, T. (1983). Implementation and applications of the spectral transformation Lanczos algorithm. In *Lecture Notes in Mathematics,* vol.973, Springer, New York, 177-188.

Ericsson, T. (1983b). Algorithms for large sparse symmetric generalized eigenvalue problems. *Report UMINF-108.83,* U. Umeä.

Erisman, A. M., Neves, K. W., and Dwarakanath, M. H. (Eds.) (1980). Electric Power Problems: The Mathematical Challenge. SIAM, Philadelphia.

Fan, K. (1949). On a theorem of Weyl concerning eigenvalues of linear transformations. *Proc. Nat. Acad. Sci. USA* **35**, 652-655.

Feler, M. G. (1974). Calculation of eigenvectors of large matrices. *J. Comput. Phys.* **14**, 341-349.

Fiedler, M. and Ptak, V. (1962). On matrices with non-positive off-diagonal elements and positive principal minors. *Czech. Math. J.* **12**, 382-400.

Fisher, N. J., and Howard, L. E. (1980). Gravity interpretation with the aid of quadratic programming. *Geophysics* **45**, 403-419.

Forsythe, G. E. (1968). On the asymptotic directions of the s-dimensional conjugate gradients method. *Numer. Math.* **11**, 57-76.

Forsythe, G. E., Malcolm, M. A., and Moler, C. B. (1977). Computer Methods for Mathematical Computations. Prentice-Hall, Englewood Cliffs, New Jersey.

Francis, J. G. F. (1961/1962). The QR transformation - a unitary analogue to the LR transformation. *Comput. J.* **4**, 1: 265-271, 2: 332-345.

Freudenberg, J. S., Looze, D. P., and Cruz, J. B. (1982). Robust analysis using singular value decomposition. *Internat. J. Control* **35**, 95-116.

Frik, G., and Lunde Johnsen, T. (1975). Note on the ill-conditioned eigenvalue problem in elastic vibration. *Comput. Methods. Appl. Mech. Engrg.* **6**, 65-77.

Gantmacher, F. R. (1959). The Theory of Matrices. I, II. Chelsea, New York.

Gehring, G. A., and Gehring, K. A. (1975). Co-operative Jahn-Teller Effects. *Reports on Progress in Physics* **38**, 1-89.

Geogakis, C., Aris, R., and Amundson, N. R. (1977). Studies in the control of tubular reactors - 2. Stabilization by modal control. *Chem. Eng. Sci.* **32**, 1371-1379.

George, A., and Liu, J. W. H. (1980). A fast implementation of the minimum degree algorithm using quotient graphs. *ACM TOMS* **6**, 337-358.

George, A., and Liu, J. W. H. (1981). Computer Solution of Large Sparse Positive Definite Systems. Prentice-Hall, Englewood Cliffs, NJ.

George, A., and Ng, E. (1983). On row and column orderings for sparse least square systems. *SIAM J. Numer. Anal.* **20**, 326-344.

Gerlakh, N. I., Zueva, N. M., and Solovyey, L. S. (1978). Linear theory of a helical MHD instability. *Magnetohydrodynamics* **14**, 431-435.

Gill, P. E., Murray, W., and Wright, M. H. (1981). Practical Optimization. Academic Press, New York.

Givens, W. (1954). Numerical computation of the characteristic values of a real symmetric matrix. *Report ORNL-1574*, Oak Ridge, Tennessee.

Godunov, S. K., and Prokopov, G. P. (1970). A method of minimal iterations for evaluating the eigenvalues of an elliptic operator. *USSR Comp. Math. Math. Phys.* **10**, 141-154.

Golub, G. H., and Kahan, W. (1965). Calculating the singular values and pseudoinverse of a matrix. *SIAM J. Numer. Anal.* **2**, 205-224.

Golub, G. H. (1968). Least squares, singular values and matrix approximations. *Apl. Mat.* **13**, 44-51.

Golub, G. H., and Reinsch, C. (1970). Singular value decomposition and least squares solutions. *Numer. Math.* **14**, 181-195.

Golub, G. H., Underwood, R., and Wilkinson, J. H. (1972). The Lanczos algorithm for the symmetric $Ax = \lambda Bx$ problem. *Report Stan-CS-72-270*, Stanford.

Golub, G. H. (1973). Some modified eigenvalue problems. *SIAM Rev.* **15.**, 318-334.

Golub, G. H. (1973b). Some uses of the Lanczos algorithm in numerical linear algebra. In *Topics in Numerical Analysis*. Miller, J. J. H. (Ed.), Academic Press, New York, 173-184.

Golub, G. H. (1974). Methods for computing eigenvalues of sparse matrix equations. In *Actas Del Seminario Sombre Methodos Numericos Modernos,* Vol. 1, Pereyra V. (Ed.), Univ. Central de Venezuela, 125-148.

Golub, G. H., and Wilkinson, J. H. (1976). Ill-conditioned eigensystems and the computation of the Jordan canonical form. *SIAM Rev.* **18**, 578-619.

Golub, G. H., and Underwood, R. (1977). The block Lanczos method for computing eigenvalues. In *Mathematical Software III.* Rice, J. R. (Ed.), Academic Press, New York, 361-377.

Golub, G. H., Luk, F. T., and Overton, M. L. (1981). A block Lanczos method for computing the singular values and corresponding singular vectors of a matrix. *ACM TOMS* **7**, 149-169.

Gruber, R. (1975). HYMMIA - band matrix package for solving eigenvalue problems. *Comput. Phys. Comm.* **10**, 30-41.

Gupta, K. K. (1973). On a combined Strum sequence inverse iteration technique for eigenproblem solution of spinning structures. *Internat. J. Numer. Methods Engrg.* **7**, 509-518.

Gupta, K. K. (1974). Eigenproblem solution of damped structural systems. *Internat. J. Numer. Methods Engrg.* **8**, 877-911.

Gupta, K. K. (1976). On a numerical solution of the supersonic panel flutter eigenproblem. *Internat. J. Numer. Methods Engrg.* **10**, 637-645.

Gupta, K. K. (1978). On a numerical solution of the plastic buckling problem of structures. *Internat. J. Numer. Methods Engrg.* **12**, 941-947.

Gupta, K. K. (1978b). Development of a finite dynamic element for free vibration analysis of two-dimensional structures. *Internat. J. Numer. Methods Engrg.* **12**, 1311-1327.

Gupta, K. K. (1979). Finite dynamic element formulation for a plane triangular element. *Internat. J. Numer. Methods Engrg.* **14**, 1431-1448.

Gustavson, F. G. (1972). Some basic techniques for solving sparse systems of linear equations. In *Sparse Matrices and Their Applications.* Rose, D. J. and Willoughby, R. A., (Eds.), Plenum, New York, 41-52.

Hamming, R. W. (1973). Numerical Methods for Scientists and Engineers. McGraw-Hill, New York.

Hamming, R. W. (1974). The frequency approach in numerical analysis. In *Studies in Numerical Analysis,* Academic Press, New York, 151-175.

Hanson, R. J., and Phillips, J. L. (1975). An adaptive numerical method for solving linear Fredholm equations of the first kind. *Numer. Math.* **24**, 291-307.

Hanson, R. J., and Norris, M. J. (1981). Analysis of measurements based on the singular value decomposition. *SIAM J. Sci. Statist. Comput.* **2**, 363-373.

HARWELL Subroutine Library (1981). Compiled by M.J. Hopper. Computer Science and Systems Division, Atomic Energy Research Establishment, Harwell, Oxfordshire, England.

Haydock, R. (1980). The recursive solution of the Schrodinger equation. *Solid State Phys.* **35**, 215-294.

Haydock, R. and Thorpe, R. (Eds.) (1981). Excitations in Disordered Systems. Plenum, New York.

Henrici, P. (1962). Bounds for iterates, inverses, spectral variation and fields of values of non-normal matrices, *Numer. Math.* **4**, 24-40.

Hestenes, M. R., and Karush, W. (1951). A method of gradients for the calculation of the characteristic roots and vectors of a real symmetric matrix. *J. Res. Nat. Bur. Standards Sect. B.* **47**, 471-478.

Hestenes, M. R., and Stiefel, E. (1952). Method of conjugate gradients for solving linear systems. *J. Res. Nat. Bur. Standards Sect. B.* **49**, 409-436.

Hestenes, M. R. (1956). The conjugate-gradient method for solving linear equations. In *Proc. Sixth Symposium Appl. Math.*, 83-102.

Hestenes, M. R. (1980). Conjugate Direction Methods in Optimization. Springer-Verlag, New York.

Hodges, D. H. (1979). Aeromechanical stability analysis for bearingless rotor helicopters. *J. Amer. Helicopter Soc.* **24**, 2-9.

Hoffman, A. J., and Wielandt, H. W. (1953) The variation of the spectrum of a normal matrix. *Duke Math. J.* **20**, 37-40.

Hori, J. (1968). Spectral Properties of Disordered Chains and Lattices. Pergamon Press, New York.

Householder, A. S. (1958). Generated error in rotational tridiagonalization. *J. Assoc. Comput. Mach.* **5**, 335-338.

Householder, A. S. (1964). The Theory of Matrices in Numerical Analysis. Blaisdell, New York.

Huang, T. C. and Parrish, W. (1978). Qualitative analysis of complicated mixtures by profile fitting x-ray diffractometer patterns. *Adv. x-Ray Anal.* **21**, 275-288.

Huang, T. C., Parrish, W. and Ayers, G. L. (1980). A rapid and precise computer method for qualitative x-ray fluorescence analysis. IBM San Jose Research Center, RJ 2912.

Jacobs, D. (Ed.) (1977). The State of the Art in Numerical Analysis. Academic Press, New York.

Jacobs, D. (Ed.) (1978). Numerical Software - Needs and Availability. Academic Press, New York.

Jennings, A., and Orr, D. R. L. (1971). Application of the simultaneous iteration method to undamped vibration problems. *Internat. J. Numer. Methods Engr.* **3**, 13-24.

Jennings, A., and Stewart, W. J. (1975). Simultaneous iteration for partial eigensolution of real matrices. *J. Inst. Math. Appl.* **15**, 351-361.

Jennings, A. (1977). Matrix Computations for Engineers and Scientists, Wiley, New York.

Jennings, A., and Agar, T. J. S. (1979). Progressive simultaneous inverse iteration for symmetric eigenvalue problems. *Comput. & Structures* **14**, 51-61.

Jennings, A. (1980). Eigenvalue methods for vibration analysis. *Shock Vibr. Digest* **12**, 3-19.

Jennings, A. (1981). Eigenvalue methods and the analysis of structural vibrations. In [Duff 1981] 109-138.

Jennings, A. (1982). Bounds for the singular values of a matrix. *IMA J. Numer. Anal.* **2**, 459-474.

Jensen, P. S. (1972). The solution of large eigenproblems by sectioning. *SIAM J. Numer. Anal.* **9**, 534-545.

Jensen, P. S. (1973). An inclusion theorem related to inverse iteration. *Linear Algebra Appl.* **6**, 209-215.

Kahan, W. (1966). When to neglect off-diagonal elements of symmetric tridiagonal matrices. *Report CS-42,* Stanford.

Kahan, W. (1967). Inclusion theorems for clusters of eigenvalues of Hermitian matrices. *Report,* U. Toronto.

Kahan, W. (1972). Private communication.

Kahan, W., and Parlett, B. N. (1974). An analysis of Lanczos algorithms for symmetric matrices. *Report ERL-M467,* UC Berkeley.

Kahan, W., and Parlett, B. N. (1976). How far should one go in the Lanczos process. In [Bunch 1976], 131-144.

Kahan, W., Parlett, B. N., and Jiang, E. (1982). Residual bounds on approximate eigensystems of nonnormal matrices. *SIAM J. Numer. Anal.* **19**, 470-484.

Kaniel, S. (1966). Estimates for some computational techniques in linear algebra. *Math. Comp.* **20**, 369-378.

Karush, W. (1951). An iterative method for finding characteristic vectors of a symmetric matrix. *Pacific J. Math.* **1**, 233-248.

Kirkpatrick, S., and Eggarter, T. P. (1972). Localized states of a binary alloy. *Phys. Rev. B* **6**, 3589-3600.

Kirkpatrick, S. (1979). Models of disordered materials. In *Eight lectures in the Proc. of the Ecole d'Ete de Physique.* Balion, R., Maynard, R. and Toulouse, G. (Eds.) North Holland, Amsterdam, 324-403.

Kreuzer, K. G., Miller, and Berger, W. A. (1981). The Lanczos algorithm for self-adjoint operators. *Phys. Lett. A* **81**, 429-432.

Kublanovskaya, V. N. (1961). On some algorithms for solving the complete problem of proper values. *Soviet Math. Dokl.* **2**, 17-19.

Kupchinov, I. I., and Ivanova, M. M. (1973). The application of eigenvalues in goedesy. *Geod. Mapp. Photogramm.* **15**, 40-43.

Lanczos, C. (1950). An iterative method for the solution of the eigenvalue problem of linear differential and integral operators. *J. Res. Nat. Bur. Standards, Sect. B* **45**, 255-282.

Lanczos, C. (1952). Solution of systems of linear equations by minimized iterations. *J. Res. Nat. Bur. Standards, Sect. B* **49**, 33-53.

Lanczos, C. (1961). Linear Differential Operators. Van Nostrand, New York.

Landauer, R. (1970). Electrical resistance of disordered one-dimensional lattices. *Philos. Mag.* **21**, 863-867.

Lawson, C., and Hanson, R. (1974). Solving Least Squares Problems. Prentice-Hall, Englewood Cliffs, New Jersey.

Lewis, J. G. (1977). Algorithms for sparse matrix eigenvalue problems. *PhD Thesis and Report CS-77-595,* Stanford.

Lewis, J. G., and Grimes, R. G. (1981). Practical Lanczos algorithms for solving structural engineering problems. In [Duff 1981], 349-355.

Licciardello, D. C., and Thouless, D. J. (1975). Conductivity and mobility edges for two-dimensional disordered systems. *J. Phys. C* **8**, 4157-4170.

Licciardello, D. C., and Thouless, D. J. (1978). Conductivity and mobility edges in disordered systems. II. Futher calculations for the square and diamond lattices. *J. Phys. C* **11**, 925-936.

LINPACK User's Guide. (1979). Dongarra, J. J., Bunch, J. R., Moler, C. B., and Stewart, G. W. SIAM Press, Philadelphia.

Lonquet-Higgens, H. C., Öpik, U., Pryce, M. H. L., and Sack, R. A. (1958). Studies of the Jahn-Teller effect. II. The dynamical problem. *Proc. Roy. Soc.* **244A**, 1-16.

Luenberger, D. G. (1973). Introduction to Linear and Nonlinear Programming. Addison-Wesley, Reading, Massachusetts.

Luk, F. T. (1980). Computing the singular-value decomposition on the ILLIAC IV. *ACM TOMS* **6**, 524-539.

Marcus, M. (1960). Basic theorems in matrix theory. *Natl. Bur. Standards Appl. Math. Ser.* **57**.

Martin, R. S., and Wilkinson, J. H. (1968). Reduction of the symmetric eigenproblem $Ax = \lambda Bx$ and related problems to standard form. *Numer. Math.* **11**, 99-110.

Martin, R. S., and Wilkinson, J. H. (1968b). The implicit QL algorithm. *Numer. Math.* **12**, 377-383.

McCormick, S. F. (1981). A mesh refinement method for $Ax = \lambda Bx$. *Math. Comp.* **36**, 485-498.

Meirovitch, L. (1980). Computational Methods in Structural Dynamics. Sijthoff & Noordhoff, Rockville, MD

Moler, C. B., and Stewart, G. W. (1973). An algorithm for generalized matrix eigenvalue problems. *SIAM J. Numer. Anal.* **10**, 241-256.

Moler, C. B., and Shavitt, I. (Eds.) (1978). Numerical Algorithms in Chemistry: Algebraic Methods. NRCC, Lawrence Berkeley Labs.

Moore, B. C. (1981). Principal component analysis in linear systems: Controllability, observability, and model reduction. *IEEE Trans. AC*-**26**, 17-32.

Moro, G., and Freed, J. H. (1981). Calculation of ESR spectra and related Fokker-Planck forms by the use of the Lanczos algorithm. *J. Chem. Phys.* **74**, 3757-3773.

NASTRAN Users Manual (1977). NASA Langley Research Center, Hampton, Virginia.

Nesbet, R. K. (1981). Large matrix techniques in quantum chemistry and atomic physics. In [Duff 1981], 161-174.

Newman, M., and Pipano, A. (1977). Fast model extraction in NASTRAN via the FEER computer program, NASTRAN Users Manual, NASA Langley Research Center, Hampton, Virginia, 485-506.

Nikolai, P. J. (1979). Algorithm 538. Eigenvectors and eigenvalues of real generalized symmetric matrices by simultaneous iteration. *ACM TOMS* **5**, 118-125.

Nour-Omid, B., Parlett, B. N., and Taylor, R. L. (1983). Lanczos versus subspace iteration for solution of eigenvalue problems. *Internat. J. Numer. Methods Engrg.* **19**, 859-871.

Nour-Omid, B., Parlett, B. N., and Taylor, R. L. (1983b). A Newton-Lanczos method for solution of nonlinear finite element equations. *Comput. & Structures* **16**, 241-252.

O'Brien, M. C. M., and Evangelou, S. N. (1980). The calculation of absorption band shapes in dynamic Jahn-Teller systems by the use of the Lanczos algorithm. *J. Phys. C* **13**, 611-623.

Ojaive, I. U., and Newman, M. (1970). Vibration modes of large structures by an automatic matrix-reduction method. *AIAA J.* **8**, 1234-1239.

O'Leary, D. P., Stewart, G. W., and Vandergraft, J. S. (1979). Estimating the largest eigenvalue of a positive definite matrix. *Math. Comp.* **33**, 1289-1292.

O'Leary, D. P. (1980). The block conjugate gradient algorithm and related methods. *Linear Algebra Appl.* **29**, 293-322.

O'Leary, D. P. (1980b). Estimating matrix condition numbers. *SIAM J. Sci. Statist. Comput.* **1**, 205-209.

O'Leary, D. P., and Simmons, J. A. (1981). A bidiagonalization-regularization procedure for large scale discretization of ill-posed problems. *SIAM J. Sci. Statist. Comput.* **2**, 474-489.

Ortega, J. M. (1960). On Sturm sequences for tridiagonal matrices. *J. Assoc. Comput. Mach.* **7**, 260-263.

Ostrowski, A. M. (1937). On determinants with dominant diagonal. *Comment. Math. Helv.* **10**, 69-96. (German).

Ostrowski, A. M. (1958-59). On the convergence of the Rayleigh quotient iteration for the computation of the characteristic roots and vectors. *Arch. Rational Mech. Anal.* **1**, 233-241, **2**, 423-428, **3**, 325-340, 341-347, 472-481, **4**, 153-165.

Paige, C. C. (1971). The computation of eigenvalues and eigenvectors of very large sparse matrices. *Ph.D. Thesis*, U. London.

Paige, C. C. (1972). Computational variants of the Lanczos method for the eigenproblem. *J. Inst. Math. Appl.* **10**, 373-381.

Paige, C. C. (1974). Bidiagonalization of matrices and solution of linear equations. *SIAM J. Numer. Anal.* **11**, 197-209.

Paige, C. C. (1974b). Eigenvalues of perturbed Hermitian matrices. *Linear Algebra Appl.* **8**, 1-10.

Paige, C. C. and Saunders, M. A. (1975). Solution of sparse indefinite systems of linear equations. *SIAM J. Numer. Anal.* **12**, 617-629.

Paige, C. C. (1976). Error analysis of the Lanczos algorithms for tridiagonalizing a symmetric matrix. *J. Inst. Math. Appl.* **18**, 341-349.

Paige, C. C. (1980). Accuracy and effectiveness of the Lanczos algorithm for the symmetric eigenproblem. *Linear Algebra Appl.* **34**, 235-258.

Paige, C. C. and Saunders, M. A. (1982). LSQR: an algorithm for sparse linear equations and sparse least squares. *ACM TOMS* **8**, 43-71.

Pandey, K. C. (1983). Theory of semiconductor surface reconstruction: Si(111)-7x7, Si(111)-2x1, and GaAs(110). *Physica* **117B&118B**, 761-766.

Parlett, B. N. (1968). Global convergence of the basic QR algorithm on Hessenberg matrices. *Math. Comp.* **22**, 803-817.

Parlett, B. N., and Reinsch, C. (1969). Balancing a matrix for calculation of eigenvalues and eigenvectors. *Numer. Math.* **13**, 293-304.

Parlett, B. N., and Poole, W. G. (1973). A geometric theory for the QR, LU, and power iterations. *SIAM J. Numer. Anal.* **10**, 389-412.

Parlett, B. N., and Scott, D. S. (1977). The Lanczos algorithm with implicit deflation. *Report ERL-M77/70,* UC Berkeley.

Parlett, B. N., and Scott, D. S. (1979). The Lanczos algorithm with selective reorthogonalization. *Math. Comp.* **33**, 217-238.

Parlett, B. N. (1980). The Symmetric Eigenvalue Problem. Prentice-Hall, Englewood Cliffs, NJ.

Parlett, B. N. (1980b). How to solve $(K - \lambda M)z = 0$ for large K and M. *Numer. Methods Engrg.* **1**, 97-106.

Parlett, B. N., and Reid, J. K. (1981). Tracking the progress of the Lanczos algorithm for large symmetric matrices. *IMA J. Numer. Anal.* **1**, 135-155.

Parlett, B. N., Simon, H., and Stringer, L. M. (1982). On estimating the largest eigenvalue with the Lanczos algorithm. *Math. Comp.* **38**, 153-165.

Parlett, B. N. (1982). Two monitoring schemes for the Lanczos algorithm. In *Computing Methods in Applied Sciences and Engineering. V.* Glowinski, R. and Lions, J. L., (Eds.) North-Holland, Amsterdam, 27-34.

Parlett, B. N., Taylor, D., and Liu, Z-S (1983). The look ahead Lanczos algorithm for large unsymmetric eigenproblems. In *Sixth International Conference on Computing Methods in Applied Science and Engineering,* Versailles, France.

Parrish, W. (1980). Advances in computerized x-ray diffractometry and x-ray analysis. IBM San Jose Research Center, RJ 2774.

Peters, G., and Wilkinson, J. H. (1969). Eigenvalues of $Ax = \lambda Bx$ with band symmetric A and B. *Comput. J.* **12**, 398-404.

Peters, G., and Wilkinson, J. H. (1970). $Ax = \lambda Bx$ and the generalized eigenvalue problem. *SIAM J. Numer. Anal.* **7**, 479-492.

Peters, G., and Wilkinson, J. H. (1974). In Accuracy of computed eigensystems and invariant subspaces. *Studies in Numerical Analysis,* Academic Press, New York, 115-135.

Peters, G., and Wilkinson, J. H. (1979). Inverse iteration, ill-conditioned equations and Newton's method. *SIAM Rev.* **21**, 339-360.

The PFORT Verifier (1974). Ryder, B. G. *Software-Practice and Experience* **4**, 359-377.

PFORT Verifier (1981). Ryder, B. G., and Hall, A. D. *Bell Laboratory Computer Science Technical Report 12.*

Platzman, G. W. (1972). North Atlantic ocean: Preliminary description of normal modes. *Science* **178**, 156-157.

Platzman, G. W. (1972b). Two dimensional free oscillations in natural basins. *J. Phys. Oceanogr.* **2**, 117-138.

Platzman, G. W. (1975). Normal modes of the Atlantic and Indian oceans. *J. Phys. Oceanogr.* **5**, 201-221.

Platzman, G. W. (1978). Normal modes of the world ocean. Part I. Design of finite-element barotropic model. *J. Phys. Oceanogr.* **8**, 323-343.

Platzman, G. W. (1979). A Kelvin wave in the eastern north Pacific ocean. *J. Geophys. Res.* **84**, 2525-2528.

Reid, J. K. (1971). On the method of conjugate gradients for the solution of large sparse systems of linear equations. In *Large Sparse Sets of Linear Equations.* Reid, J. K. (Ed.), Academic Press, New York, 231-254.

Reid, J. K. (1980). A survey of sparse matrix computation. In [Erisman 1980], 41-69.

Ruhe, A. (1974). SOR-methods for the eigenvalue problem with large sparse matrices. *Math. Comp.* **28**, 695-710.

Ruhe, A. (1974b). Iterative eigenvalue algorithms for large symmetric matrices. In *Numerische Behandlung von Eigenwertaufgaben.* L. Collatz, ed. Birkhauser, Basel, 97-115.

Ruhe, A. (1975). On the closeness of eigenvalues and singular values for almost normal matrices. *Linear Algebra Appl.* **11**, 87-93.

Ruhe, A. (1976). Computation of eigenvalues and eigenvectors. In *Sparse Matrix Techniques.* Barker, V. A. (Ed.), Springer, New York,130-184.

Ruhe, A. (1979). Implementation aspects of band Lanczos algorithms for computation of eigenvalues of large sparse symmetric matrices. *Math. Comp.* **33**, 680-687.

Rutishauser, H. (1959). Theory of gradient methods. In *Refined Iterative Methods for Computation of the Solution and the Eigenvalues of Self-Adjoint Boundary Value Problems.* Engeli, M., Ginsburg, J., Rutishauser, H., and Stiefel, E. L. Eds. Birkhauser, Basel.

Rutishauser, H. (1969). Computational aspects of F. L. Bauer's simultaneous iteration method. *Numer. Math.* **13**, 4-13.

Rutishauser, H. (1970). Simultaneous iteration methods for symmetric matrices. *Numer. Math.* **16**, 205-223.

Saad, Y. (1980). On the rates of convergence of the Lanczos and the block-Lanczos methods. *SIAM J. Numer. Anal.* **17**, 687-706.

Saad, Y. (1982). The Lanczos biorthogonalization algorithm and other oblique projection methods for solving large unsymmetric systems. *SIAM J. Numer. Anal.* **19**, 485-506.

Sahasrabudhe, S. C., and Vaidya, P. M. (1979). Estimation of singular values of an image matrix. *IEEE Trans. ASSP-27*, 434-436.

Schwarz, H. R. (1968). Tridiagonalization of a symmetric band matrix. *Numer. Math.* **12**, 231-241.

Schwarz, H. R., Rutishauser, H. and Stiefel, E. (1973). Numerical Analysis of Symmetric Matrices. Prentice-Hall, Englewood Cliffs, New Jersey.

Schwarz, H. R. (1974). The eigenvalue problem $(A-\lambda B)x = 0$ for symmetric matrices of high order. *Comput. Methods Appl. Mech. Engrg.* **3**, 11-28.

Schwarz, H. R. (1974b). The method of coordinate overrelaxation for $(A-\lambda B)x = 0$. *Numer. Math.* **23**, 135-151.

Schwarz, H. R. (1977). Two algorithms for treating $Ax = \lambda Bx$. *Comput. Methods Appl. Mech. Engrg.* **12**, 181-199.

Scott, D. S. (1978). Analysis of the symmetric Lanczos algorithm. *Ph.D. Thesis,* UC Berkeley.

Scott, D. S. (1979). Block Lanczos software for symmetric eigenvalue problems. *ORNL Report CSD-48,* Oak Ridge, Tennessee.

Scott, D. S. (1979b). How to make the Lanczos algorithm converge slowly. *Math. Comp.* **33**, 239-247.

Scott, D. S. (1980). The advantages of inverted operators in Rayleigh-Ritz approximations. *SIAM J. Sci. Statist. Comput.* **3**, 68-75.

Scott, D. S. (1981). The Lanczos algorithm. In *Sparse Matrices and Their Uses,* Duff, I. S. (Ed.), Academic Press, New York, 139-159.

Scott, D. S. (1981b). Solving sparse symmetric generalized eigenvalue problem without factorization. *SIAM J. Numer. Anal.* **18**, 102-110.

Scott, D. S., and Ward, R. C. (1982). Solving symmetric-indefinite quadratic lamda-matrix problems without factorization. *SIAM J. Sci. Statist. Comput.* **3**, 58-67.

Sears, T., Miller, T. A., and Bondybey, V. E. (1981). The Jahn-Teller effect in $C_6F_6^{+}$. *J. Chem. Phys.* **74**, 3240-3248.

Shavitt, I. (1970). Modification of Nesbet's algorithm for the iterative evaluation of eigenvalues and eigenvectors of large matrices. *J. Comput. Phys.* **6**, 124-130.

Shavitt, I., Bender, C. F., Pipano, A., and Hosteny, R. P. (1973). The iterative calculation of several of the lowest or highest eigenvalues and corresponding eigenvectors of very large symmetric matrices. *J. Comput. Phys.* **11**, 90-108.

Simon, H. D. (1982). The Lanczos algorithm for solving symmetric linear systems. *PhD Thesis UC Berkeley.*

Simon, H. D. (1984). The Lanczos algorithm with partial reorthogonalization. *Math. Comp.* **42**, 115-142.

Slonczewski, J. C., Müller, K. A., and Berlinger, W. (1970). Dynamic Jahn-Teller effect of an impurity in a spontaneously distorted crystal. *Phys. Rev. B* **1**, 3545-3551.

Smithies, F. (1958). Integral Equations. Cambridge University Press.

SPARSPAK User Guide (1979). George, A., Liu, J. W. H., and Ng, E. *CS Dept. Tech. Report,* U. Waterloo.

Stein, J., and Krey, U. (1979). Numerical studies on the Anderson localization. *Z. Phys. B* **34**, 287-296.

Stewart, G. W. (1973). Introduction to Matrix Computations. Academic Press, New York.

Stewart, G. W. (1973b). Error and perturbation bounds for subspaces associated with certain eigenvalue problems. *SIAM Rev.* **15**, 727-764.

Stewart, G. W. (1974). The numerical treatment of large eigenvalue problems. In *Information Processing 74* North Holland, Amsterdam 666-672.

Stewart, G. W. (1975). Methods of simultaneous iteration for calculating eigenvectors of matrices. In *Topics in Numerical Analysis II.* Miller, J. J. H. (Ed.), Academic Press, New York, 185-196.

Stewart, G. W. (1976). A bibliographical tour of the large, sparse generalized eigenvalue problem. In [Bunch, 1976], 113-130.

Stewart, G. W. (1977). On the perturbation of pseudo-inverses, projections and linear least squares problems. *SIAM Rev.* **9**, 634-662.

Stewart, G. W. (1978). Perturbation theory for the generalized eigenvalue problem. In *Recent Advances in Numerical Analysis,* Academic Press, New York, 193-206.

Stewart, W. J., and Jennings, A. (1981). A simultaneous iteration algorithm for real matrices. *ACM TOMS* 7, 184-198.

Stiefel, E. (1958). Kernel polynomials in linear algebra and their numerical applications. *Nat. Bur. Standards Appl. Math. Ser.* **49**, 1-22.

Symm, H. J., and Wilkinson, J. H. (1980). Realistic error bounds for a simple eigenvalue and its associated eigenvector. *Numer. Math.* **35**, 113-126.

Szyld, D. B., and Widlund, O. B. (1979) Applications of conjugate gradient type methods to eigenvalue calculations. In *Advances in Computer Methods for Partial Differential Equations. III.* Proc. Third IMACS Internat. Symposium, Lehigh U., Bethlehem, Pennsylvania, 167-173.

Takahashi, H. (1981) On the Lanczos method in eigenvalue problems. In *Proc. Kyoto U. Symp.,* 119-139.

Taussky, O., and Marcus, M. (1962). Eigenvalues of finite matrices. In *Survey of Numerical Analysis,* Todd, J. A. (Ed.), McGraw-Hill, New York, 279-297.

Taussky, O. (1963). On the variation of the characteristic roots of a finite matrix under various changes of its elements. In *Recent Advances in Matrix Theory,* Schneider, H. (Ed.), U. Wisconsin Press, 125-138.

Thouless, D. J. (1974). Electrons in disordered systems and the theory of localization. *Physics Reports (Sect. C of Phys. Letters)* 13, 93-142.

Tikhonov. A. N., and Arsennin, V. Y. (1977). Solution of Ill-posed Problems. Wiley, New York.

Underwood, R. (1975). An iterative block Lanczos method for the solution of large sparse symmetric eigenproblems. *Ph. D. Thesis and Report, STAN-CS-75-496,* Stanford.

van der Vorst, H. A. (1982). A generalized Lanczos scheme. *Math. Comp.* 39, 559-561.

van Kats, J. M., and van der Vorst. H. A. (1976). Numerical results of the Paige-style Lanczos method for the computation of extreme eigenvalues of large sparse matrices. *Acad. Comp. Centrum Report TR-3.* U. Utrecht.

van Kats, J. M., and van der Vorst. H. A. (1977). Automatic monitoring of Lanczos schemes for symmetric or skew symmetric generalized eigenvalue problems. *Acad. Comp. Centrum Report TR-7*, U. Utrecht.

Van Loan, C. F. (1976). Generalizing the singular value decomposition. *SIAM J. Numer. Anal.* **13**, 76-82.

Van Ness, J. E., Brasch, F. M., Jr., Landaren, G. L. and Naumann, S. T. (1980). Analytical investigation of dynamic instability occurring at Powerton station. *IEEE Trans. PAS-99*, 1386-1395.

Varah, J. M. (1970). Computing invariant subspaces of a general matrix when the eigensystem is poorly conditioned. *Math. Comp.* **24**, 137-149.

Varah, J. M. (1973). On the numerical solution of ill-conditioned linear systems with applications to ill-posed problems. *SIAM J. Numer. Anal.* **10**, 257-267.

Varah, J. M. (1975). A lower bound for the smallest singular value of a matrix. *Linear Algebra Appl.* **11**, 3-5.

Varah, J. M. (1979). A practical examination of some numerical methods for linear discrete ill-posed problems. *SIAM Rev.* **21**, 100-111.

Varah, J. M. (1983). Pitfalls in the numerical solution of linear ill-posed problems. *SIAM J. Sci. Statist. Comput.* **4**, 164-176.

Varga, R. S. (1962). Matrix Iterative Analysis. Prentice-Hall, Englewood Cliffs, New Jersey.

Varga, R. S. (1976). M-matrix theory and recent results in numerical linear algebra. In [Bunch 1976], 375-387.

Ward, R. C., and Gray, L. J. (1978). Eigensystem computation for skew-symmetric matrices and a class of symmetric matrices. *ACM TOMS* **4**, 278-285.

Weaire, D., and Hodges, C. (1978). Anderson localization and the recursion method. *J. Phys. C.* **11**, L685-L689.

Weaver, W., Jr., and Yoshida, D. M. (1971). The eigenvalue problem for banded matrices. *Comput. & Structures* **1**, 651-664.

Webster's Third New International Dictionary Unabridged, G. & C. (1971). Merriam Co., Springfield, MA.

Weingarten, V. I., Ramanathan, R. K., Chen, C. N. (1983). Lanczos eigenvalue algorithm for large structures on a minicomputer. *Comput. & Structures* **16**, 253-257.

Weyl, H. (1949). Inequalities between the two kinds of eigenvalues of a linear transformation. *Proc. Nat. Acad. Sci. USA* **35**, 408-411.

Whitehead, R. R. (1972). A numerical approach to nuclear shell-model calculations. *Nucl. Phys.* **A182**, 290-300.

Whitehead, R. R., Watt, A., Cole, B. J., and Morrison, I. (1977). Computational methods for nuclear shell-model calculations. In *Advances in Nuclear Physics*, 9. Baranger, M., and Vogt, E. (Eds.), Plenum Press, New York, 123-176.

Wiberg, T. (1973). A combined Lanczos and conjugate gradient method for the eigenvalue problem of large sparse matrices. *Report UMINF-42.73*, U. Umeå.

Wiberg, T. (1979). A convergence theorem for the Lanczos algorithm for computing eigenvalues of large sparse matrices. *Report UMINF-77.79*, U. Umeä.

Wilkinson, J. H. (1958). The calculation of eigenvectors of codiagonal matrices. *Comput. J.* **1**, 90-96.

Wilkinson, J. H. (1958b). The calculation of eigenvectors by the method of Lanczos. *Comput. J.* **1**, 148-152.

Wilkinson, J. H. (1960). Householder's method for the solution of the algebraic eigenproblem. *Comput. J.* **3**, 23-27.

Wilkinson, J. H. (1962). Calculation of the eigenvalues of a symmetric tridiagonal matrix by the method of bisection. *Numer. Math.* **4**, 362-367.

Wilkinson, J. H. (1962b). Calculation of the eigenvectors of a symmetric tridiagonal matrix by inverse iteration. *Numer. Math.* **4**, 368-376.

Wilkinson, J. H. (1965). The Algebraic Eigenvalue Problem. Oxford University Press, New York.

Wilkinson, J. H. (1968). Global convergence of tridiagonal QR with origin shift. *Linear Algebra Appl.* **1**, 409-420.

Wilkinson, J. H., and Reinsch, C. (1971). Handbook for Automatic Computation. Vol. II. Linear Algebra. Springer-Verlag, New York.

Wilkinson, J. H. (1972). Inverse iteration in theory and practice. *Symposia Mathematica* 10, 361-379.

Wilkinson, J. H. (1977). Some recent advances in numerical linear algebra. In [Jacobs 1977], 3-53.

Wilkinson, J. H. (1978). Singular-value decomposition - basic aspects. In [Jacobs 1978], 109-135.

Winant, C. D., Inman, D. L., and Nordstrom, C. E. (1975). Description of seasonal beach changes using empirical eigenfunctions. *J. Geophys. Res.* **80**, 1979-1984.

Winkelman, J. R., Chow, J. H., Bowler, B. C., and Avramovic, B. (1981). An analysis of interarea dynamics of multi-machine systems. *IEEE Trans PAS* **400**, 754-763.

Yale Sparse Matrix Package. (1982). Eisenstat, S. C., Gursky, M. C., Schultz, H. M., and Sherman, A. H. I-Symmetric Codes. *Inter. J. Numer. Methods Engrg.* **18**, 1145-1151.

Yale Sparse Matrix Package. (1982b). Eisenstat, S. C., Gursky, M. C., Schultz, H. M., and Sherman, A. H. II-Nonsymmetric Codes. *CS Report 114, Yale.*

AUTHOR INDEX

A

Abo-Hamd, M. and Utku (1978)
Anderson, N. and Karasalo (1975) 168
Anderson, P. J. and Liozou (1973) 196
Anderssen, R. S. and Bloomfield (1974) 175
Andrews, H. C. and Hunt (1977) 28, 175, 178
Arsenin, V. Y. (1977) (See Tikhonov, A. N.)

B

Backus, G. and Gilbert (1970) 175
Barakat, R. and Buder (1979) 174
Barnes, E. (1982) 18, 243
Barth, W., Martin and Wilkinson (1967) 91
Bathe, K. J. and Wilson (1976) 25
Bauer, F. L. and Fike (1960) 198
Björck, Å. (1976) 13
Björck, Å. and Duff (1980) 179
Bloomfield, P. (1974) (See Anderssen, R. S.)
Bondybey, V.E. (1981) (See Sears, T.)
Buder, T.E. (1979) (See Barakat, R.)

C

Chow, J. H., Cullum, and Willoughby (1984)
     18-19, 30
Cline, A. K., Golub, and Platzman (1976) 19
Cline, A. K., Moler, Stewart, and Wilkinson
     (1979) 168
Cole, B. J. (1977) (See Whitehead, R. R.)
Cullum J. and Donath (1974) 59, 64-5, 212-3,
     232-3, 242-4
Cullum, J., Donath and Wolfe (1975) 221, 243
Cullum, J. (1978) 231, (1979, 1980) 175,
     (1980b) (See Chow, J. H.)
Cullum, J. and Willoughby, (1979, 1980, 1981)
     59, 74-5, 107, 139
Cullum, J., Willoughby and Lake (1983)
     93, 128, 181
Cullum J. and Willoughby (1984) 84

D

Davidson, E. R. (1983) 18, 24
Davis, C. and Kahan (1970) 19, 38, 240
Dean, P. (1972) 18
Donath, W. E. and Hoffman (1973) 243-244
Donath, W. E. (1974, 1975) (See Cullum, J.)
Duff, I. S. (1980) (See Bjorck, A.)

E

Eberlein, P. J. (1971) 195, 206
Eckart, C. and Young (1936) 177
Edwards, J. T., Licciardello, and Thouless
     (1979) 59, 70-2, 125
Eggarter, T. P. (1972) (See Kirkpatrick, S.)
EISPACK (1976, 1977) 5, 10. 13, 34, 76,
     89-90, 195, 205, 247
Ekstrom, M. P. and Rhoads (1974) 174, 177
Ericsson, T. and Ruhe (1980) 29, 59,
     61-3, 163
Ericsson, T. (1983) 59

F

Fan, K. (1949) 221
Fiedler, M. and Ptak (1962) 169
Fike, C. T. (1960) (See Bauer, F. L.)
Fisher, N. J. and Howard (1980) 174
Freed, J. H. (1981) (See Moro, G.)

G

Gantmacher, F. R. (1959) 196-7
Gilbert, F. (1970) (See Backus, G.)
Givens, W. (1954) 42
Godunov, S. K. and Prokopov (1970) 103
Golub, G. H. and Kahan (1965) 28, 174,
     178-9, 181-2
Golub, G. H. (1968) 178, (1976)
     (See Cline, A. K.)
Golub, G. H. and Reinsch (1970) 178
Golub, G. H. and Underwood (1975, 1977) 41,
     59, 63-4, 212-3, 220, 227, 237
Golub, G. H., Luk, and Overton (1981) 181, 185
Gustavson, F. G. (1972) 28

H

Hamming, R. W. (1973) 16
Hanson, R. J. (1974) (See Lawson, C.)
Hanson, R. J. and Phillips (1975) 28
HARWELL Subroutine Library (1981)
Haydock, R. (1980) 18
Hestenes M. R., and Karush (1951) 214
Hestenes, M. R., and Stiefel (1952) 95-6
Hestenes, M.R. (1956) 95, (1980) 97
Hoffman, A. J. and Wielandt (1953) 20
Hoffman, A. J. (1973) (See Donath, W. E.)
Howard, L. E. (1980) (See Fisher, N. J.)
Householder, A. S. (1958) 42, (1964) 95, 170

Huang, T. C. and Parrish (1978) 175
Hunt, B. R. (1977) (See Andrews, H. C.)

J

Jennings, A. (1977) 18, 22-3, 25, 70, 89, 128
Jiang, E. (1982) (See Kahan, W.)

K

Kahan, W. (1965) (See Golub, G.H.),
    (1970) (See Davis, C.)
Kahan, W. and Parlett (1976) 67
Kahan, W., Parlett and Jiang (1982) 198
Kaniel, S. (1966) 40, 220
Karasalo, I. (1975) (see Anderson, N.)
Karush, W. (1951) 40-1, 214-6, 220 (1951)
    (See Hestenes, M. R.)
Kirkpatrick, S. and Eggarter (1972) 142
Kirkpatrick, S. (1979) 18, 145
Krey, U. (1979) (See Stein, J.)

L

Lake, M. (1983) (See Cullum, J.)
Lanczos, C. (1950) 32 (1961) 17, 27, 179
Lawson, C. and Hanson (1974) 28, 164,
    172-4
Lewis, J. G. (1977) 53, 55, 59, 65-6,
    68-69, 212
Licciardello, D. C. (1979) (See Edwards, J. T.)
LINPACK User's Guide (1979) 28, 173-4,
    176, 178, 185
Liozou, G. (1973) (See Anderson, P. J.)
Luenberger, D. G. (1973) 117, 216-8, 245
Luk, F. T. (1981) (See Golub, G. H.)

M

Marcus, M. (1960) 167
Martin, R. S. and Wilkinson (1968) 205-6,
    (1967) (See Barth, W.)
Miller, T. A. (1981) (See Sears, T.)
Moler, C. (1979) (See Cline, A. K.)
Moro, G. and Freed (1981) 194, 199
Morrison, I. (1977) (See Whitehead, R. R.)

N

NASTRAN User's Guide (1977) 18, 25, 59
Nesbet, R. K. (1981) 18, 24
Newman, M. and Pipano (1977) 59-61

O

O'Leary, D. P. (1980) 222
O'Leary, D. P. and Simmons (1981) 179
Ostrowski, A. M. (1937) 169
Overton, M. L. (1981) (See Golub, G. H.)

P

Paige, C. C. (1971) 43, 45, 50, 52, 56, 64,
    71, 102, 128 , (1972) 45-7, 103, (1974) 179,
    181, (1976) 45, 48-9, 101-2, 105-8, (1980) 45,
    49-53, 102-3, 105-8, 123, 130
Paige, C. C. and Saunders (1975) 95
Pandey, K. C. (1983) 18, 24
Parlett, B. N. and Scott (1979) 34, 59,
    62, 67-8
Parlett, B. N. (1976) (See Kahan, W.)
    (1980) 1, 16-7, 21, 37-40, 57, 67, 89, 93,
    (1982) 59, 67
Parlett, B. N. and Reid (1981) 59, 70,
    72-3, 125
Parrish, W. (178) (See Huang, T. C.)
Peters, G. and Wilkinson (1969) 26
Phillips, J. L. (1975) (See Hanson, R.J.)
Pipano, A. (1977) (See Newman, M.)
Platzman, G. W. (1978) 19, (1976)
    (See Cline, A.K.)
Prokopov, G. P. (1970) (See Godunov, S.K.)
Ptak, V. (1962) (See Fiedler, M.)

R

Reid, J. K. (1971) 95, (1981)
    (See Parlett, B. N.)
Reinsch, C. (1970) (See Golub, G. H.)
Rhoads, R. L. (1974) (See Ekstrom, M. P.)
Ruhe, A. (1975) 168, (1979) 212, (1980)
    (See Ericsson, T.)
Rutishauser (1959) 104

S

Saad, Y. (1980) 37-8, 41-2, 227
Saunders, M. A. (1975) (See Paige, C.C.)
Scott, D. S. (1979) 53, 55, 68-9, 212,
    (1979) (See Parlett, B. N.), (1981) 59, 67-9
Sears, T., Miller and Bondybey (1981) 18
Simmons, J. A. (1981) (See O'Leary, D. P.)
Smithies, F. (1958) 175
SPARSPAK User's Guide (1979) 30, 157, 163
Stein, J. and Krey (1979) 18
Stewart, G. W. (1973) 1, 8, 17, 19-20, 23-24,
    43, 76, 87, 164, 166, 177, 195-6, 198,
    (1977) 173, (1979) (See Cline, A. K.)
Stiefel, E. (1952) (See Hestenes, M. R.)

T

Thouless, D. J. (1979) (See Edwards, J. T.)
Tikonov, A. N. and Arsenin (1977) 174

U

Underwood, R. (1975)
Underwood, R. (1977) (See Golub, G. H.)

V

van Kats, J. N. and van der Vorst (1976, 1977)
    59, 70-2, 125, 142
van der Vorst, H. A. (1976, 1977)
    (See van Kats, J. M.)
Varga, R. S. (1962) 169, 171, (1976) 169
Varah, J. M. (1973) 176, (1975) 168-9,
    (1979) 176

W

Watt, A. (1977) (See Whitehead, R. R.)
Weyl, H. (1949) 167
Whitehead, R. R., Watt, Cole and Morrison

(1977) 18, 24
Wielandt, H. W. (1953) (See Hoffman, A. J.)
Wilkinson, J. H. (1960) 43, (1965) 43, 79,
    83, 88, 90, 157-8, 166, 195-6, 199, (1967)
    (See Barth, W.), (1968) 206, (1968) (See
    Martin, R. S.), (1969) (See Peters, G.)
    (1978) 178, (1979) (See Cline, A. K.)
Willoughby, R. A. (1979, 1980, 1981, 1983,
    1984), (See Cullum, J.)
Wilson, E. L. (1976) (See Bathe, J.K.)
Wolfe, P. (1975) (See Cullum, J.)

Y

Young, G. (1936) (See Eckart, C.)

SUBJECT INDEX

Applications, 18-9, 243-4
Chebyshev polynomials, 15-6
Complex symmetric matrices, 3, 8, 194-209
    definitions, 3, 8, 194-5
    Eberlein Jacobi algorithm, 195-6
    Lanczos procedure, eigenvalues, 199-204
    Lanczos procedure, eigenvectors, 204
    perturbation theory, 198
    properties, 196-8, 201
    QL algorithm for tridiagonal, 204-9
Conjugate gradients
    block, 226-7
    definitions, 95-6
    equivalence with Lanczos tridiagonalization
        exact arithmetic, 95-101
        finite arithmetic, 101-118
    quadratic functions, 217-8
    residual bounds, 104-7
    s-step conjugate gradients, 218-20
Eigenvalues
    definition, 7
    interlacing, 22
    Lanczos procedures, 59-75, 136-52,
        157-60, 162-3, 199-204
    Sturm Sequencing, 22-3
Eigenvectors
    definition, 7
    Lanczos procedures, 59-75, 136-52,
        157-60, 162-3, 199-204
Factorizations
    Cholesky, 12-3
    sparse, 29-31
    triangular, 12
Generalized eigenvalue problems, 7, 24-6, 59-63,
    162-3
Hermitian matrices, 3, 23-4
    Lanczos procedures, 157-62
Invariant subspaces, 9
Inverse iteration, 9
Krylov Subspaces, 14
Lanczos phenomenon, 102, 119-21
Lanczos procedures
    basic iterative single-vector, 41
    block, 63-9, 181-2, 213, 232-41
    hybrid, 244-7
    limited reorthogonalization, 64-9, 232-41
    no reorthogonalization, 69-75, 92-5, 136-52,
        186-7, 201-4
    practical procedures, 53-8
    single-vector, 59-63, 67-8, 70-5, 136-52,
        186-7, 201-4
Lanczos procedures, block
    an application, 243-4
    basic block recursion, 210-2
    basic iterative block, 213

block conjugate gradients, 226
block Rayleigh quotient, 221-31
block steepest ascent, 224-5
comparison with single-vector, 250
computation of second block, 232-3
computing both extremes of spectrum, 228-31
convergence estimates, 239-41
Cullum, J. and Donath, W. E., 64-5, 232-44
dropping 'zero' Lanczos vectors, 233-37
Golub, G. H. and Underwood, R., 63-4, 227-8,
    237
hybrid procedure, 244-50
Krylov subspaces, 210-1, 225
Lewis, J. G., 65-9
losses of orthogonality, 238-9
operations counts, 242
optimization interpretation, 220-8
practical implementation, 232-41
Scott, D. S., 67-9
Lanczos procedures, single-vector
No reorthogonalization
    basic procedure, 33-4
    complex symmetric matrices, 199-204
    Cullum, J. and Willoughby, R. A., 74-5,
        92-5, 136-52, 201-4
    Edwards, J. T. et al., 71-2
    error estimates (exact arithmetic), 140-1
    generalized problems, 162-3
    Hermitian matrices, 157-62
    Identification test, 121-30
    introduction, 92-5
    numerical results, 142-50, 155-7
    orthogonality (local), 139
    Parlett, B. N. and Reid, J. K., 72-4
    practical procedures, 53-8
    real symmetric matrices, 136-55
    singular values/vectors, 178, 186-7, 190-3
    spurious eigenvalues, 130-6
    test problems, 131, 142
    Van Kats, J. M. and van der Vorst, H. A., 70-1
Lanczos tridiagonalization
    block version, 55
    computational variants, 46-8
    definition, 95
    equivalence with conjugate gradients
        exact arithmetic, 95-101
        finite arithmetic, 107-118
    error analysis, exact arithmetic, 35-40
    error analysis, finite arithmetic, 44-53
        error analysis, 48-9
        error estimates, 51
        losses of orthogonality, 48, 50
        Ritz vectors, norms, 50-1
        Spectrum of Lanczos matrices, 52-3
    Paige, C. C., 44-53

Matrices, definitions
  adjoint of, 6
  Cholesky Factorization, 12-3
  column permutation, 11
  complex plane rotations, 12
  complex symmetric, 3
  diagonal, 5
  diagonalizable, 8
  factorization, 12
  Hermitian, 3, 23-4
  Hessenberg, 5
  Householder transformation, 13
  ill-conditioned, 10
  nondefective, 8, 194-5
  orthogonal, 4, 10
  permutation, 4-5
  plane rotation, 11-2
  positive definite, 7
  principal submatrices, 6
  real orthogonal, 4
  real symmetric, 3, 17
  row permutation, 11
  similarity transformations, 11
  skew symmetric, 6
  sparse, 3, 28-9
  sparse factorizations, 29-31
  triangular, 5
  tridiagonal, 4, 76-91
  unitary, 4, 10
Matrix transformations
  column permutation, 11
  factorizations, 29-31
  Householder, 13
  orthogonal, 10
  plane rotations, 11-2
  reorderings, 30
  row permutation, 11
  similarity, 11
  unitary, 10
Norms
  vector, 1
  matrix, 2
Orthogonalization
  Gram-Schmidt, 13, 34
  modified Gram-Schmidt, 13-14
  orthogonal projections, 14
  reorthogonalization
    limited, 64-9
    total, 59-64
Perturbation theory

  complex symmetric, 198
  Hermitian matrices, 19
  real symmetric matrices, 19-20
Rayleigh quotients, 8
Residual estimates
  Hermitian, 20-2
  real symmetric, 20-2
Ritz vectors, 15, 57-8
Real rectangular matrices
  See singular values/vectors
Real symmetric matrices, 17-23
  perturbation theory, 19-20
  error estimates, 20-2
  Lanczos procedures, 59-75, 232-250,
    136-157
Singular values and vectors
  applications, 28, 172-7
  block Lanczos procedure, 181-2
  bounds on, 168-71
  data reduction, 177-8
  definitions, 9-10, 26-8
  equivalent symmetric problem, 27-8, 179-81
  Golub, G. H., Luk, F. T. and Overton, M. L.,
    181-2
  ill-posed problems, 174-7
  Lanczos recursion, 181-4
  matrix scalings, 171-2
  numerical examples, 187-93
  properties of Lanczos matrices, 184-5
  pseudo inverses, 172-4
  relationships with eigenvalues, 166-8
  single-vector Lanczos procedure, 178,
    186-7, 190-3
  singular value decomposition, 26-7, 164-6
  solving systems of equations, 172-4
Spurious eigenvalues
  example, 130-6
  identification test, 121-30
Spectral decomposition, 8
Test matrices, 131, 142
Tridiagonal matrices, 76-91, 204-9
  adjoint of, 79-82
  complex symmetric, 84-5, 204-9
  definitions, submatrices, 78-9
  determinant recursions, 79, 83
  eigenvectors, 82-3
  fundamental identity, 84
  Hermitian, 85-6
  inverse iteration, 86-9
  QL algorithm, 204-9
  Sturm sequencing, 89-91
  tridiagonal, general, 85